Annual Festival on Quantum Phenomena, Quantum Control and Geometry of Quantum States (Quantum Fest 2015)

Journal of Physics: Conference Series Volume 698

Mexico City, Mexico
19 - 23 October 2015

ISBN: 978-1-5108-2281-8
ISSN: 1742-6588

Printed from e-media with permission by:

Curran Associates, Inc.
57 Morehouse Lane
Red Hook, NY 12571

Some format issues inherent in the e-media version may also appear in this print version.

Printed by Curran Associates, Inc. (2016)

For permission requests, please contact the Institute of Physics
at the address below.

Institute of Physics
Dirac House, Temple Back
Bristol BS1 6BE UK

Phone: 44 1 17 929 7481
Fax: 44 1 17 920 0979

techtracking@iop.org

Additional copies of this publication are available from:

Curran Associates, Inc.
57 Morehouse Lane
Red Hook, NY 12571 USA
Phone: 845-758-0400
Fax: 845-758-2633
Email: curran@proceedings.com
Web: www.proceedings.com

Table of contents

Volume 698

Quantum Fest 2015
19–23 October 2015, Mexico City, Mexico

Accepted papers received: 29 February 2016
Published online: 7 April 2016

Preface

Papers

Quantum Fest 2015

Preface

The *Quantum Fest* is a periodic annual festival on Quantum Phenomena, Quantum Control and Geometry of Quantum States, organized by the Center for Research and Advanced Studies (Cinvestav by its acronym in Spanish) and Unidad Profesional Interdisciplinaria en Ingeniería y Tecnologías Avanzadas del I.P.N. (UPIITA-IPN) in México City, Mexico. The aim of this meeting is to bring together students and researchers which are engaged in the subjects of the festival, from both theoretical and experimental approaches, in order to get lively discussions and to enable a closer contact between them.

The festival was celebrated for the first time in the Physics Department of Cinvestav (2010), since then it has been alternatively hosted in Cinvestav and UPIITA. This is the sixth *Quantum Fest* and the first time that it is formally hosted in the Tecnológico de Monterrey (ITESM).

The *Quantum Fest 2015* took place from October 19 to 23 in *Aulas VI* of the Tecnológico de Monterrey, Campus Estado de México (CEM). We would like to thank the willing of the ITESM-CEM to offer its facilities as the venue of the festival; all its help provided to simplify the logistics and organization of the conference has been welcomed and is acknowledged.

This volume is dedicated to the memory of our dear friend and colleague Sujeev Wickramasekara who passed away suddenly on December 28th 2015 at the age 48. Sujeev participated several times in the festival, he was an excellent speaker and, as a part of the audience, he used to pay special attention to the students' presentations. His comments were always addressed to improve the students' work, so that Sujeev was very popular among the young participants of the festival. We lost not only a great colleague and promissory scientist, but also a beautiful person who we shall always remember.

Professor Manuel Gadella kindly agreed to write the tribute to Sujeev Wickramasekara. The obituary is included in these Proceedings as the opening article.

The topics addressed at the short courses of the *Quantum Fest 2015* were geometry of quantum states, quantum entanglement, fractional calculus with applications in physics (mainly in optics), operator theory of evolution equations and quantum optics.

The Lecturers of Quantum Fest 2015 were:

Karol Życzkowski (Jagiellonian University, Kraków, Poland)

Karol Penson (Sorbonne Universités, Université Paris VI, LPTMC, France)

Vladimir I. Man'ko (Lebedev Physical Institute, Moscow, Russia)

Julio C. Gutiérrez-Vega (Tecnológico de Monterrey, Campus Monterrey, México)

Besides the courses lectured by Profs. Życzkowski, Penson and Gutiérrez-Vega, there were 11 plenary presentations, 12 contributed talks and 25 posters. This time, 48 participants from Tecnológico de Monterrey, Instituto Politécnico Nacional, Cinvestav, and Universidad Nacional Autónoma de México came together to attend the courses, talks and poster sessions delivered during the festival.

We would like to mention that the plenary as well as the contribution talks are presented only by invitation of the festival Organizing Committee.

The Plenary Speakers of Quantum Fest 2015 were:

Mauricio Carbajal (Physics Department, Cinvestav, México)

Octavio Castaños (ICN-UNAM, México)

David Fernández (Physics Department, Cinvestav, México)

Eduardo Gómez (Insituto de Física, UASLP, México)

Véronique Hussin (Department of Mathematics, University of Montréal, Canada)

Rocío Jáuregui (IF-UNAM, México)

Máximo López-López (Physics Department, Cinvestav, México)

Margarita A. Man'ko (Lebedev Physical Institute, Moscow, Russia)

Bogdan Mielnik (Physics Department, Cinvestav, México)

José Récamier (ICF-UNAM, México)

Eric Rowell (Department of Mathematics, Texas A& M University, USA)

Sujeev Wickramasekara (Department of Physics and Astronomy, The University of Iowa, USA)

Unfortunately, the participation of Profs. M. Mank'o and V.I. Mank'o was canceled due to unexpected drawbacks. Nonetheless, they submitted a contribution for the proceedings of the conference, the one that we are happy to include in this volume.

The informal and lively discussions at *Aulas VI* of Tecnológico de Monterrey, CEM, during either the coffee breaks or the poster sessions, have been a breeding ground for scientific collaboration between the participants of the conference. We are certain that this is crucial to strength the rapprochement between students and researchers.

The best poster expositor, M.Sc. Zurika Blanco-Garcia, was the recipient of an academic recognition and received the book *Geometry of Quantum States. An Introduction to Quantum Entanglement* of I. Bengtsson and K. Życzkowski, autographed by Prof. Życzkowski, as a gift from the organizers. The figure in the cover of this volume has been borrowed from the cartel presented by Miss Blanco-Garcia.

The members of the 2015 Organizing Committee were:

Sara Cruz y Cruz (UPIITA-IPN)

Francisco Delgado (Tecnológico de Monterrey, CEM)

Oscar Rosas-Ortiz (Physics Department, Cinvestav)

The *Quantum Fest 2015* would not have been possible without the valuable financial support of the sponsor institutions: Cinvestav and UPIITA-IPN.

Our thanks to M.Sc. Rubén Santiago, *Director del Departamento de Física y Matemáticas*, and to the academic authorities of Tecnológico de Monterrey, Campus Estado de México, for endless support and host during the festival.

Special thanks to Mmes Miriam Lomelí, Maribel de la Cruz and Mónica Otero, Secretaries of the festival, for invaluable help in many aspects of the organization, including the design of the posters and distribution of announcements, hotel and flight reservations, and the scheduling of the event. Miriam, Maribel and Mónica: the *Bar Quantum* is very grateful to you.

The publication of this volume of *Journal of Physics: Conference Series* has been supported by the Academic Affairs Office of Cinvestav, the Physics Department of Cinvestav, Instituto Politécnico Nacional (project SIP 20160527), and Tecnológico de Monterrey Campus Monterrey (Grupo de Enfoque de Óptica y Láseres de la Escuela de Ingeniería y Ciencias). We would like to express our gratitude to Profs. Ricardo Félix Grijalva (Provost of Cinvestav), Máximo López López (Chair of the Physics Department, Cinvestav) and Julio C. Gutiérrez-Vega (Tecnológico de Monterrey, Campus Monterrey) for facilitating the administrative duties related with the publication of these Proceedings.

The present volume includes only the positively refereed contributions that were submitted by the participants of the festival.

Finally, we would like to announce that the seventh *Quantum Fest* will take place in the facilities of Unidad Profesional Interdisciplinaria en Ingeniería y Tecnologías Avanzadas del I.P.N., México City, during the third week of October 2016.

Oscar Rosas-Ortiz
Physics Department
Cinvestav

México City, February 2016

Peer review statement

All papers published in this volume of *Journal of Physics: Conference Series* have been peer reviewed through processes administered by the proceedings Editors. Reviews were conducted by expert referees to the professional and scientific standards expected of a proceedings journal published by IOP Publishing.

Journal of Physics: Conference Series **698** (2016) 012001 doi:10.1088/1742-6596/698/1/012001

In Memoriam: Sujeev Wickramasekara (1967-2015)

M Gadella

Departmento de Física Teórica, Atómica y Optica. Universidad de Valladolid. Paseo Belén 7, 47011. Valladolid Spain.

E-mail: manuelgadella1@gmail.com

Abstract. This is an obituary in Memoriam of Sujeev Wickramasekara who died suddenly on December 28th 2015.

Sujeev Wickramasekara 1967-2015

Sujeev Wickramasekara passed away suddenly on December 28th 2015 at the age 48. He was an excellent scientist with widespread knowledge in the fields of theoretical and mathematical physics. He was a young spirit, hard working and passionate physicist and mathematician; great teacher, highly devoted to his students; and above all an outstanding human being. He had novel and interesting ideas in his field of research, some of which were being developed at the time of his unexpected death.

Sujeev was born in Colombo (Sri Lanka) on September 3rd, 1967. He is survived by his wife, Tammy Nyden, his step sons, Cole and Jonah, his parents Walter Keerthiraja Wickramasekara and Gurukande Wedaarachchige Chandrani, his sister Chamindi and his brother Neshan.

After completing his secondary studies at the Royal College in Colombo, he went to the USA. He earned his Bachelors degree in physics (Summa Cum Laude) at the University of Southern California. Then, he moved to Austin (Texas), where he completed his Ph. D. under Professor Arno Bohm. During this period, Sujeev published 15 scientific papers in leading journals.

After teaching as the Wiess Professor of Physics and Astronomy at Rice University (Texas) and as an Assistant Professor at St. Olaf College (Minnesota), he received a permanent position at Grinnell College (Iowa) in 2005, where he received tenure in 2010 and earned a Full Professorship in 2015. At Grinnell, he displayed a tremendous range from challenging majors with advanced seminars in quantum theory, to teaching sections of general physics and tutorials on the growth and convergence of scientific knowledge and "The Isolation and Connectedness of Human Life". He was a supportive and thoughtful colleague and mentor to his peers and a generation of students.

In addition, Sujeev had a permanent collaboration with the Math-Phys groups of the University of Iowa in Iowa City and the University of Valladolid in Spain. He was a winner of the prestigious Cottrell College Science Award and the Harris Faculty Fellowship.

Among his non-scientific personal interest, he had a personal taste for classical music, in particular the choral works of Schubert and Beethoven, as well as poetry and literature. Above of all, he had a particular dedication to family life. He was an excellent husband and a great step father.

This rings us to the qualities of Sujeev as human being. He was a great friend of all his scientific collaborators, always ready to help others in all aspects of life, either professional or personal. I have witnessed his dedication and patience with students. Sujeev was a great loss for all of us who had the great fortune to know him.

As mentioned earlier, Sujeev earned his Ph.D. at the University of Texas at Austin. He started with some studies on the applications of the de Sitter group $SO(3,2)$ as a spectrum generating group. Then, the application of these ideas in the study of internal symmetries on a model for hadrons [1].

The main subject of his dissertation dealt with semigroup representations of the Poincaré group and their applications to the quantum theory of decay of relativistic unstable particles. This came after a collaboration with his adviser A. Bohm and his fellow graduate students Nathan Harshman and Hani Kaldass [2–7].

At the same time, he also began with one of his deepest scientific interests: the Einstein theory of relativity and its consequences [8].

Sujeev scientific quest covered a variety of subjects. With Nathan Harshman, he explored the intricacies of several aspects of quantum entanglement and their consequences in the theory of quantum information [9, 10]. At the same time, he continued with his analysis of quantum symmetries [11–13], in particular those induced by the Poincaré group [14] as well as on the consequences of representations of Poincaré semigroups in the theory of decay of relativistic particles [15–18]. He also worked in black hole thermodynamics [19].

His concerns about the very fundamentals of quantum physics drove him to an investigation of the result of changes, either inertial or not, of reference systems in quantum physics. I like to believe that these concerns are a result of both, classroom thoughts shared with advanced students and his own experience with symmetry transformations in quantum mechanics, through group theory and group representations. This idea came after the consideration of some of his publications [12–14, 20, 21].

During the last few years and with the collaboration of WH Klink, he was investigating how the same quantum phenomena may look like for different observers which are accelerated with respect to each other. This implies the use of non-inertial transformations of reference frames. Furthermore, this can be seen from three different points of view. In the former, one considers inertial transformations governed by the Galilei group. This yields to the usual non-relativistic quantum mechanics. In this case, a rather general set of non-inertial transformations is governed by the so called Galilei line group. This is a non-trivial generalization of the Galilei group and has an infinite countable number of generators that form a Kac-Moody algebra. Central extensions of the Galilei line group with physical meaning make sense as *loops*, which are a generalization of groups in which the associativity condition has been dropped [22, 23]. It is quite interesting to see that, from the perspective of quantum dynamics, non-inertial transformations may produce additional terms in the Hamiltonian of a single particle, which are not obtained by the canonical quantization of equivalent classical transformations [22–25].

The second point of view was to extend the above analysis to special relativity using the Poincaré group instead of the Galilei group. The third one, was even more ambitious as it included curved space-time, and extending the work to general relativity using the de Sitter group. In some sense, this last step would have been an attempt towards a study of quantum gravity.

In addition to these projects, Sujeev was writing a book, with other authors, on the mathematical foundations of quantum decaying systems and continuing his research in quantum decaying states.

Sujeev was not only an active collaborator with the groups in Iowa City or Valladolid. He also was an active participant in the Quantum Fest organized each year in Mexico City. In fact, his talk in 2015 was his last one to an international scientific meeting. His contribution appears in this volume.

I wish to emphasize that the list of references below is far from being complete. Among his scientific collaborators, let me mention the following (I hope the list be essentially complete): A Bohm, BK Button, V Colussi, M Gadella, F Gómez-Cubillo, A Guneratne, H Kaldass, P Kielanowski, WH Klink, BR MacGregor, AE McCoy, W Mogavero, NL Harshman, HT Reich, L Rodríguez and T Yildirim.

Acknowledgments
I am sincerely grateful to Tammy Nyden, Sujeev's wife, for her help with the present obituary.

References

[1] Wickramasekara S 1997 *Int. J. Theor. Phys.* **36** 2409
[2] Bohm A, Harshman NL, Kaldass H and Wickramasekara S 2000 *Europ. J. Phys* **18** 333
[3] Bohm A, Harshman NL, Wickramasekara S and Kielanowski P 2000 *Phys. Lett.* A **264** 425
[4] Bohm A, Kaldass H and Wickramasekara S 2003 *Fort. Phys.* **51** 569
[5] Bohm A, Kaldass H and Wickramasekara S 2003 *Fort. Phys.* **51** 604
[6] Wickramasekara S and Bohm A 2003 *J. Math. Phys.* **44** 930
[7] Bohm A, Kielanowski P and Wickramasekara S 2006 *Ann. Phys.* **321** 2299
[8] Wickramasekara S 2001 *Class. Quant. Grav.* **18** 5353
[9] Harshman NL and Wickramasekara S 2007 *Phys. Rev. Lett.* **98** 080406
[10] Harshman NL and Wickramasekara S 2007 *Open Sys. Inf. Dyn.* **14** 3421
[11] Wickramasekara S 2009 *Ann. Phys.* **324** 2506
[12] Colussi V, Reich HT and Wickramasekara S 2009 *Ann. Phys.* **324** 2490
[13] Colussi V and Wickramasekara S 2008 *Ann. Phys.* **323** 3020
[14] Riech HT and Wickramasekara S 2010 *Ann. Phys.* **325** 1081
[15] Wickramasekara S 2009 *J. Math. Phys.* **50** 042305
[16] Wickramasekara S 2009 *J. Math. Phys.* **50** 072302
[17] Wickramasekara S and Harshman NL 2009 *Rep. Math. Phys.* **64** 123
[18] Gadella M, Gómez-Cubillo F, Rodríguez L and Wickramasekara S 2013 *J. Math. Phys.* **54** 072303
[19] Button BK, Rodríguez L and Wickramasekara S 2013 *J. High Ener. Phys.* **10** 144
[20] Mogavero W and Wickramasekara S 2011 *Int. J. Theor. Phys.* **50** 2278
[21] MacGregor BR, McCoy AE and Wickramasekara S 2012 *Ann. Phys.* **327** 2310
[22] Klink WH and Wickramasekara S 2015 *Relativity, Symmetry and the Structure of Quantum Theory I. Galilean quantum theory* (Bristol, IOP Concise Physics)
[23] Klink WH and Wickramasekara S 2013 *Ann. Phys.* **336** 261
[24] Klink WH and Wickramasekara S 2013 *Phys. Rev. Lett.* **111** 160404
[25] Klink WH and Wickramasekara S 2014 *Ann. Phys.* **340** 94

Non-Inertial Quantization: Truth or Illusion?

S Cruz y Cruz

UPIITA, Instituto Politécnico Nacional
Av. Instituto Politecnico Nacional, 2580 CP 07340 CDMX, Mexico

E-mail: sgcruzc@ipn.mx

B Mielnik

Physics Department, Cinvestav, AP 14-740 07000 CDMX, Mexico

E-mail: bogdan@fis.cinvestav.mx

Abstract. The quantum Hamiltonians in non inertial frames, while proper to describe the time evolution, might give a false information about the radiation of physical systems. The formal peculiarities observed for the uniformly rotating and accelerating frames might be caused by an excess of verbal approaches in the present day quantum theories.

1. Introduction

While the relativistic theories create increasingly flexible geometries, their quantum counterparts stick persistently to the same abstract scheme, based on the "crystalline rigid" Hilbert space geometry, with the state evolution maintained linear at the cost of increasing the number of variables (in form of tensor products, in a sense, the modern equivalents of Ptolemy's epicycles). On more sophisticated level, the branches of the perturbative graphs are reinterpreted as the *virtual quanta*, propagating endlessly in their underworld, the *polarized vacuum* (comparable to *Matmos* of R. Vadim [1], where some creatures wait for an occasion to emerge in the real world). In spite of the well known successes of this allegoric image, the solutions of some elementary problems remain incomplete.

2. Moving frames

Curiously, this includes also the quantum evolution operators $U(t, t_0)$ driven by variable Hamiltonians $H(t)$:

$$i\frac{d}{dt}U(t,t_0) = H(t)U(t,t_0). \tag{1}$$

If $H(t)$ are real, then $U(t_1, t_0)$ are unitary and they can be represented as

$$U(t_1,t_0) = e^{-i(t_1-t_0)F(t_1,t_0)} \tag{2}$$

where the (non unique) self-adjoint $F(t_1, t_0)$ are the *effective Hamiltonians* for the time intervals $[t_0, t_1]$. An important case are the periodic systems with $H(t+T) = H(t)$. If the evolution within one period $[t_0, t_0 + T]$ is given by $U(T) = U(t_0 + T, t_0) = \exp(-iTF)$, $F = F^\dagger$, its repetitions in the next periodicity intervals define the *Floquet process*

$$U(nT) = U(T)^n = e^{-itF}, \quad t = nT, \quad n \in \mathbb{Z} \tag{3}$$

where the (non unique) $F = F(t_0+T, t_0)$ is known as the *quasi-energy operator*. The hypothesis of Zeldovich [2] tells that the spectral frequencies emitted/absorbed by a periodic system are given by the differences of $F-$levels, modulo multiples of $\hbar\omega$ (where $\omega = \frac{2\pi}{T}$):

$$\Delta E = F_n - F_m + l\hbar\omega, \quad l \in \mathbb{Z}. \tag{4}$$

In spite of its naturally, the Zeldovich hypothesis has some physical limits. It does not permit to distinguish the emission and absorption events. It becomes practically useless if the periods T of the Hamiltonians are too large. In fact, if T would be of order of magnitude of days instead of nanoseconds, then how the radiating system will know that within one day the external interactions will be exactly repeated? The same about the trajectories of 2-level states described as *splins* [3]. Suppose, the experimentalist generates an exact, periodic spin trajectory: how the radiating particle will know that he will not change his mind after a while? (Hence, the doubts whether the Zeldovich hypothesis can determine the short term radiative response).

An interesting case are the rotating systems, where the simple unitary transformation reduces the evolution to a constant generator (the "Hamiltonian in the rotating frame"). In the simplest cases, this happens for the initial Hamiltonians $H(t) = H_0 + H_\mathrm{I}(t)$, where H_0 is a basic, rotation invariant part and $H_\mathrm{I}(t)$ is an interaction term containing some rotating vector (*e.g.*, an external electric field $\mathbf{E} = \mathcal{E}(\cos\omega t, \sin\omega t, 0)$). Assuming that the time dependence of $H_\mathrm{I}(t)$ can be expressed by: $H_\mathrm{I}(t) = e^{-i\omega(t-t_0)\mathbf{Ms}} H_\mathrm{I}(t_0) e^{i\omega(t-t_0)\mathbf{Ms}}$, where $\mathbf{M}$ is the orbital angular momentum, $\mathbf{s}$ a fixed unit vector and the exponentials of $\mathbf{Ms}$ rotate the internal parameters of $H_\mathrm{I}(t)$, the whole Hamiltonian $H(t)$ can be written as

$$H(t) = e^{-i\omega(t-t_0)\mathbf{Ms}} H(t_0) e^{i\omega(t-t_0)\mathbf{Ms}}, \quad \omega \in \mathbb{R}. \tag{5}$$

If so, the evolution operator (2) can be conveniently factorized. The substitution

$$U(t, t_0) = e^{-i\omega(t-t_0)\mathbf{Ms}} W(t, t_0) \tag{6}$$

leads to

$$W(t, t_0) = e^{-i(t-t_0)G}, \tag{7}$$

where

$$G = G(t_0) = H(t_0) - \omega\mathbf{Ms}. \tag{8}$$

Since in the time moments $t - t_0 = nT$ the evolution (6) reduces to (7), the generator $G = G(t_0)$ is one of the quasi-energy operators (adequate to test Zeldovich hypothesis [2]); but since (7) holds for all $t \in \mathbb{R}$, G is simultaneously interpretable as the "*Hamiltonian in the rotating frame*" (with some tempting suggestions about the fundamental quantization problems [4,5]). Note that G does not depend on t, though it depends on the "anchorage moment" t_0. To fix attention, put $t_0 = 0$ and consider $G = G(0)$. The t-independence of (8) suggests that the system described by (7) obeys the "quantum theory in non inertial frames", with some formal hints that it should emit/absorb the G-quanta. The idea, though, is not free of dangers.

Some of them are of structural nature. In fact, one can notice that the operators (8) can differ substantially from the traditional Hamiltonians. Thus e.g. if the initial $H(0)$ is a positive quadratic form of the canonical variables $\mathbf{q} = (x, y, z)$ and $\mathbf{p} = (p_x, p_y, p_z)$, it might easily happen that G has no lower bound. This was indeed detected for the charged Schrödinger particle in rotating magnetic fields [6, 7] where G becomes a non-positive combination

$$G = \omega_0 H_0 - \omega_1 H_1 + \omega_2 H_2 \tag{9}$$

of 1D oscillators H_0, H_1, H_2 (with $\omega_j \geq 0, j = 0, 1, 2$). The analogous results hold for the rotating electric fields, rotating oscillators [8–10] etc. Independently of any deeper problems, the non-positive character of (9) can awake a lot of doubts against too literal understanding of G as a "true Hamiltonian", i.e., as the energy analogue in the rotating frame.

Yet, an attempt of incorporating the indefinite structures (9) into the radiation theory has been carried in the well known papers on the atomic or molecular states analogous to the "Trojan asteroids" circulating around the Lagrange points of the rotating potentials [8–12]. The indefinite evolution generators G in the rotating frames were then introduced into the spontaneous emission formulae (as it seems, in hope to reduce the problem of the radiation theory to the time-independent Hamiltonians). Yet, some doubts did not disappear. While the "Hamiltonian" (8) explains reasonably well the motion of the rotating states, the problem remains whether one can extrapolate still further, treating G as a key to determine the quantum jumps, emission rates, lifetimes, etc [6, 7, 13–15].

The formal experiment adopting this idea turned even more interesting than intended, since it was a step to test the Quantum Field Theory (QFT) "in the rotating frames" [8, 10, 12]. The results at this stage were not yet conclusive, since the quadratic forms of the Hamiltonians around the "Lagrange wells" were just an approximation, so the effects of the "inverted oscillators" could not be checked for arbitrary field intensities. Indeed, the rotating field $\mathcal{E}$ in the "Trojan models" can be neither too strong nor too weak, otherwise the system goes out of the stability zone and the bound states disappear. Moreover, even for adequate $\mathcal{E}$ the tunnel effect presents a notable competition, making it difficult to check the exact contribution of G to the supposed radiative decay. However, the problem admits some simple toy models, which facilitate conclusions.

To avoid complications, we applied the same methods [8–10, 12] to a more elementary case which shares some properties of Trojans [10] but is free of the ionization and tunnel effects. We refer simply to the 2D harmonic oscillator in the rotating electric field

$$H(t) = H_0 - e\mathcal{E}\mathbf{n}(t)\mathbf{x}, \tag{10}$$

where $H_0 = \frac{\mathbf{p}^2}{2} + \nu^2 \frac{\mathbf{x}^2}{2}$ and $\mathbf{n}(t) = (\cos\omega t, \sin\omega t, 0)$ is a unit vector rotating in the x, y-plane. Quite obviously, $H(t)$ has the form (5), so if $t_0 = 0$, the transformation to the rotating frame leads to a constant generator:

$$G = H_0 - e\mathcal{E}x - \omega(xp_y - yp_x). \tag{11}$$

The peculiarity of (11) shows up already at this point. Indeed, an elementary algebraic argument (c.f. [6]) reduces G to a pair of 1D oscillators:

$$G = \omega_+ A_+^\dagger A_+ + \omega_- A_-^\dagger A_- + \nu + \frac{e^2\mathcal{E}^2}{2(\omega^2 - \nu^2)}, \tag{12}$$

where $\omega_\pm = \nu \pm \omega$ and $A_\pm$ are given by

$$A_\pm = \frac{1}{2\sqrt{\nu}}\left(\nu x \pm i\nu y + ip_x \mp p_y \mp \frac{e\mathcal{E}}{\omega \pm \nu}\right), \tag{13}$$

satisfying the commutation rules

$$[A_-, A_+] = [A_-^\dagger, A_+] = 0, \quad [A_\pm, A_\pm^\dagger] = 1, \tag{14}$$

(compare [6, 10, 12]). The generator G is singular at $\omega = \nu$, due to the interaction between the rotating frame and quantum trajectories rotating in two opposite directions. Suppose, we use it for $\omega < \nu$ and $\omega > \nu$. For $\omega < \nu$, both ladders are just softly modified. However, if $\omega > \nu$, then $\omega_- < 0$ and the second oscillator is turned upside down, producing the top state instead of the ground state, together with an infinite ladder of negative eigenvalues. By adopting G as a criterion for the radiation (see [8, 9, 12, 16]) and the standard rules, one might be tempted to infer that the *ground-top* state of (12) is unstable and must fall down spontaneously to the first negative level. Such is indeed the conclusion in the works on the rotating systems [8, 10, 11]. Yet, the story does not end up so easily. If the argument is right, then the spontaneous transitions to the subsequent negative levels should occur, producing a radiative avalanche described humorously by Roy Glauber: "*To be in actual possession of an inverted oscillator would be a fantastic thing: it would solve world energy problem*" [17].

A tempting hypothesis is that (10-13) can indeed produce a radiative cascade at the cost of the rotating field (10) pumping the energy to the system (compare [18]). However, the idea does not apply. To show this it is enough to take the limit $\mathcal{E} \to 0$ (a test counterproductive in the "Trojan case" [8, 10], but relevant in ours). One therefore has

$$A_\pm = \frac{1}{2\sqrt{\nu}} \left(\nu x \pm i\nu y + ip_x \mp p_y\right). \tag{15}$$

So, after an elementary calculation

$$A_+^\dagger A_+ = \frac{1}{2\nu}(H_0 - \nu) - \frac{1}{2}M_z, \tag{16}$$

and

$$A_-^\dagger A_- = \frac{1}{2\nu}(H_0 - \nu) + \frac{1}{2}M_z. \tag{17}$$

Hence

$$G = \omega_+ A_+^\dagger A_+ + \omega_- A_-^\dagger A_- = (H_0 - \nu) - \omega M_z. \tag{18}$$

Here the singularity no longer exists, but if $\omega > \nu$, one of the oscillators is still inverted, with the descending ladder of eigenvalues suggesting the spontaneous system jumps to $-\infty$. If treating seriously the G-levels, it would mean: *NOTHING, if rotates fast enough, produces radiation!*

However, the supposed effect cannot be the result of the energy pumping (no energy can be pumped by vanishing $\mathcal{E}$!) but rather an illusion caused by over-interpreting G.

This is indeed one of few cases where the "formal story" of QFT becomes visibly unreal. In fact, the puzzle disappears after a simple redefinition. It is enough to notice that for $\mathcal{E} = 0$ the eigenstates of the "negative ladder" are simultaneously the eigenstates of the initial H_0 with the increasing, positive energies in eyes of the inertial observer. In fact, the simple calculation for $\mathcal{E} = 0$ with $A_\pm$ given by (15-17) yields

$$H_0 - \nu = \nu \left(A_+^\dagger A_+ + A_-^\dagger A_-\right). \tag{19}$$

So, G and H_0 commute and the "positive" and "negative" eigenstates of G are just a part of the ordinary, positive state ladder of H_0. More explicitly, if

$$|n_+, n_-\rangle = \frac{1}{\sqrt{n_+! n_-!}} (A_+^\dagger)^{n_+} (A_-^\dagger)^{n_-} |0, 0\rangle, \tag{20}$$

then one has

$$G|n_+, n_-\rangle = (n_+\omega_+ + n_-\omega_-)\,|n_+, n_-\rangle, \tag{21}$$

but simultaneously

$$(H_0 - \nu)|n_+, n_-\rangle = \nu\,(n_+ + n_-)\,|n_+, n_-\rangle. \tag{22}$$

The question arises up to what extend one can still adopt the concept of the *rotational frequency shift* [16], supposing that while for the inertial observer (19) the system states fall down to the oscillator ground state, the same process in eyes of the rotating observer looks as the system descent down the negative ladder emitting some modified frequencies? Such seems the hypothesis in [16], where the transitions down the negative ladder is considered as one of stability challenges.

This is certainly not the case in our oscillator model (15-17) where the same process can be observed simultaneously from the inertial and rotating frames for the same (though rearranged) ladders. For the inertial observer all oscillator states must *shrink* falling to the ground state. So, can the same states, in eyes of the rotating observer, expand in one of their canonical pair of variables, when descending down to $-\infty$? This turns impossible: the space extension of the atomic states are exactly the same in the rotating and in the inertial frames! Hence, the quantum states in the rotating frame cannot expand going down, but must also shrink *going up*. This is what indeed happens, as the "negative ladder" is an inverted part of the positive one. This inverts also the conclusion of [16], where the spontaneous transition from the most compact state ψ_0 to the less localized ψ_1 (cf. [16], Eq. (22), p.56), is considered as one of stability dangers. Our hypothesis is that the "negative levels" are not the stability danger but inversely, offer some little help to the system stability.

We heard some comments that the QFT in the rotating frame remains valid for $\omega > \nu$, but then the quantum states jumping on the negative ladder would get rid of their negative energies, by *emitting* the negative energy quanta $E = \nu - \omega$, meaning that they indeed go up, and everything is correct!

Yet, the *negative energy photons* are a significant sacrifice of the common sense, just to preserve the formal shape of the theory. Worse, because if one wants to use the Hamiltonian G for $\mathcal{E} = 0$ and $\omega = \nu$, one thus obtains the rotating frame plagued by zero-energy quanta (a strange concept, tempting the peculiar questions, e.g., how many *0-frequency photons* do we need to describe the static electric field?). It might mean that the QFT in non-inertial frames is only a sort of verbal exercise whose correct understanding requires always the return to the inertial frames. This was indeed noticed by Delande and Zakrzewski [12]. They write: "*Hence, emission of the spontaneous σ^+ polarized photon increases the energy in the rotating frame*" [12], p. 473, col. 2, lines 3-5 from bottom. With our only comment, that by "energy" they mean the G-eigenvalue.

Of course, our toy model is non-relativistic. Already in special relativity the rotating coordinates would loose their sense beyond $2\pi r \geq \frac{c}{\omega}$, see the discussions by Padmanabhan [19] and Earman [20]. However, it is precisely our oversimplified case, permitting to observe simultaneously the behaviour of quantum system in the inertial and non-inertial frames, which may illustrate some other interpretational problems.

3. Rotating $\mathcal{E}$ as seen in the inertial frame

Hence, what about the rotating $\mathcal{E} \neq 0$ observed from the inertial frame? Basically, it is a member of a wide family of the quadratic systems described by the Floquet Hamiltonians [21]. Yet, in our case, the evolution equation (1) can be exactly solved for any time limits, giving explicitly the unitary operators $U(t_1, t_0)$ and the effective Hamiltonians $F(t_1, t_0)$. This can be conveniently done in the interaction instead of rotating frame:

$$U(t_1, t_0) = e^{-i(t_1 - t_0)H_0} V(t_1, t_0) \tag{23}$$

obtaining $V(t_1, t_0) = \exp[-iK(t_1, t_0)]$, where $K(t_1, t_2)$ is linear in $\mathbf{x}, \mathbf{p}$-operators, with coefficients expressed through elementary trigonometric integrals. Now, if $\omega \neq \nu$, the traditional

Baker-Campbell-Hausdorff formula with the "polarization derivative" [22], or else, the techniques of Gilmore [23] permit to write down $U(t_1, t_0)$ explicitly in form (2) with $F(t_1, t_0)$ given in terms of $\tau = (t_1 + t_0)/2$ and $\delta t = t_1 - t_0$ [24]:

$$F(t_1, t_0) = \frac{\mathbf{p}^2}{2} + \nu^2 \frac{\mathbf{x}^2}{2} - \frac{\xi}{\sin \frac{\nu}{2}\delta t}\left[a\nu(\cos\omega\tau x + \sin\omega\tau y) + b(\sin\omega\tau p_x - \cos\omega\tau p_y)\right] \tag{24}$$

where $\xi = e\mathcal{E}/(\omega^2 - \nu^2)$ and

$$a = \nu \sin\frac{\nu}{2}\delta t \cos\frac{\omega}{2}\delta t - \omega\cos\frac{\nu}{2}\delta t \sin\frac{\omega}{2}\delta t, \qquad b = \nu\cos\frac{\nu}{2}\delta t \sin\frac{\omega}{2}\delta t - \omega \sin\frac{\nu}{2}\delta t\cos\frac{\omega}{2}\delta t.$$

For $\delta t \to 0$, (24) reproduces the instantaneous Hamiltonian $H(\tau)$, while for $\delta t = T$, it yields the effective Hamiltonian for the entire period T:

$$\begin{aligned} F(t_0) &= \tfrac{1}{2}\left[(p_x + \omega\xi\sin\omega t_0)^2 + (p_y - \omega\xi\cos\omega t_0)^2 \right.\\ &\qquad \left. + \nu^2\left((x - \xi\cos\omega t_0)^2 + (y - \xi\sin\omega t_0)^2\right) - f_0\right], \end{aligned} \tag{25}$$

with $f_0 = \xi^2(\omega^2 + \nu^2)$. Differently than $G(t_0)$, the $F(t_0)$ is a *continuous extension* of $H(t)$. The divergence of ξ for $\mathcal{E} \neq 0$ and $\omega \to \nu$ reflects the fact that the electric field rotating with the oscillator frequency ν must cause the resonance of the quantum mechanical system. Besides, all illusions disappear. There are no negative oscillators, no ladders of eigenvalues descending to $-\infty$, no radiative avalanches precipitated by violently rotating but vanishing fields. Note that the motion integral (24) was known for the forced oscillators (see Brewer and Holthouse [25]), but apparently, without appreciating its possibilities to construct the alternative Floquet Hamiltonians for the rotating case.

It seems to confirm the hypothesis that for the non inertial frames some concepts split: while the "non inertial Hamiltonians" describe well the unitary evolution, they might falsify the radiation effects which require different generators, in our case (25) rather than (12-13); bad news for too formal radiation theories though good news for the stability of the trapped states.

In fact if such problems exist even for the uniformly rotating frames, this implies a warning about more general cases of quantum theory in non-inertial frames and/or relativistic space-times. One of them is still intensely discussed.

4. Unruh radiation

In his inspired 1973 study Fulling [5] wondered about the unclearly defined concepts of QFT leading to the peculiar forms of the creation and annihilation operators in non-inertial reference frames. If one deduces the annihilation and creation operators of quantum fields from the positive and negative frequency components of their classical equivalents (the definition automatically adopted in all QFT versions), then the creation/annihilation operators in the non-inertial frame do not coincide with those of the inertial Minkowski frame, but are altered by linear transformations of Bogolubov type. As it seems it exhibited one of weak points of the too verbal formulation of QFT. Indeed, why the "quantization" should follow the calligraphic formulae of field energy in terms of "frequencies"? Moreover, why the physical observer whose space-time trajectory is not temporal straight line, comparing the frequencies of the hypothetical external fields to his proper time, should feel surrounded by the clouds of real (unreal?) quanta?

The authors interested in [5] started to analyse the (physically peculiar) case of quantum systems undergoing a constant, never-ending acceleration with trajectories limited to one of Rindler sectors in the Minkowski space-time. In his interesting paper Davies [26] wondered about the capacity of the mirror in this universe to produce the real quanta and about the questionable aspects of Hawking theory on the black hole evaporation. However, it seems that the initial doubts almost disappeared in an intense trend of publications.

In his 1976 study Unruh adopts *bona fide* the non inertial quantization in Rindler coordinates [27]. While considering his elementary example of an endlessly accelerated detector, c.f. [27], Eq. (3.1a), he presents a kind of allegoric picture (see [27], p. 885):

"What the detector regards as the detection (and thus absorption) of a Φ-quantum, the Minkowski observer sees as the emission by the detector of such a quantum. The energy of this emission (...) comes from the external field accelerating the detector".

In his further arguments, assuming a coupling constant ϵ between the quantum detector and the massless field Φ in the polarized Minkowski vacuum, Unruh concludes that the result is *what one would expect of a detector in a thermal bath of photons of the temperature* $\frac{a}{2\pi}$. The authors interested in the idea (see, e.g. [28, 29]) expressed it in a more complete form

$$T = \frac{a\hbar}{2\pi ck}. \tag{26}$$

The consistency of units on both sides of (26) is notable! However, the provocative argument about the absorption–emission equivalence in different frames, even if shaking, is not entirely clear. At least, some elements of the hypothesis are absent and they seem essential for the precise understanding of the assumed phenomenon.

In the collection of papers, interpreting (26) a lot of authors simplify the problem introducing a hypothetical point-like Unruh-De Witt detector (in the simplest case, just a spin $\frac{1}{2}$ particle). Sharing the accelerated motion, it "perceives" the bath of virtual quanta of the polarized Minkowski vacuum invisible for any inertial observer. In what follows, we shall be interested not so much in the thermal effect (which can occur for various reasons, without excluding the macroscopic propulsion), but in the first place, in the specific doctrine of the accelerated detector which "sees" (and is bombarded by) some virtual quanta of Minkowski vacuum. Here, some details are still missing. We shall make a preliminary list, starting from apparently minor points.

(1) Can the detector be truly microscopic (e.g. a micro particle which should "click", consistently with Unruh-De Witt idea)? An idealization seems excessive, even if it might avoid some troubles. Indeed, *what does it mean* that the point-like particle "clicks"? The assumption seems to ignore the nature of quantum detection acts. We do not think that the existing QM interpretation is sacred, but if one tries to predict some effects in *frames of the existing theory* then one should respect its assumptions. They tell us that in order to detect anything, the detector should produce some macroscopic (or at least mesoscopic) effects. So, it should have some macroscopic or mesoscopic parts. If not, the detector will never click. It will just continuously evolve, obeying the quantum mechanical unitary propagators.

(2) Yet, the avalanche of papers is dominated by the maximally simplified model of the point-like detectors. Is this unavoidable? Can the supposed Unruh radiation affect the macroscopic objets? Assuming for a moment that such radiation indeed exists, the answer must be: yes!. A shortcut statement was offered by Unruh himself: "it is real enough to roast a steak" (see also the polemic discussions of Scully et al [30–32]) So, the steak (or at least its molecules) is the correct detector of the hypothetical Unruh bath. All this does not yet explain everything about the macroscopic detectors; compare the description of the "thermometer" in [28, 29], or else, the critical study of Narozhny et al [33–35] who write: "*...its treatment in the literature has led to contradictory results. (...) This is because different points of a finite size body rigid with respect to Rindler coordinate frame, move actually with different accelerations in MS*". So, was it the reason to restrict the story to the point-like detectors?.

Suppose, however, one is ready to accept some slight errors of the acceleration a in (26). The discussion, anyhow, cannot escape the next questions.

(3) One of the original Unruh models of an extended detector is a cabin moving with a constant, linear acceleration. But if cabin, then (forgive us this extreme vision) why not the spaceship? Moreover, to protect the story from being too narrow, one should also ask, whether the detection must consist in particle *absorption*? In the real labs (inertial or not) it is certainly not so. The particle can run in some sensitive medium like the bubble chamber or a block of photographic emulsion, leaving a visible trace but without being absorbed. Suppose then that a cabin of an accelerated spaceship (the laboratory) contains a sensitive blocks of photographic emulsion. Is it forbidden to expect that the particles of the Unruh bath (which can roast the steak) will mark some traces in the block of emulsion? If not forbidden, then some more questions inevitably follow.

(4) In fact, suppose in an accelerated lab there is one detector ready to *absorb* particles, inside of a block of photographic emulsion, ready to *register* the trajectories. Suppose now, there are two observers watching the same phenomenon. One is a physicist of the laboratory, sharing the accelerated motion. Another one is an inertial observer in the second spaceship accompanying the accelerating one in a little fragment of its trajectory. Suppose moreover, that both spaceships have the transparent windows: so the inertial observer, (who cannot see directly the virtual Minkowski quanta), will anyhow notice the macroscopic traces in the accelerated laboratory. According to the doctrine, he must see them running in the opposite direction: not *absorbed* but *emitted*, not *to* but *out* of the detector! This ambivalence, though, is not easy to sustain. It could be true only if the *time* of an accelerated system was running backward with respect to the inertial time, which is not the case neither for the rotating nor for accelerating frames.

A chance to mask the difficulty (saving the inspiration of [27]) would be to assume that the photographic emulsion shares the fate of the beefsteak, absorbing such an avalanche of the bath particles that any "trajectories" disappear in the radiation chaos. However, we do not think that this is a good answer. The formula (26) shows an extremely delicate effect, with bath particle seldom achieving high energies sufficient to paint the macroscopic traces, so should such effect occur, it is difficult to assume that it will be masked by some general trajectory chaos.

(5) Moreover, must the Unruh particles be massless? The original work of Fulling [5] on Klein-Gordon fields does not exclude particles with non-vanishing rest mass. Now, while in our present day experiments there may be still some doubts about the emission moments of some very tiny particles (e.g. neutrinos) in our earthly (non-inertial!) laboratories there seem to be no danger of mistaking the emission and absorption acts for heavy bodies. Yet, one of characteristic aspects of the publication trend on Unruh radiation is that (with some exceptions) it focuses attention on the massless scalar fields. May it be some (subconscious?) resistance against slow and massive particles which do not obey the image of delicate, almost not registrable trajectories? If so, even this cannot help, since the high energy photons mark also the macroscopic traces! Moreover, on p. 886, top of col. 2, in his paper [27] Unruh writes: "*...one could also say that the accelerated proton has detected one of many high energy neutrinos which are present in the Minkowski vacuum in the proton's accelerated frame of reference*". The model of particle bath containing neutrinos is repeated in all papers explaining proton decay observed in the accelerated frame. However, if neutrinos, then why not other massive particles, like nucleons, atomic nuclei, or even tennis balls? Can their flight direction be opposite in different frames? Certainly not in the Rindler's frame used to formulate the bath hypothesis.

(6) We must conclude, that the supposed change of the flight direction of the hypothetical Unruh particles under the transition from the inertial to non-inertial frame might be allegoric but not real. It is simply not possible that the bath particle falls *into* the detector in eyes of one observer, but is *emitted* in eyes of another. So, if the detector reacts at all, both observers must either see the absorption, or the emission of the detected particle. Now, since for the inertial observer the virtual particle of Minkowski vacuum are undetectable, both can only

detect the same particle emission by the accelerated detector, though looking slightly different in coordinate of both observers. (Similarly, as in our naive example of the rotating frames, to distinguish between *emission* and *absorption*, only the inertial observer has a good key) As it seems, this simple argument makes stronger the conclusion of Earman, who writes carefully: "*Of three approaches to the Unruh effect* (...) *the least helpful is* (...) *of thermal flux of Fulling quanta*", see [20], Ch.8, p.93. See also the transparent analysis by Padmanabhan [19, 36].

However, in some works the subject acquires an almost esoteric sense. So, the caloric bath particles are no longer virtual Minkowski quanta (as initially suggested by Unruh), but they are something else (Fulling particles?). If massive, they would be anyhow invisible, since they remain sharply localized, below the Planck distance, etc... However, a careful analysis of some interpretational dangers typical for quantum theory in sub-manifolds of Minkowski spacetime were already discussed by Padmanabhan who notices: "*...we extend our analysis to non uniformly accelerated frames* (...) *Here we reach the surprising conclusion that no ambiguity arises in the particle definition. Thus, physically realizable detectors do respond to the Minkowski definition of particles*" [36], p.248. See also the short and transparent notes in [37]. As if it was not enough, some descriptions of the supposed "thermal bath" reveal that the detector will be affected by an infinite numbers of 0-energy photons. In spite of all efforts, this is a sign that the formalism entered into an illusionary domain. The applied "regularization" deserves perhaps some artistic prize, but we doubt that it will convince the detector! (reported in [38], p. 20).

5. The thermal effect?

Our critical remarks still do not explain the sense of the formula (26). In the stream of publications an implicit idea seems to be that if only the existence of (an extremely weak) thermal effect in the accelerating bodies can be experimentally proved, it will justify the Unruh discovery, even if not interpreted exactly according to his hypothesis. Indeed, as one can notice, the two parts of the original Unruh argument [27] "*What the detector regards as the detection (and thus absorption) of a Φ-quantum, the Minkowski observer sees as the emission by the detector of such a quantum*", and "*The energy of this emission* (...) *comes from the external field accelerating the detector*" are rather in narrative than logical connection. They express a kind of provocative idea, which made the hypothesis so widely discussed, but... it can still happen, that a questionable theory might help to guess the true result! In what follows, we henceforth try to review various points of view about the caloric effect (26) without rejecting a priori the "detector mysteries", as the unknown coupling with the vacuum scalar field Φ or with forces maintaining the acceleration.

As far as we could find, the hypothesis about the thermal effect appeared originally for the physical bodies moving with a constant linear acceleration, on the trajectories restricted to one of Rindler wedges in the Minkowski space-time. The information signals available on such trajectory are as well restricted by the Rindler horizons. The most natural hypothesis was that the response of the detector depends on its internal structure. The simplest assumption about an interaction between the detector with an outside field Φ was described in the first Unruh paper [27], in terms of a coupling constant ϵ, which after the 2nd order perturbation steps appears in the expressions for the transition probabilities (see e.g. formulae (3.2-4) on p. 884, then (3.10), (3.12), (3.14) p.p. 885-6). As stated by Unruh, if $\epsilon = 0$, the detector is not coupled with Φ and the caloric effect will not materialize. However, Unruh skips some details. He only explains that (if $\epsilon \neq 0$), the detector reacts "as if it was submerged in the particle bath" of temperature T. Quite similarly, Grove and Ottewill introduce the coupling constant λ [39]. Yet, they do not explain why the final thermal effect (26) does not depend on the coupling constants.

The only explanation one can imagine is that the temperature (26) is independent of any detector, it is an abstract aspect of the Minkowski polarized vacuum limited to the Rindler wedge, though the consequences will materialize only when the detector persists moving with a constant acceleration a and will finally achieve the thermodynamical equilibrium state, which may happen after shorter or longer time depending on the value of ϵ or λ (see alfo Ford and Connel [40]). In some cases, though, the calculation reveals that the equilibrium requires the time comparable with the age of the universe (see [38], p. 823, col. 2, down). How nice: so, shall we have a chance to observe the detector still before the thermal equilibrium?

This explanation was critically analysed in ample studies by Earman [20], and by Narozny et al [33,35]. What cannot be understood is that a detector in any instant of its existence, with an acceleration a (which at any moment can change), should "feel" the properties of the abstract *polarized vacuum* in an entire, infinite Rindler zone. This may look more natural, if the detector follows forever one of killing trajectories from $-\infty$ to $+\infty$. Even so, it is incomprehensible, how in any moment of its existence, it can know, that his acceleration will not change in the future? As humorously expressed by Earman, it is an assumption "*...which requires in turn a knowledge of entire past and future of the observer; but barring backward causation, no laboratory registration taken at a finite time is sensitive to what the instrument does in the future*". Quite obviously, the assumption about a detector moving with a constant, never-ending acceleration is of Science-Fiction type, and *cannot be realized as a trajectory of any physical object* [33].

Some attempts of obtaining more credible scenario were made by assuming that the detector was switched on in the past and switched off in the future. If switched on/off violently, the discontinuous jump was falsifying the expected result (26) (see ample reports in [38]). The desired formula (26) for T was obtained in the papers assuming the very slow, adiabatic switching on the detector in the distant past and switching off in the distant future. However, if the "detector" is an electron, can somebody explain, e,g., what does it mean that the electron is switched off adiabatically? Does it mean that its coupling constants ϵ, λ, etc, will die out? And again, how the electron will know that it will be switched off, not so soon, but in a distant future? As it seems, what can be switched is not the detector itself (electron?) but simply the accelerating mechanism. This is, however, a different story, which leads us naturally to the spaceship model.

6. The cosmic travel?

A realistic trajectory of a space-ship in the Minkowski spacetime can have indeed some intervals of constant acceleration but they do not need and cannot continue indefinitely. In some moments the traveling laboratory might change its acceleration or even recover an inertial (straight) world line e.g. to save the fuel (we apologize for using so elementary argument!) The lines $x = \pm ct$ then cease to be horizons, there is nothing special in the coordinate center $x = t = 0$ and the fine mathematical problems in the Rindler wedge no longer matter. Hence, the collections of papers with attention focused on the "distinguished point", inviolable horizons, etc, seem to be trapped in too narrow and unphysical visions. In the ample literature on the subject one can also find some efforts to define the correct "asymptotic conditions" in the past and in the future, under which the conclusion about acceleration heating could become correct. However, why the effect (if real) should depend on any asymptotes in the future?

Of course, the travelling detector can change his trajectory in quite capricious way. One may thus wonder whether the supposed Unruh radiation, if any, should not produce some strictly local phenomena, bursts of radiation from the "detector's eye" caused by instants of accelerating force, having not much to do with the asymptotic behaviour, and neither responding to the universal "radiation bath"?

A mathematically elegant attempt to describe the finite-time analogue of Unruh phenomenon was undertaken by Rovelli and Martinetti [41] but apparently, succeeded only to stress the difficulty. The authors describe a hypothetical detector conserving his constant acceleration in a certain final interval of its proper time. By applying a global view, they describe the "detector" perceptions during his entire *lifetime* (meaning the constant acceleration interval between certain time moments τ_0 and τ_1).Using the theory of local observables [42] they assume that the detector, in principle, interacts with the events localized in a "diamond", i.e. the spacetime region (τ_0, τ_1) between the future cone of the initial point $P(\tau_0)$ and the past cone of the end point $P(\tau_1)$.

As the result, they obtain the graphic of the"inverse temperature" $\beta(\tau)$, symmetric with respect to the center of the diamond's time. Suppose, we take the derivation *bona fide*. The value of the parameter $\beta(\tau)$ is symmetric and maximal (corresponding to the Unruh effect) in the center of the interval, i.e. in the middle of the accelerated trajectory (see [41], Figure 1). Let us suppose that β is measurable. The problem with causality is then quite evident. Even if not reading exactly his future, the intelligent detector approaching the middle of the $\beta(\tau)$-graphic could discriminate at least same *scenarios* which can never happen: for instance, that the acceleration of his vehicle just in the next few seconds will come to an abrupt end, e.g., due to an accidental defect of the accelerating motor. This is not everything, however.

In all papers on the non-inertial detectors registering the virtual quanta invisible in the inertial frame, we never noticed an inverse problem: that this can happen only at the cost of overlooking some *real quanta*, visible for the inertial observers. In fact, if the traditional annihilation operators a_j are given by the linear (Bogolubov) transformations of their Rindler equivalents, i.e.,

$$a_j = \int_{-\infty}^{\infty} dkU(j,k)b_k + \int_{-\infty}^{+\infty} dkV(j,k)b_k^\dagger$$

see [5], formulae (30) and (40)), where $V((j,k)$ is a non-trivial function, describing the probability of observing the Minkowski vacuum as the bath of Unruh particles. Then in the equivalent formula

$$a_j^\dagger = \int_{-\infty}^{+\infty} dk\bar{U}(j,k)b_k^\dagger + \int_{-\infty}^{+\infty} dk\bar{V}(j,k)b_k$$

the same function $\bar{V}(j,k)$ will express the probability of perceiving the physical $a^\dagger|0>$ particle as the physical vacuum. So, should the detector be bombarded by a sequence of real particles with the creation operators $a_j^\dagger$ there will be always the probability (however little) that they will all be unperceived by an accelerating detector! However, could such a chance exist also if the detector (spaceship) collides with massive beam of heavy, charged particles (e.g. protons), which must destroy it in eyes of of any inertial observer... but could it remains invisible in the accelerated frame? What about the macroscopic charge and mass values under the transition from an inertial to the Rindler frame?

All this may indicate that, in spite of so many successes (the anomalous magnetic moments, etc.), the "verbal saga" of QFT, at least this time, is overdone. The "non inertial Hamiltonians", while offering a part of the truth, create also some mythologies which cannot be literately taken. Though we are outsiders in the subject, we cannot abstain from citing a fragment of the original Fulling work [5] (even if Fulling himself changed later his point of view). He writes about the supposed particle bath: "... *One must accept the fact, however, that the quanta of this Fock representation do not correspond to physical particles. One must find something else to play the role of the basic observables of the theory...*" ([5], p. 2857, down the col. 1). Indeed, if there is any vacuum friction opposing the acceleration of microsystems, then even the Nernst ideas of the "zero point radiation" [40, 43–45], in spite of all doubts, are easier to accept than the vision of the particle bath, real in one reference frame and absent in another. Does it mean that an ample trend of inspired publications, indeed an almost superhuman effort of so many authors was on vane? Certainly not. Even if questionable, it has an enormous success, precisely by approaching the applicability limit of the present day theories. After all, Cristobal Columbus did not found his dreamed way to India. Yet, he achieved something else. He just discovered an obstacle.

Acknowledgments
The financial support of Project SIP 20160527 (Instituto Politécnico Nacional, Mexico) is acknowledged.

References

[1] Vadim R 1968 Barbarella
[2] Zeldovich Ya B 1967 *JETP* **24** 1006
[3] Brody D C, Holm C D and Meier D M 2012 *Phys. Rev. Lett.* **109** 100501
[4] Schmutzer E and Plebański J 1977 *Fortsch. Phys.* **25** 37
[5] Fulling S A 1973 *Phys. Rev.* D **7** 2850
[6] Mielnik B and Fernández C D J 1989 *J. Math. Phys.* **30** 537
[7] Fernández C D J 1990 *Acta Phys.Polon.* **B 21** 589
[8] Białynicki-Birula I and Białynicka-Birula Z 1996 *Phys. Rev. Lett.* **77** 4298
[9] Białynicki-Birula I and Białynicka-Birula Z 1997 *Phys. Rev. Lett.* **78** 2539
[10] Białynicki-Birula I, Kaliński M and Eberly J H 1995 *Phys. Rev. Lett.* **75** 973
[11] Białynicki-Birula I, Kaliński M and Eberly J H 1994 *Phys. Rev. Lett.* **73** 1777
[12] Delande D and Zakrzewski J 1998 *Phys. Rev.* A **58** 466
[13] Farelly D, Lee E and Uzer T 1995 *Phys. Rev. Lett.* **75** 972
[14] Ford G W and O'Connell R F 1997 *Physica* A **243** 377
[15] Delgado F and Mielnik B 1998 *Phys. Lett.* A **249** 369
[16] Białynicka-Birula Z and Białynicki-Birula I 1997 *Phys. Rev.* A **56** 3623
[17] Glauber R J 1986 *Proc. Int. School of Coherent Optics (Ustroń, Poland 1985)*, ed Kujawski A (Ossolineum Wrocław) p 9-30
[18] Cohen-Tannoudji C, Dupont-Roc J and Grynberg G 1992 *Atom-Photon Interactions* (New York: J Wiley & Sons)
[19] Padmanabhan T 1985 *Class. Quantum Grav.* **2** 117
[20] Earman J 2011 *Studies Hist. Philos. Mod. Phys.* **42** 81
[21] Hagedorn G A, Loss M and Slawny J 1986 *J. Phys. A: Math. Gen.* **19** 521
[22] Mielnik B and Plebanski J 1970 *Ann. Inst. H. Poincare* A **12** 215
[23] Gilmore R 1974 *J. Math. Phys* **15** 2090
[24] Cruz y Cruz S 2005 Ph.D. Thesis Cinvestav Mexico
[25] Breuer h P and Holthaus M 1991 *Ann. Phys.* **211** 249
[26] Davies P C W 1975 *J. Phys.* A **8** 609
[27] Unruh W G 1975 *Phys. Rev.* D **14** 870
[28] Bell J S and Leinaas J M 1983 *Nucl. Phys.* B **212** 131
[29] Bell J S and Leinaas J M 1987 *Nucl. Phys.* B **284** 488
[30] Scully M O, Kocharovsky V V, Belyanin A, Fry E and Capasso F 2003 *Phys. Rev. Lett* **91** 243004
[31] Hu B L and Roura A 2004 *Phys. Rev. Lett* **93** 129301
[32] Scully M O, Kocharovsky V V, Belyanin A, Fry E and Capasso F 2004 *Phys. Rev. Lett* **93** 129302
[33] Narozhny N B, Fedotov A M, Karnakov B M, Mur V D and Belinski v A 2001 *Phys. Rev.* D **65** 025004
[34] Fulling S A and Unruh w G 2004 *Phys. Rev.* D **70** 04870
[35] Narozhny N B, Fedotov a M, Karnakov B M, Mur V D and Belinski V A 2004 *Phys. Rev.* D **70** 048702
[36] Padmanabhan T 1982 *Astr. Space Sci.* **83** 247
[37] Padmanabhan T 1990 *Phys. Rev. Lett* **64** 2471
[38] Crispino L C B, Higuchi A and Matsas G E A 2008 *Rev. Mod. Phys.* **80** 787
[39] Grove P G and Ottewill A C 1983 *Phys. Rev.* D **16** 3905
[40] Ford G W and O'Connell R F 2006 *Phys. Lett.* A **350** 17
[41] Martinetti P and Rovelli C 2003 *Class. Quantum Grav.* **20** 4919
[42] Haag R 1992 *Local Quantum Physics* (New York: Springer-Verlag)
[43] de la Peña L, Valdes-Hernandez A and Cetto A M 2008 *Am. J. Phys.* **76** 947
[44] de la Peña L, Valdes-Hernandez A and Cetto A M 2009 *Found. Phys.* **39** 1240
[45] de la Peña L, Valdes-Hernandez A and Cetto A M 2010 *Phys.* E **42** 308

Maximally Entangled Multipartite States: A Brief Survey

M Enríquez[1,2], I Wintrowicz[3,4] and K Życzkowski[3,5]
[1]UPIITA, Instituto Politécnico Nacional, Av. IPN 2580 CP 07340 México D.F., México
[2]Departamento de Física, Cinvestav, AP 14-740, 07000 México DF, Mexico
[3]Institute of Physics, Jagiellonian University, Kraków, Poland
[4]Department of Theoretical Physics, University of Lodz, Łódź, Poland
[5]Center for Theoretical Physics, Polish Academy of Sciences, Warszawa, Poland
E-mail: menriquezf@fis.cinvestav.mx, iwona@merlin.phys.uni.lodz.pl, karol@tatry.if.uj.edu.pl

Abstract. The problem of identifying maximally entangled quantum states of a composite quantum systems is analyzed. We review some states of multipartite systems distinguished with respect to certain measures of quantum entanglement. Numerical results obtained for 4-qubit pure states illustrate the fact that the notion of maximally entangled state depends on the measure used.

1. Introduction

Multipartite entangled states represent a potential tool in several tasks of quantum information, quantum teleportation, error correction codes, among others [1, 2]. Therefore, the study of entanglement properties of such states is essential and recently has been a field of intense research [1,3–7]. In the case of a two qubit systems the famous Bell state $|\Psi_2^+\rangle = \frac{1}{\sqrt{2}}(|00\rangle + |11\rangle)$ is *maximally entangled*, as the reduced state is maximally mixed, $\mathrm{Tr}_A|\Psi_2^+\rangle\langle\Psi_2^+| = \mathbb{1}_2/2$. The same property is typical to the entire orbit of locally equivalent states, $(U_A \otimes U_B)|\Psi_2^+\rangle$, with $U_A, U_B \in U(2)$. In a similar way, for any bipartite quantum system consisting of two subsystems with N levels each there exists a generalized Bell state,

$$|\Psi_N^+\rangle = \frac{1}{\sqrt{N}} \sum_{i=1}^{N} |i\rangle_A \otimes |i\rangle_B, \tag{1}$$

such that the reduced density matrix is maximally mixed, so its entanglement entropy is maximal, $E(|\Psi_N^+\rangle) = \log N$. It is also well justified to call it the *maximally entangled* state, as also other measures of bipartite entanglement achieve its maximum for this very state.

The situation changes for systems consisting of three or more subsystems. There exist several measures of multipartite entanglement and it happens that a given state is most entangled with respect to one measure, but other entanglement measure achieves its maximum for some other state. For instance, among all three-qubit pure states the genuine three-party entanglement measured by three tangle [8] is largest for the state $|GHZ\rangle = \frac{1}{\sqrt{2}}(|000\rangle + |111\rangle)$

while the two-tangle τ_2 and the persistence of entanglement [9] is has its maximum for the state $|W\rangle = \frac{1}{\sqrt{3}}(|001\rangle + |010\rangle + |100\rangle)$.

The aim of this work is to present a brief review of multipartite entanglement and to discuss possible options of selecting distinguished states which maximize some entanglement measures. The problem of identifying multipartite maximally entangled states can be approached from different perspectives. For instance, we can analyze the entanglement entropy of a multipartite state decomposition in a product basis minimized over all local unitary transformations. Another way to describe entanglement in a composite systems is considering some of its bipartitions and then take the average of certain entanglement measure. This last notion is related to concepts like maximally multipartite-entangled state (MMES) [10], k-uniform states [2] and absolutely maximally entangled (AME) states [6, 7].

This contribution is organized as follows, in Section 2 the minimal decomposition entropy is introduced as a multipartite entanglement measure. Besides, a numerical search for three-qubit maximally entangled states with respect this measure is accomplished. Section 3 is devoted to describe entanglement properties by analyzing some bipartite reductions of a composite system. The notion of absolutely maximally entangled states is given in Section 4. Its connection with orthogonal arrays, multiunitary matrices, mutually orthogonal Latin squares is briefly discussed as well. In Section 5 a list of some maximally entangled states and their minimal decomposition entropy is presented. Besides, we discuss some numerical results obtained for four-qubit random pure states and compare several entropy-based measures of entanglement. Finally, some conclusions are given in Section 6.

2. Multipartite quantum states and its decomposition entropy

Consider the general case of a pure state $|\psi\rangle$ of a quantum system composed of K subsystems with N levels each. Let us write it in a product basis

$$|\psi\rangle = \sum_{i_1=1}^{N} \cdots \sum_{i_K=1}^{N} C_{i_1,i_2,\dots,i_K} |i_1\rangle \otimes |i_2\rangle \otimes \cdots \otimes |i_K\rangle \tag{2}$$

and discuss, how to quantify its entanglement? The tensor C is described by N^K complex entries normalized as usual, $\sum_{i_1=1}^{N} \cdots \sum_{i_n=1}^{N} |C_{i_1,i_2,\dots,i_K}|^2 = 1$. If all of them but one are equal to zero, the state is separable. Of course the converse does not hold, since any separable state written in a generic product basis has several non-zero components. Thus it is natural to use local unitary transformations, $U_{\text{loc}} = V_1 \otimes \cdots \otimes V_K \in U(N)^{\otimes K}$, and to look for a distinguished product basis [11].

It will be convenient to use a multi-index $\mu = (i_1, i_2, \dots, i_K)$, with $\mu \in \{1, \dots, N^K\}$, and to introduce a normalized probability vector, $p_\mu = |C_\mu|^2 = |C_{i_1,i_2,\dots,i_K}|^2$ of length N^K. As usual it can be characterized by the Rényi entropy

$$S_q(\vec{p}) = \frac{1}{1-q} \log \left(\sum_{\mu=1}^{N} p_\mu^q \right). \tag{3}$$

For $q \to 1$ this quantity reduces to the standard Shannon entropy

$$S_1(p) = -\sum_{\mu=1}^{N} p_\mu \log p_\mu, \tag{4}$$

which in the context of the decomposition of a state $|\psi\rangle$, is called the *Ingarden–Urbanik* entropy [12, 13] and written $S^{\text{IU}}(|\psi\rangle) = S\big(p(|\psi\rangle)\big)$. From this point on log stands for the natural logarithm.

Making use of the generalized entropies S_q for any multipartite state $|\psi\rangle \in H_N^{\otimes K}$ we define the minimal *Rényi–Ingarden–Urbanik* entropy [5],

$$S_q^{\mathrm{RIU}}(\psi) \;=\; \min_{U_{\mathrm{loc}}} S_q\left[p(U_{\mathrm{loc}}|\psi\rangle)\right], \tag{5}$$

where the minimum is taken over all local unitaries U_{loc}. Such an approach with $q = 1$ was applied by Bravyi [14], who showed that the *minimal decomposition entropy* $S_1^{\mathrm{RIU}}(\psi)$ characterizes entanglement of any pure state $|\psi\rangle$ and determines the minimal information gained by performing projective measurements in each local basis. For any separable state one can find a product basis such that the probability vector p_μ is pure, so one has $S_q^{\mathrm{RIU}}(\psi_{sep}) = 0$ for any $q \geq 0$. In the case of a bipartite state the optimal basis is given by the singular value decomposition of the matrix C, so the familiar Schmidt vector λ determines the Rényi entropy of entanglement, $S_q^{\mathrm{RIU}}(\psi) = S_q(\lambda)$. However, for $K \geq 3$ subsystems one works with a K-index tensor (2) and in general there are no analytic techniques to find the minimum (5), so one has to rely on numerical methods [5].

The minimal decomposition entropy S_q^{RIU} is by construction constant along the orbit of locally unitary states,

$$|\psi\rangle \underset{LU}{\longleftrightarrow} |\phi\rangle \quad \Rightarrow \quad S_q^{\mathrm{RIU}}(|\psi\rangle) = S_q^{\mathrm{RIU}}(|\phi\rangle) \tag{6}$$

for any $q \geq 0$. The problem of determining whether two multipartite pure states belong to the same LU orbit was solved for qubits by Kraus [15] and later generalized for higher dimensions [16].

Discussing the generalized decomposition entropy S_q^{RIU} let us distinguish some other values of the Rényi parameter, apart of the Shannon value, $q = 1$. The case $q = 0$, corresponding to the Hartley entropy, is related to the rank R of the tensor C –the minimal number of the components in the product decomposition [17],

$$C_{i_1,i_2,\ldots,i_K} = \sum_{\nu=1}^{R} \gamma_\nu \, a_{i_1}^{\nu} \otimes b_{i_2}^{\nu} \otimes \ldots k_{i_K}^{\nu}, \tag{7}$$

involving R products of K vectors combined with arbitrary coefficients γ_ν. The minimal Hartley entropy, $S_0^{\mathrm{RIU}}(\psi) = \log R$, was introduced in [18] to characterize multipartite entanglement and called *Schmidt measure*. The name refers to the bi-partite case, as then R is equal to the Schmidt rank – the number of positive components of the Schmidt vector.

It is known that for any three-qubit state a five term representation exists [19, 20], so in this case $R \leq 5$. In the general case of $|\psi\rangle \in H_N^{\otimes K}$ Carteret, Higuchi and Sudbery have shown [11], that a suitable choice of each of K local unitary matrices $V_j \in U(N)$ allows one to bring $N(N-1)/2$ entries of the tensor C to zero. This implies a simple upper bound for the rank of the tensor,

$$R \leq R_{\max} = N^K - KN(N-1)/2, \tag{8}$$

and for the minimal Rényi decomposition entropy,

$$S_q^{\mathrm{RIU}}(\psi) \;\leq\; S_0^{\mathrm{RIU}}(\psi) \;\leq\; \log R_{\max}. \tag{9}$$

The first inequality is a consequence of monotonicity of the Rényi entropy with respect to the Rényi parameter q [21].

Minimal decomposition entropy S_2^{RIU}, characterizing the maximal purity of the vector p_μ, was studied in [22] in the context of the coding theory. However, even more important is the

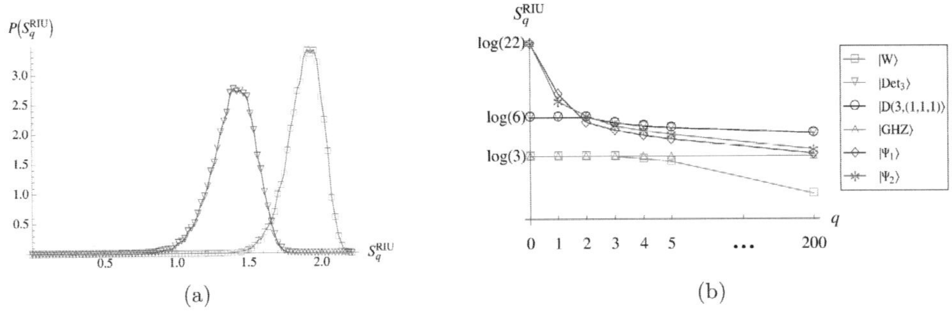

Figure 1: (a) Distributions of the minimal decompositon entropy for random states with $q = 1$ (□) and $q = 2$ (▽) (b) The minimal RIU entropy computed for representative three-qutrit states.

limiting case $q \to \infty$ of the Chebyshev entropy, $S_\infty = -\log p_{\max}$. This entropy characterizes the size of the largest product component, $S_\infty^{\mathrm{RIU}}(|\psi\rangle) = -\log F_{\max}$, where

$$F_{\max}(\psi) = \max_{U_{\mathrm{loc}}} |\langle\psi|U_{\mathrm{loc}}|0\rangle^{\otimes K}\rangle|^2 \tag{10}$$

denotes the maximal fidelity of $|\psi\rangle$ with respect to any separable state. Thus its function $D_{\mathrm{FS}} = \arccos(\sqrt{F_{\max}})$ represents the Fubini–Study distance to the closest separable state $|\psi_{\mathrm{sep}}\rangle$, often used to quantify entanglement [23, 24]. For this purpose one uses also related quantities [25–29], including infidelity, $E_G = 1 - F_{\mathrm{sep}}$, and $E_L = -\log(F_{\mathrm{sep}}) = S_\infty^{\mathrm{RIU}}$, respectively called linear and logarithmic *geometric measures* of quantum entanglement.

Note also that the closest separable state $|\psi_{\mathrm{sep}}\rangle$, provides the principal component [17, 30] of the analyzed tensor (7) distinguished by the coefficient γ_ν with the largest modulus.

For any bipartite case the maximal fidelity (10) is explicitly given by the largest coefficient $\lambda_{\max}$ of the Schmidt vector [26, 31], and can be also calculated for a large class of three-qubit states [32–34]. In other cases one has to rely on numerical optimization, which becomes cumbersome if the number K of subsystems becomes large. Hence it is not easy to to identify the K-party state for which the minimal decomposition entropy S_q^{RIU} is maximal for $q = 1$ or any other value of the Rényi parameter q.

2.1. *Searching for three-qutrit maximally entangled states*

We are interested in identifying three-qutrit states for which the minimal decomposition entropy (5) is the largest. Some results in the case of three and four-qubit states were obtained by analyzing the distributions of S_q^{RIU} for an ensemble of random states [5]. For three-qutrits the corresponding distributions are shown in Figure 1 a) with $q = 1, 2$. Moreover, we take $q = 200$ as a bound for the case $q = \infty$. To optimize the quantity (5) we perform a random walk over the space of unitary matrices. We found two candidate random states to be maximally entangled with respect the minimal RIU entropy, which in the canonical basis read

$$|\Psi_1\rangle = \sum_{j=0}^{26} c_j^1 |j\rangle, \quad |\Psi_2\rangle = \sum_{j=0}^{26} c_j^2 |j\rangle, \tag{11}$$

where the index j is written in base 3 and the coefficients are give by

$$c^1 = (0.193e^{1.7i}, 0.323e^{-2.01i}, 0.16e^{-2.16i}, 0.229e^{-2.22i}, 0.232e^{-3.12i}, 0.186e^{-2.5i}, 0.239e^{-2.34i}, 0.141e^{-0.411i}, 0.159e^{-0.512i}, 0.099e^{1.54i}, 0.144e^{-2.43i}, 0.148e^{2.13i}, 0.263e^{-1.62i}, 0.322e^{0.475i}, 0.216e^{-1.95i}, 0.068e^{-1.39i}, 0.030e^{-2.89i}, 0.171e^{1.91i}, 0.253e^{-2.82i}, 0.022e^{-0.225i}, 0.06e^{-1.2i}, 0.003e^{2.64i}, 0.133e^{-1.52i}, 0.202e^{2.2i}, 0.194e^{1.08i}, 0.207e^{1.13i}, 0.274e^{-2.29i})$$

$$c^2 = (0.245e^{0.074i}, 0.024e^{2.49i}, 0.248e^{1.66i}, 0.069e^{1.55i}, 0.256e^{0.114i}, 0.118e^{-2.88i}, 0.313e^{-1.24i}, 0.076e^{2.77i}, 0.149e^{0.208i}, 0.208e^{2.56i}, 0.227e^{-2.88i}, 0.157e^{2.27i}, 0.072e^{3.08i}, 0.2e^{-1.07i}, 0.199e^{-1.87i}, 0.13e^{-1.95i}, 0.133e^{1.5i}, 0.218e^{-1.68i}, 0.244e^{-1.84i}, 0.191e^{-3.05i}, 0.049e^{2.61i}, 0.144e^{1.22i}, 0.226e^{2.14i}, 0.278e^{-2.46i}, 0.227e^{0.773i}, 0.186e^{-2.11i}, 0.218e^{-1.52i})$$

One can compare the minimal decomposition entropy of the former states with some representative three-qutrit states. For instance, the K-qudit Dicke state, invariant with respect to permutations, reads

$$|D[K, \vec{k}]\rangle = \sqrt{\frac{\prod_i k_i!}{K!}} \sum_{\pi \in S_K} |\underbrace{0\cdots 0}_{k_0}\underbrace{1\cdots 1}_{k_2}\ldots\underbrace{(d-1)\cdots(d-1)}_{k_{d-1}}\rangle. \tag{12}$$

The well-known K-qudit W-state and GHZ-state can be represented as linear combinations of some particular cases of the Dicke states (12)

$$\begin{aligned} |\mathrm{W_K}\rangle &= |D[K, (1, K-1, 0, \ldots, 0)]\rangle, \\ \sqrt{K}|\mathrm{GHZ}_K^d\rangle &= |D[K, (K, \ldots, 0)]\rangle + |D[K, (0, K, \ldots, 0)]\rangle + \ldots + |D[K, (0, \ldots, K)]\rangle. \end{aligned} \tag{13}$$

For the states (12) the maximal fidelity with respect the closest separable state (10) is known [29], hence their decomposition entropy with $q = \infty$ can be computed straightforward

$$S_\infty^{\mathrm{RIU}}(D[K, \vec{k}]) = -\log\left[\frac{K!}{\prod_i k_i!}\prod_{i=0}^{d-1}\left(\frac{k_i}{K}\right)^{k_i}\right]. \tag{14}$$

On the other hand, for the determinant state of N subsystems with N levels each,

$$|\mathrm{det}_N\rangle = \frac{1}{\sqrt{N!}}\sum_{i_1\cdots i_N=1}^{N} \epsilon_{i_1,\ldots,i_N}|i_1\cdots i_N\rangle, \tag{15}$$

Bravyi [35] has reported that $S_1^{\mathrm{RIU}}(\mathrm{det}_N) = \log N!$. On the other hand, it is not so difficult to show that the states $|\mathrm{det}_3\rangle$ and $|D[3, (1, 1, 1)]\rangle$ are LU-equivalent by making the transformations $|2\rangle_3 \to -|2\rangle_3$, $|0\rangle_1 \to -|0\rangle_1$, $|2\rangle_1 \to -|2\rangle_1$ and $|1\rangle_3 \to -|1\rangle_3$, and hence $S_1^{\mathrm{RIU}}(D[3, (1, 1, 1)]) = \log 6$. Fig. 1 b) depicts a comparison between the states (11) with some other three-qutrit states.

3. Partition into two subsystems

Is there any way to describe entanglement of a multipartite state an an explicit way, without performing any awkward optimizations? One simple option is to analyze, under what possible transformations a given state remains entangled. The *persistence* p of entanglement of a K-party pure state $|\phi\rangle$ is defined [9] as the minimal number of local measurements which have to be performed to destroy the entanglement completely. As any measurements of a single qubit transforms the K qubit GHZ state to a product state we have $p(|GHZ_K\rangle) = 1$. On the other hand after a measurement of single qubit the W state reamins entangled, so $p(|W_3\rangle) = 2$ and $p(|W_K\rangle) = K - 1$. Hence from the persistence perspective, the W-state is more entangled than GHZ.

Another manageable way to handle multipartite entanglement is to restrict our attention to the much simpler case of bipartite entanglement. For instance, one can split the K-partite system into M and $K - M$ parties, and use standard techniques to characterize bipartite entanglement, and possibly average over different splittings. This approach enables us to write down a reasonable definition of multipartite entanglement,

$$E_x(|\psi_K\rangle) = \frac{1}{L_K} \sum_A S_x(\rho_A), \tag{16}$$

where $\sigma_A = \text{Tr}_{\bar{A}}|\psi\rangle\langle\psi|$ and the summation goes over all L_K selected partitions of the system into two parties $\{A, \bar{A}\}$. The measure S_x of the bipartite entanglement can be arbitrary, and one often takes the standard von Neumann entropy of the reduced state $S_1(\sigma)$, the Rényi entropy of order two, $S_2 = -\log \text{Tr}\sigma^2$, its function called linear entropy, $S_2^{HC} = 1 - \text{Tr}\sigma^2$, equivalent to the Tsallis entropy of order two, or the Chebyshev (minimal) entropy S_∞.

The simplest choice is to fix the number M of subsystems in the distinguished set A to unity, so then the number of possible bipartite splittings taken into account scales linearly, $L_K = K$. Such a quantity, characterizing the average entanglement of a single subsystem with the remaining part of the system, is called the Mayer–Wallach measure, as these authors [36] applied (16) for any K-qubit state. The measure S_x they used to quantity entanglement occurred to be equivalent [14] to the linear entropy S_2^{HC}. This observation allows us to use this measure also for N-level subsystems and write $Q_1(\phi) = \frac{N}{N-1}\langle S_2^{HC}\rangle$, where the average is taken over all K different splittings, for which $K-1$ subsystems are traced out and the reduced state describes a single subsystem. A suitable prefactor assures that this quantity is bounded by unity, $Q_1^{\max} = 1$, and the maximum is achieved e.g. for the generalized GHZ state,

$$|GHZ_K^N\rangle = \frac{1}{\sqrt{N}}\Big(|1\rangle^{\otimes K} + |2\rangle^{\otimes K} + \cdots + |N\rangle^{\otimes K}\Big). \tag{17}$$

To take into account also entanglement between larger subsystems it is convenient to fix the size of the smaller subsystem, writting $M = |A| \leq K - M$, and to average a suitable entanglement measure E_x over all $L_{K,M} = \binom{K}{M}$ splittings into bipartitions of this size. This approach leads, for instance, to a natural generalization of the Mayer–Wallach measure,

$$Q_M(|\phi\rangle) \ := \ \frac{N^M}{N^M - 1}\left\langle 1 - \left(\text{Tr}_{\bar{A}}|\phi\rangle\langle\phi|\right)^2\right\rangle_M, \tag{18}$$

where the average is taken over all $\binom{K}{M}$ partial traces with respect to environments $\bar{A}$ consisting of $K - M$ parties. By definition this quantity achieves the upper bound $Q_M \leq 1$ if there exists a pure state $|\psi_K\rangle$ such that its M-party reduced density matrix $\text{Tr}_{\bar{A}}|\psi_K\rangle\langle\psi_K|$ with respect to *any* such partition $\bar{A}$ is maximally mixed. Such states are called *M-uniform* [2] or maximally multipartite entangled [10].

An alternative way to characterize the degree of mixing of reduced density matrices can be based on quantum analogue of the classical Rényi entropy (3),

$$S_q(\sigma) := \frac{1}{1-q} \log \text{Tr}(\sigma^q), \tag{19}$$

defined for $q \geq 0$ and $q \neq 1$. It can be considered as a natural generalization of the von Neumann entropy, $-\text{Tr}\sigma \log \sigma$, obtained for $q \to 1$. We are going to distinguish three cases, $q = 1$, $q = 2$

and $q \longrightarrow \infty$, and rewrite the corresponding expressions in terms of the eigenvalues of λ_i of the reduced state σ,

$$S_1(\sigma) = -\sum_i \lambda_i \log \lambda_i \qquad \text{the von Neumann entropy,} \tag{20a}$$

$$S_2(\sigma) = -\log \mathrm{Tr}(\sigma^2) \qquad \text{the Rényi entropy with q=2,} \tag{20b}$$

$$S_\infty(\sigma) = -\log \lambda_{\max} \qquad \text{the Chebyshev entropy.} \tag{20c}$$

Here $\lambda_{\max}$ denotes the largest eigenvalue.

In particular, to characterize the degree of mixing of reductions of states consisting of four subsystems we will use a quantity which depends on the Rényi parameter q and characterizes reductions to one and two-party systems

$$\widetilde{S}_q(\rho) := \sum_{j=1}^{4} \frac{S_q(\rho_j)}{\log 2} + \sum_{k=2}^{4} \frac{S_q(\rho_{1k})}{2\log 2}. \tag{21}$$

Here ρ_j and ρ_{1k} denote reduced density matrices containing one and two particles respectively. Thus the above quantity represents a weighted average over all relevant splittings of the four-party system, and the prefactors $\log 2$ and $2\log 2$ are added to assure that both terms have the same weight, so the value of $\widetilde{S}_q(\rho)$ for any 4-qubit state belongs to the interval $[0, 7]$. For any separable pure state one has $\widetilde{S}_q(|\psi_{\rm sep}\rangle\langle\psi_{\rm sep}|) = 0$, while the extremal value $\widetilde{S}_q = 7$ would correspond to a (non-existing) absolutely maximally entangled state of four qubits. Note that for $q = 2$ the quantity (21) is related to (16) as it corresponds to the sum $Q_1 + Q_2$ of quantities describing one- and two-party reductions defined in (18).

Let us return back to the general case of system composed of an arbitrary number of K systems. Any M-uniform state with $M \geq 2$ is by construction $(M-1)$-uniform, since partial trace of a maximally mixed state is maximally mixed, but the converse does not hold. We already know that the generalized GHZ state (17) is 1-uniform as it maximizes Q_1, but it is not two-uniform. Thus families of M-uniform states form a hierarchy of entanglement. Note that an M-uniform state of a system with K parties may exist if $M \leq K/2$, since any partial trace on a larger subsystem cannot be maximally mixed [2]. It is thus justified to distinguish M-uniform states of a system with $K = 2M$ subsystems, which are called *absolutely maximally entangled* (AME) [37].

To understand implications of these conditions on the general tensor form (2), let us first consider a bipartite state $|\psi\rangle = \sum_{\mu,\nu} C_{\mu,\nu}|\mu,\nu\rangle$ of $N \times N$ system. This state is maximally entangled, and thus 1-uniform, if the partial trace of the projector is equal to the maximally mixed state, $\rho_* = \frac{1}{N}\mathbb{1}$,

$$\mathrm{Tr}_B|\psi\rangle\langle\psi| = \sum_\nu C_{\mu,\nu}C^\dagger_{\nu,\mu'}|\mu\rangle\langle\mu'| = \rho_* \;\Leftrightarrow\; CC^\dagger = \frac{1}{N}\mathbb{1}. \tag{22}$$

This condition is satisfied if the matrix $U = \sqrt{N}C$ is unitary.

4. Arbitrary number of partitions

Consider now a general state (2) of $K = 2M$ subsystems and use composite indices, $\mu = (i_1, \ldots, i_M)$ and $\nu = (i_{M+1}, \ldots, i_{2M})$. Repeating the above argument for the splitting A versus $\bar{A}$ we see that the partial trace with respect to the partition $\bar{A}$ consisting of subsystems $i_{M+1}, \ldots, i_{2M}$ is maximally mixed, if the matrix $C_{\mu,\nu}$ obtained by a suitable reshaping of the rescaled tensor C is unitary. The same argument works for any other choice of M indices defining the composite index μ and the corresponding partition A, so we arrive at the following conclusion [6].

Proposition. A state (2) of $K = 2M$ subsystems with N levels each is M-uniform, and thus *absolutely maximally entangled*, if matrices $C_{\mu,\nu}$ obtained by reshaping the tensor $N^{M/2}C$ into a square matrix of size N^M are unitary for all $\binom{2M}{M}$ choices of M indices out of $2M$, which determine the composite indices μ and ν.

This is the case if the unitary matrix U obtained by reshaping the analyzed tensor remains unitary after performing suitable reorderings of their elements. Such matrices are called *multiunitary*. The corresponding tensors, called *perfect tensors*, are used to construct holographic codes and propose toy models for the bulk/boundary correspondence (anti de Sitter space and conformal field theory) [38].

In the case of a four-party system one has to check condition (22) for three splittings, 12, 13 and 14, which is equivalent to verifying if three matrices, $C_{(1,2)(3,4)}$, $C_{(1,3)(2,4)}$ and $C_{(1,4)(2,3)}$ multiplied by a constant are unitary. Denoting the first matrix U we need to check that its partial transpose U^{T_2} and the reshuffled matrix U^R – see [21] – are unitary. Although there are $\binom{4}{2} = 6$ choices of two subsystems out of four, three other cases will also be covered since as U is unitary so is U^T.

Any familiar Bell state $|\psi_2^+\rangle$ is one-uniform, so it forms the simplest AME state of two qubits. Interestingly, there are no two-uniform, AME states for four qubits – see [39, 40]. In short, the assumption that partions AB, AC and AD of the system $ABCD$ are maximally mixed implies so many constaints that no 4-qubit state can satisfy all of them. In other words there are no multiunitary matrices of order four: if U and $U^{T_2} \in U(4)$ then $U^R \notin U(4)$. This effect can be compared to the *frustration* in a spin system [41]. Larger systems offer more parameters to play with and indeed, for system composed of 4 subsystems with $N = 3$ levels each there exists an AME state.

$$3|\Psi_3^4\rangle = |0000\rangle + |0112\rangle + |0221\rangle + |1011\rangle + |1120\rangle + |1202\rangle + |2022\rangle + |2101\rangle + |2210\rangle. \quad (23)$$

Such a state can be constructed with help of orthogonal arrays [7] or mutually orthogonal Latin squares [6]. The corresponding multiunitary matrix U_9 of order 3^2 is related to a sudoku design, see [6]. There exist also 2-uniform state of 5 qubits [42] and 3-uniform AME state of six qubits [4] and they can be used to design quantum error correction codes. It would be also interesting to study such states for larger systems, as a general question, for what K-partite systems with N levels M-uniform state exists remains open [1, 6, 41].

5. A list of selected highly entangled states

In Table 1 we present a list of some multipartite quantum states that are identified as maximally entangled with respect certain entanglement measures. The catalog itemizes states related to the entanglement criteria previously discussed. In addition, the states are organized with respect their minimal decomposition entropy (5).

5.1. *Two qubits*

In the case of two qubits each one of the Bell states

$$|\Psi_2^{\pm}\rangle = \frac{1}{\sqrt{2}}(|00\rangle \pm |11\rangle), \quad |\Phi_2^{\pm}\rangle = \frac{1}{\sqrt{2}}(|01\rangle \pm |10\rangle), \quad (24)$$

is 1-uniform as the reduced density matrix with respect any of the subsystems is maximally mixed. Moreover, it is not difficult to show that these states are mutually LU-equivalent. On the other hand, for two-qudits the maximally entangled state is given by (1) (including the LU-equivalent states). The minimal RIU entropy (5) in both cases can be computed by means the singular value decomposition of the corresponding coefficient matrices.

State	S_1^{RIU}	S_2^{RIU}	S_∞^{RIU}
Two qubits			
Bell states: $\|\Psi^\pm\rangle$, $\|\Phi^\pm\rangle$	$\log 2$	$\log 2$	$\log 2$
Two qudits			
GHZ-state: $\|\mathrm{GHZ}_2^d\rangle$	$\log d$	$\log d$	$\log d$
Three qubits			
GHZ-state: $\|\mathrm{GHZ}_2^3\rangle$	$\log 2$	$\log 2$	$\log 2$
W-state: $\|\mathrm{W}_3\rangle$	$\log 3$	$\log 3$	$\log(9/4)$
Random "maximal" state: $\|\Phi_1^{\max}\rangle$	1.277	1.020	0.610
Random "maximal" state: $\|\Phi_2^{\max}\rangle$	1.141	1.108	0.727
Four qubits			
Higuchi and Sudbery state: $\|\mathrm{HS}\rangle$	* $\log 6$	* $\log 6$	* $\log(9/2)$
Cluster states: $\|\mathrm{C}_k\rangle$	* $\log 4$	* $\log 4$	* $\log 4$
Random "maximal" state: $\|\Psi_1^{\max}\rangle$	1.934	1.573	0.934
Symmetric state: $\|D[4,(2,0)]\rangle$	$\log(8/\sqrt{3})$	$\log(24/7)$	$\log(8/3)$
$BSSB_4$ state: $\|BSSB_4\rangle$	1.733	* $\log(16/3)$	1.393
Ψ_4 state: $\|\Psi_4\rangle$	1.733	* $\log(16/3)$	* $\log 4$
Hyperdeterminant state: $\|HD\rangle$	1.561	* $\log(9/2)$	1.110
Yeo and Chua state: $\|YC\rangle$	1.602	1.410	1.002
L state: $\|L\rangle$	1.561	* $\log(9/2)$	1.110
GHZ state: GHZ_4	$\log 2$	$\log 2$	$\log 2$
Five qubits			
1-uniform state $\|\Psi_5\rangle$	* $\log 8$	1.716	1.199
AME state $\|\Omega_{5,2}\rangle$	* $\log 8$	* $\log 8$	2.03
Symmetric state: $\|D[5,(2,0)]\rangle$	2.263	* $\log 5$	$\log(216/625)$
Three qutrits			
Random "maximal" state: $\|\Psi_1\rangle$	2.205	1.707	1.1452
Random "maximal" state: $\|\Psi_2\rangle$	2.072	1.82	1.221
Symmetric state $\|D[3,(1,1,1)]\rangle$	* $\log 6$	* $\log 6$	$\log(9/2)$
Four qutrits			
AME state: $\|\Omega_{4,3}\rangle$	* $\log 9$	* $\log 9$	* $\log 9$
Symmetric state: $\|D[4,(2,1,1,0)]\rangle$	* $\log 12$	* $\log 12$	$\log(16/3)$

Table 1: Some multipartite maximally entangled states with respect different measures and their minimal RIU entropy (5). The star indicates that the value has been conjectured from numeric calculations.

5.2. Three qubits

It is known that the 1-uniform 3-qubit GHZ-state is maximally entangled with respect the three-tangle but it does not maximizes the minimal decomposition entropy (5) for none value of the Rényi parameter q [5]. In fact, there exist two candidate random states to be maximal with respect this measure with $q=1$ and $q=2$, which in the computational basis read

$$|\Phi_1^{\max}\rangle = 0.27|000\rangle + 0.377|100\rangle + 0.326|010\rangle + 0.363|001\rangle + 0.740e^{-0.79\pi i}|111\rangle, \tag{25a}$$
$$|\Phi_2^{\max}\rangle = 0.438|000\rangle + 0.29|100\rangle + 0.371|010\rangle + 0.316|001\rangle + 0.698e^{-0.826\pi i}|111\rangle. \tag{25b}$$

The minimal decomposition entropy S_∞^{RIU} attains its largest value value for the W-state, however such a state is not 1-uniform.

5.3. Four qubits

It has been conjectured that for the Rényi parameter $q \geq 2$ the 1-uniform state

$$|\mathrm{HS}\rangle = \frac{1}{\sqrt{6}}\left[|0011\rangle + |1100\rangle + w(|0101\rangle + |1010\rangle) + w^2(|0110\rangle + |1001\rangle)\right], \quad w = \exp(2i\pi/3), \tag{26}$$

is maximally entangled with respect the minimal decomposition entropy [5]. This state also gives the maximal average von Neumann entropy of partial traces averaged over all possible 3 splittings of 4 qubit system into two bipartite systems [40]. For the special case of $q = 1$ the following state obtained numerically with a random walk procedure,

$$|\Psi_1^{\max}\rangle = 0.630|0000\rangle + 0.281|1100\rangle + 0.202|1010\rangle + 0.24|0110\rangle + 0.232e^{0.494\pi i}|1110\rangle$$
$$+0.059|1001\rangle + 0.282|0101\rangle + 0.346e^{-0.362\pi i}|1101\rangle + 0.218e^{0.626\pi i}|1011\rangle \quad (27)$$
$$+0.304|0011\rangle + 0.054e^{-0.725\pi i}|0111\rangle + 0.164e^{0.372\pi i}|1111\rangle,$$

is a candidate to be maximally entangled with respect the minimal RIU entropy [5]. On the other hand, the three cluster states [43]

$$|C_1\rangle = \frac{1}{2}\left(|0000\rangle + |0011\rangle + |1100\rangle - |1111\rangle\right), \quad (28a)$$
$$|C_2\rangle = \frac{1}{2}\left(|0000\rangle + |0110\rangle + |1001\rangle - |1111\rangle\right), \quad (28b)$$
$$|C_3\rangle = \frac{1}{2}\left(|0000\rangle + |0101\rangle + |1010\rangle - |1111\rangle\right), \quad (28c)$$

maximize the Rényi entropy of partial trace for $q \geq 2$ [44]. Numerical computations support the conjecture that for these states the RIU entropy does not depend on the Rényi parameter q [5]. As another example we consider the ground state of the 4-qubit XXX Heisenberg model. This state is locally equivalent to the symmetric Dicke state $D[4,(2,0)]$ and yields $Q_1 = 1$ [36]. Finally, we emphasize the fact that there exist no 4-qubit AME states [6, 39, 40].

The state introduced by Gour and Wallach [44]

$$|L\rangle = \frac{1}{2\sqrt{3}}((1+\omega)(|0000\rangle + |1111\rangle) + (1-\omega)(|0011\rangle + |1100\rangle) + \quad (29)$$
$$\omega^2(|0101\rangle + |0110\rangle + |1001\rangle + |1010\rangle)), \quad \omega = \exp(2i\pi/3),$$

maximizes the measure Q_2 defined in (18), i.e. the linear entropy $\langle S_2^{HC}\rangle$ of the reduced state averaged over three different splittings into two pairs of bi-partite states. Another state introduced in [44] and called $|M\rangle$ there, which is conjectured to yield the maximal value of the average von Neumann entropy of partial traces, is shown [45] to be locally equivalent to the HS state (26).

Another distinguished state

$$|HD\rangle = \frac{1}{\sqrt{6}}(|1000\rangle + |0100\rangle + |0010\rangle + |0001\rangle + \sqrt{2}|1111\rangle), \quad (30)$$

was called the 'hyperdeterminant state', as it maximizes the four-qubit hyperdeterminant $\mathrm{Det}_4(|\varphi\rangle)$ [45]. This important invariant with respect to local transformations is maximized for the above two states, $\mathrm{Det}_4(|L\rangle) = \mathrm{Det}_4(|HD\rangle) = 1$. However, there exist highly entangled states, $|D_{4,1}\rangle, |D_{4,2}\rangle, |C_k\rangle, |HS\rangle, |GHZ_4\rangle$, for which $\mathrm{Det}_4(|\varphi\rangle) = 0$.

The state $|BSSB_4\rangle$ found by Brown et.al [46] by a numerical search,

$$|BSSB_4\rangle = \frac{1}{2\sqrt{2}}(|0110\rangle + |1011\rangle + i(|0010\rangle + |1111\rangle) + (1+i)(|0101\rangle + |1000\rangle)), \quad (31)$$

is highly entangled with respect to measure based on the average linear entropy of bi-partite reductions. The following state introduced by Yeo and Chua [47] $|YC\rangle$ can be used to perform a faithful teleportation of an arbitrary two-qubit entangled state,

$$|YC\rangle = \frac{1}{2\sqrt{2}}(|0000\rangle - |0011\rangle - |0101\rangle + |0110\rangle + |1001\rangle + |1010\rangle + |1100\rangle + |1111\rangle). \quad (32)$$

It is conjectured to maximize entanglement measures based on Rényi entropy of partial traces for $q \geq 2$. Let us also distinguish the tensor product of the Bell states,

$$|\psi_4\rangle = |\Psi_+\rangle \otimes |\Psi_+\rangle = \frac{1}{2}(|0000\rangle + |0011\rangle + |1100\rangle + |1111\rangle). \tag{33}$$

It is worth mentioning that Verstraete et al. [48] divide all 4-qubit pure states in nine different classes, to which any state can be transformed by SLOCC operations. One of these classes called *generic* contains states of the form [44],

$$|\varphi\rangle = z_0|\Phi^+\rangle|\Phi^+\rangle + z_1|\Phi^-\rangle|\Phi^-\rangle + z_2|\Psi^+\rangle|\Psi^+\rangle + z_3|\Psi^-\rangle|\Psi^-\rangle, \tag{34}$$

where $|\Phi^{\pm}\rangle$ and $|\Psi^{\pm}\rangle$ denote the standard Bell states (24), while $z_0, z_1, z_2, z_3 \in \mathbf{C}$ represent complex coefficients normalized as $\sum_{i=0}^{3} |z_i|^2 = 1$. A state belonging to this class represented in the computational basis reads

$$\begin{aligned}|\psi\rangle =& \frac{z_0+z_1}{2}|0000\rangle + \frac{z_0-z_1}{2}|0011\rangle + \frac{z_2+z_3}{2}|0101\rangle + \frac{z_2-z_3}{2}|0110\rangle + \\ & \frac{z_2-z_3}{2}|1001\rangle + \frac{z_2+z_3}{2}|1010\rangle + \frac{z_0-z_1}{2}|1100\rangle + \frac{z_0+z_1}{2}|1111\rangle. \end{aligned} \tag{35}$$

It is easy to see that several of the states mentioned above belong to this class. For instance, the states $|\psi_4\rangle$, $|GHZ_4\rangle$ and $|L\rangle$ correspond to real vectors of coefficients, equal to $(1, 0, 0, 0)$; $(1, 1, 0, 0)/\sqrt{2}$ and $(1, \omega, \omega^2, 0)/\sqrt{3}$ respectively.

To illustrate how the above states are distinguished with respect to the measures of entanglement based on degree of mixing of reduced density matrices we use the weighted sum (21) of Renyi entropies, which takes into account all possible one and two-party reductions. Fig. 2 presents numerical data of average Rényi entropies of reduced density matrices for random pure states of four-qubit systems. The data are shown in the planes spanned by two Rényi entropies chosen from the triple $\widetilde{S}_1$, $\widetilde{S}_2$, $\widetilde{S}_\infty$. In other words we see three projections of the 3D body formed in the space spanned by these three quantities. The data obtained for the ensemble of states belonging to the generic class (34) presented in the right panels show that these 1-uniform states are characterized by a high degree of entanglement.

Our numerical results confirm that the states listed above are distinguished indeed: the state HS (26) gives the maximal average entropy $\widetilde{S}_1$, while the entropies $\widetilde{S}_2$ and $\widetilde{S}_\infty$ are maximized by the YC state (32) and the cluster states (28a). It is worth to note that the state $|\Psi_1^{max}\rangle$, defined in (11) and distinguised by the minimal decomposition entropy, is not characterized by a high value of the entropies $\widetilde{S}_q$ of reductions, as it falls outside the area covered by the states from the generic class. Let us emphasize once more that there are no 2-uniform states for 4-qubit systems: the dots corresponding to states analyzed are located far away from the bounds represented in the plots by red dotted lines.

5.4. *Five qubits*

We first consider the 1-uniform state

$$2\sqrt{2}|\Psi_5\rangle = |00000\rangle + |10011\rangle + |00101\rangle + |11001\rangle + |10110\rangle + |01111\rangle + |11100\rangle, \tag{36}$$

that was introduced by Brown *et al.* as a highly entangled state with respect the negative partial transpose criterion [46]. This can be also obtained by means orthogonal arrays [7]. On the other hand, the 5-qubit AME state

$$2\sqrt{2}|\Omega_{5,2}\rangle = |00000\rangle + |00011\rangle + |01100\rangle - |01111\rangle + |11010\rangle + |11001\rangle + |10110\rangle - |10101\rangle, \tag{37}$$

fulfils $Q_1 = 1$ and has applications in quantum teleportation, quantum-state sharing, superdense coding and orthogonal arrays, as well [7, 49]. Interestingly, numerical computations show that the symmetric state $|D(5,(2,0))\rangle$ has larger S_1^{RIU} than these two.

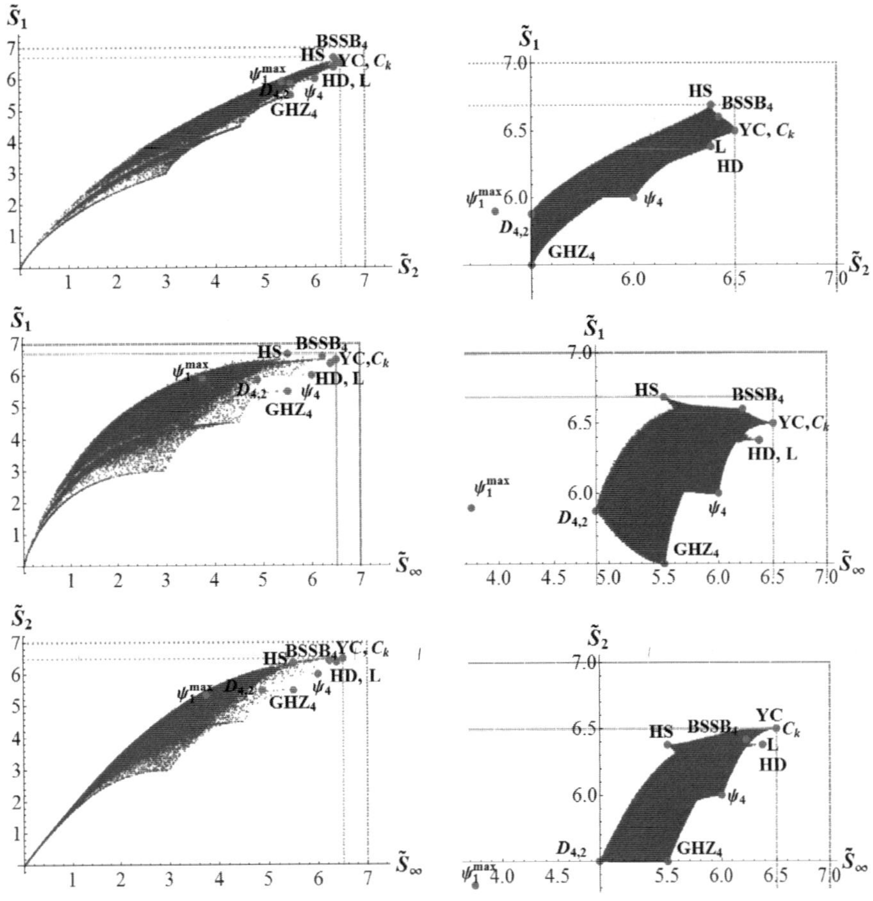

Figure 2: Average Rényi entropies $\widetilde{S}_1$, $\widetilde{S}_2$, $\widetilde{S}_\infty$ of the reduced density matrices for random four-qubit states for three choices of the axes. Left plots show values for 10^6 random states, while right plots present data for random states of the class (34). Red dotted lines indicate upper bounds for the entanglement measures, while yellow line represents the maximal values obtained numerically. Distinguished states are labeled by their names, while $D_{4,2}$ denotes the state $|D[4,(2,0)]\rangle$ defined in (12).

5.5. *Three qutrits*

We focus on the random states (11) as well as the symmetric state $|D(3,(1,1,1))\rangle$. Among several three-qutrit states this last attains the highest value for S_∞^{RIU} while the former states have the largest value of S_1^{RIU} and S_2^{RIU}, respectively.

5.6. *Four qutrits*

As an example of maximally entangled state we consider the state (23) for which our numeric calculations allow us to conjecture that the minimal decomposition entropy is the same regardless the value of the Rényi parameter. However, the symmetric state $|D(4,(2,1,1,0))\rangle$ attains a larger value of (5) with $q = 1,2$.

6. Conclusions

We have reviewed some entanglement properties of various multipartite states. In general, characterization of multipartite entanglement can be approached from different perspectives as one can study various measures of entanglement related to different aspects of non-locality. In the case of bipartite systems one identifies the class of states locally equivalent to the generalized Bell state, for which several measures of entanglement achieve their maximal values. However, for systems with more components the situation changes: in general, a highly entangled state with respect to a single measure of entanglement can display average, low or even vanishing degree of entanglement, if another measure is used. This can be seen in Fig. 2, where several four-qubit states are plotted according to the average Rényi entropy. The problem of distinguishing states for which the different entanglement measures attain their maximum value remains open even for some measures of three and four-parties entanglement. This fact is evident from Table 1 where several multipartite states are listed with respect their minimal RIU entropy.

Acknowledgements

The authors are grateful to Cinvestav, UPIITA-IPN and ITESM-CEM for an invitation to attend the conference Quantum Fest 2015 organized in México in October, 2015. Financial support by Polish National Science Center under the project number DEC-2011/02/A/ST1/00119 (K.Ż.) and by the Mexican Council for Science and Technology, Conacyt, (M.E.) is thankfully acknowledged.

References

[1] Arnaud L and Cerf N 2013 Exploring pure quantum states with maximally mixed reductions *Phys. Rev.* A **87** 012319
[2] Scott A J 2004 Multipartite entanglement, quantum-error-correcting codes, and entangling power of quantum evolutions *Phys. Rev.* A **69** 052330
[3] Amico L, Fazio R, Osterloh A and Vedral V 2008 Entanglement in many-body systems *Rev. Mod. Phys.* **80** 517
[4] Borras A, Plastino A R, Batle J, Zander C, Casas M and Plastino A 2007 Multi-qubit systems: highly entangled states and entanglement distribution *J. Phys.* A, **40** 13407
[5] Enríquez M, Puchała Z and Życzkowski K 2015 Minimal Rényi–Ingarden–Urbanik entropy of multipartite quantum states *Entropy* **17** 5063
[6] Goyeneche D, Alsina D, Latorre J I, Riera A and Życzkowski K 2015 Absolutely maximally entangled states, combinatorial designs and multi-unitary matrices *Phys. Rev.* A **92** 032316
[7] Goyeneche D and Życzkowski K 2014 Genuinely multipartite entangled states and orthogonal arrays *Phys. Rev.* A **90** 022316
[8] Coffman V, Kundu J and Wootters W K 2000 Distributed entanglement *Phys. Rev.* A **61** 052306
[9] Briegel H J and Raussendorf R 2001 Persistent entanglement in arrays of interacting particles *Phys. Rev. Lett.* **86** 910
[10] Facchi P, Florio G, Parisi G and Pascazio S 2008 Maximally multipartite entangled states *Phys. Rev.* A **77** 060304 R
[11] Carteret H A, Higuchi A and Sudbery A 2000 Multipartite generalization of the Schmidt decomposition *J. Math. Phys.* **41** 7932
[12] Ingarden R and Urbanik K 1962 Quantum informational thermodynamics *Acta Phys. Polon.* **21** 281
[13] Spekkens R W and Sipe J E 2001 Non-orthogonal preferred projectors for modal interpretations of quantum mechanics *Found. Physics* **31** 1403
[14] Brennen G K 2003 An observable measure of entanglement for pure states of multi-qubit systems *Quant. Inf. Comp.* **3** 619
[15] Kraus B 2010 Local unitary equivalence of multipartite pure states *Phys. Rev. Lett.* **104** 020504
[16] Liu B, Li J, Li X and Qiao C 2012 Local unitary classification of arbitrary dimensional multipartite pure states *Phys. Rev. Lett.* **108** 050501
[17] Lathauwer L D, Moor B D and Vandewalle J 2000 A multilinear singular value decomposition *SIAM J. Matrix Anal. Appl.* **21** 1253
[18] Eisert J and Briegel H J. 2001 Schmidt measure as a tool for quantifying multiparticle entanglement *Phys. Rev.* A **64** 022306

[19] Acín A, Andrianov A, Costa L, Jané E, Latorre J I and Tarrach R 2000 Generalized Schmidt decomposition and classification of three-quantum-bit states *Phys. Rev. Lett.* **85** 1560
[20] Acín A, Andrianov A, Jané E and Tarrach R 2001 Three-qubit pure state canonical forms *J. Phys.* A **34** 6725
[21] Bengtsson I and Życzkowski K 2006 *Geometry of Quantum States* (Cambridge University Press)
[22] Parker M G and Rijmen V 2001 The quantum entanglement of binary and bipolar sequences Ed T Helleseth *Sequences and Their Applications* (Springer- London UK) (Discrete Mathematics and Theoretical Computer Science Series)
[23] Shimony A 1995 Degree of entanglement *Ann. NY. Acad. Sci.* **755** 675
[24] Vedral V and Plenio M B 1998 Entanglement measures and purification procedures *Phys. Rev.* A **57** 1619
[25] Barnum H and Linden N 2001 Monotones and invariants for multi-particle quantum states *J. Phys.* A **34** 6787
[26] Brody D C and Hughston L P 2001 Geometric quantum mechanics *J. Geom. Phys.* **38** 19
[27] Cao Y and Wang A 2007 Revised geometric measure of entanglement *J. Phys.* A **40** 3507
[28] Chen L, Aulbach M and Hajdušek M. 2014 A comparison of old and new deffnitions of the geometric measure of entanglement *Phys. Rev.* A **89** 042305
[29] Wei T C and Goldbart P M 2003 Geometric measure of entanglement and applications to bipartite and multipartite quantum states *Phys. Rev.* A, **68** 042307
[30] Kolda T and Bader B W 2009 Tensor decompositions and applications *SIAM Rev.* **51** 455
[31] Życzkowski K and Bengtsson I 2002 Relativity of pure states entanglement *Ann. Phys. (N.Y.)* **295** 115
[32] Tamaryan L, Park D, Son J W and Tamaryan S 2008 Geometric measure of entanglement and shared quantum states *Phys. Rev.* A **78** 032304
[33] Chen L, Xu A and Zhu H 2010 Computation of the geometric measure of entanglement for pure multiqubit states *Phys. Rev.* A **82** 032301
[34] Hilling J J and Sudbery A 2010 The geometric measure of multipartite entanglement and the singular values of a hypermatrix *J. Math. Phys.* **51** 072102
[35] Bravyi S 2003 Entanglement entropy of multipartite pure states *Phys. Rev.* A **67** 012313
[36] Meyer D A and Wallach N R. 2002 Global entanglement in multiparticle systems *J. Math. Phys.* **43** 4273
[37] Helwig W, Cui W, Latorre J I, Riera A and Lo H K 2012 Absolute maximal entanglement and quantum secret sharing *Phys. Rev.* A **86** 052335
[38] Pastawski F, Yoshida B, Harlow D and Preskill J 2015 Holographic quantum error-correcting codes: toy models for the bulk/boundary correspondence *JHEP* **06** 149
[39] Gisin N and Bechmann-Pasquinucci H 1998 Bell inequality, bell states and maximally entangled states for n qubits? *Phys. Lett.* A **246** 1
[40] Higuchi A and Sudbery A 2000 How entangled can two couples get? *Phys. Lett.* A **272** 213
[41] Facchi P, Florio G, Marzolino U, Parisi G and Pascazio S 2010 Multipartite entanglement and frustration *New J. Phys.* **12** 025015
[42] Laflamme R, Miquel C, Paz J and Zurek W 1996 Perfect quantum error correcting code *Phys. Rev. Lett.* **77** 198–201
[43] Raussendorf R, Browne D E and Briegel H J 2003 Measurement-based quantum computation with cluster states *Phys. Rev.* A **68** 022312
[44] Gour G and Wallach N R 2010 All maximally entangled four-qubit states *J. Math. Phys.* **51** 112201
[45] Alsina D and Latorre J I 2014 Entanglement and Nonlocality in Many-Body Systems: a primer Preprint
[46] Brown IDK, Stepney S, Sudbery A and Braunstein S. 2005 Searching for highly entangled multi-qubit states *J. Phys.* A **38** 1119
[47] Yeo Y and Chua W K 2006 Teleportation and dense coding with genuine multipartite entanglement *Phys. Rev. Lett.* **96** 060502
[48] Verstraete F, Dehaene J, De Moor B and Verschelde H 2002 Four qubits can be entangled in nine different ways *Phys. Rev.* A **65** 052112
[49] Muralidharan S and Panigrahi P 2008 Perfect teleportation, quantum-state sharing, and superdense coding through a genuinely entangled five-qubit state *Phys. Rev.* A **77** 032321

Hidden Correlations in Indivisible Qudits as a Resource for Quantum Technologies on Examples of Superconducting Circuits

M A Man'ko[1] **and V I Man'ko**[1,2]

[1] Lebedev Physical Institute, Leninskii Prospect 53, Moscow 119991, Russia

[2] Moscow Institute of Physics and Technology (State University), Dolgoprudnyi, Moscow Region 141700, Russia

E-mail: mmanko@sci.lebedev.ru

Abstract. We show that the density-matrix states of noncomposite qudit systems satisfy entropic and information relations like the subadditivity condition, strong subadditivity condition, and Araki–Lieb inequality, which characterize hidden quantum correlations of observables associated with these indivisible systems. We derive these relations employing a specific map of the entropic inequalities known for density matrices of multiqudit systems to the inequalities for density matrices of single-qudit systems. We present the obtained relations in the form of mathematical inequalities for arbitrary Hermitian $N\times N$-matrices. We consider examples of superconducting qubits and qudits. We discuss the hidden correlations in single-qudit states as a new resource for quantum technologies analogous to the known resource in correlations associated with the entanglement in multiqudit systems.

1. Introduction

Quantum correlations associated, for example, with the phenomenon of entanglement [1] which exists in bipartite and multipartite systems, like systems of qubits or qudits, provide a resource for development of quantum technologies [2]. The properties of correlations can be characterized by different kinds of entropic and information relations known for both classical probability distributions and classical observables (see, for example, [3]) and quantum density matrices of composite systems, like bipartite and tripartite systems, where entropic inequalities given as the subadditivity and strong subadditivity conditions provide information on the presence and degree of correlations [4–6].

Among quantum systems studied in connection with possible applications in quantum technologies, the superconducting circuits based on Josephson junctions are considered from both theoretical and experimental points of view [7–21].

Recently, it was pointed out that quantum correlations known for bipartite and multipartite systems also exist in the systems without subsystems (indivisible or noncomposite systems) [22–25]. We called such correlations in noncomposite systems the hidden correlations. For example, analogs of the entropic and information inequalities like the subadditivity condition, the strong subadditivity condition, and the Araki–Lieb inequality [26] expressed in the form of matrix relations for density matrices of single qudit systems exist for indivisible systems as well.

The aim of this paper is to review the approach developed in [23, 24] for obtaining new entropic inequalities for systems without subsystems and extend this approach for obtaining new entropic inequalities not only for the state density matrices but also for Hermitian matrices of observables. The idea of the approach is based on the fact that any Hermitian matrix can be mapped on the nonnegative matrix with unit trace. The corresponding tool was applied in [27] to obtain the relation of entropy S and energy E of qudit systems, $E + S \leq \ln Z(-1)$, where $Z(\beta)$ is the system partition function, with β being the inverse temperature.

The approach formulated provides the possibility to extend all density-matrix inequalities known for multipartite systems to density-matrix inequalities for indivisible systems as well as to all observables. Here, we concentrate on particular examples of superconducting qudits and finite-dimensional systems.

This paper is organized as follows. In section 2, we review the properties of superconducting circuit states modeled by a parametric quantum oscillator with vibrating voltage and current. In section 3, we present a new inequality for the classical single qudit system in the form of subadditivity condition. In section 4, we discuss the approach to study Hermitian matrices using a map of the matrices on nonnegative matrices to derive new inequalities for Hermitian matrices. In section 5, we consider examples of inequalities for classical probabilities and classical observables. In section 6, we obtain new inequalities for quantum observables on the example of artificial atom realized by a superconducting qudit with $j = 3/2$. In section 7, we give the conclusions and prospectives.

2. Superconducting qubits

Superconducting devices based on application of Josephson junctions are discussed [7–10] as possible technical instruments for developing new technologies, for example, quantum computing. The idea of such applications is related to the fact that the Josephson junction realizes a model of the electric circuit with inductance L and capacitance C [11], i.e., the current and voltage in devices with Josephson junctions vibrate as the momentum and position of a mechanical oscillator does (see, for example, [28]). The frequency of vibrations ω in the electric circuit is determined by a factor proportional to $(LC)^{-1/2}$, and for high frequencies and low temperatures, such that $\hbar\omega \geq T$, the oscillator behaves as a quantum oscillator called the superconducting circuit, where the current and voltage satisfy the Heisenberg uncertainty relation [29], the Schrödinger–Robertson uncertainty relation [30, 31], and the purity-dependent uncertainty relation [32].

If the circuit parameters L and C are constant, the stationary states of the quantum oscillator correspond to the energy levels E_n ($n = 0, 1, 2, \ldots$). If the oscillator excitations are such that only finite number of levels N is involved, the stationary states of the superconducting circuit are identified with the qudit states of $j = (N-1)/2$. For example, if only two oscillator energy levels are involved, the superconducting circuit realizes a superconducting qubit and, in this case, a set of Josephson junctions models a multiqubit system, which can be employed to provide quantum information devices.

To realize the dynamical (nonstationary) Casimir effect, in [11–13] it was suggested to use the Josephson junction with time-dependent parameters [inductance $L(t)$ and capacitance $C(t)$], which is an analog of the oscillator with time-dependent frequency $\omega(t)$. The current and voltage in parametric superconducting circuits are generated due to temporal variations in the Josephson-junction parameters analogously to the photon in squeezed-states generation in resonators with vibrating boundaries due to the dynamical Casimir effect. The photons created due to the dynamical Casimir effect were registered in the devices where the Josephson junctions are employed [33].

The oscillator with time-dependent frequency is described by the Hamiltonian (we use dimensionless variables $\hbar = m = \omega(0) = 1$)

$$\hat{H} = \frac{\hat{p}^2}{2} + \omega^2(t)\frac{\hat{q}^2}{2}. \qquad (1)$$

In [34], it was found that the parametric oscillator has linear in the position and momentum integrals of motion called dynamical invariants (see, for example, [32, 35, 36]) of the form

$$\hat{a}(t) = -\frac{i}{\sqrt{2}}\left(\dot{\mathcal{E}}(t)\hat{q} - \mathcal{E}(t)\hat{p}\right), \qquad (2)$$

where the complex function $\mathcal{E}(t)$ satisfies the classical equation of motion for a parametric oscillator $\ddot{\mathcal{E}}(t) + \omega^2(t)\mathcal{E}(t) = 0$ under the initial conditions $\mathcal{E}(t) = 1$, $\dot{\mathcal{E}}(0) = i$, and $\hat{a}(0) = \frac{1}{\sqrt{2}}(\hat{q} + i\hat{p})$. The linear invariants $\hat{a}(t)$ and $\hat{a}^\dagger(t)$ satisfy the commutation relation $\left[\hat{a}(t), \hat{a}^\dagger(t)\right] = 1$. For $\omega(t) = 1$, the integral of motion $\hat{a}(t) = e^{it}\hat{a}(0)$.

An analog of the ground state of the stationary oscillator (superconducting circuit), being the squeezed state, has the wave function

$$\psi_0(x,t) = \left(\pi\mathcal{E}^2(t)\right)^{-1/4} \exp\frac{i\dot{\mathcal{E}}(t)x^2}{2\mathcal{E}(t)}. \qquad (3)$$

Dispersions of the position $\sigma_{xx}(t)$ and momentum $\sigma_{pp}(t)$ (current and voltage in a superconducting circuit) are determined by the function $\mathcal{E}(t)$ as follows:

$$\sigma_{xx}(t) = \frac{|\mathcal{E}(t)|^2}{2}, \qquad \sigma_{pp}(t) = \frac{|\dot{\mathcal{E}}(t)|^2}{2}. \qquad (4)$$

The correlation coefficient $r(t)$ of the current and voltage $r(t) = \frac{\sigma_{xp}(t)}{\sqrt{\sigma_{xx}(t)\sigma_{pp}(t)}}$ is given by the bound in the Schrödinger–Robertson uncertainty relation

$$\sigma_{xx}(t)\sigma_{pp}(t) \geq \left[4\left(1 - r^2(t)\right)\right]^{-1}, \qquad (5)$$

which provides the equality

$$|\mathcal{E}(t)\dot{\mathcal{E}}(t)|^2 = \left[4 - r^2(t)\right]^{-1}. \qquad (6)$$

The eigenfunctions of time-dependent invariants $\hat{n}(t) = \hat{a}^\dagger(t)\hat{a}(t)$ are analogs of stationary states of superconducting circuits with time-dependent parameters L and C. The eigenvalues of the integral of motion $\hat{n}(t)$ do not depend on time and take the values $n = 0, 1, 2, \ldots, \infty$.

For finite number of excited states $\mid n\rangle$ $(n = 0, 1, 2, \ldots, N)$ of a parametric superconducting circuit, the states can be considered as an approximation of qudit states with $j = (N-1)/2$. These states can be used analogously to the states of stationary superconducting circuits in quantum technologies. For example, if $j = 1/2$, a parametric superconducting qubit can be realized.

The symplectic tomogram (probability distribution) [37] of the squeezed vacuum state (3) of the superconducting circuit

$$w(X, t \mid \mu, \nu) = \frac{1}{2\pi|\nu|}\left|\int \psi_0(y,t) \exp\left(\frac{i\mu}{2\nu}y^2 - \frac{iX}{\nu}y\right) dy\right|^2 \qquad (7)$$

reads

$$w(X, t \mid \mu, \nu) = \frac{1}{\sqrt{2\pi\sigma_{xx}(t)}} \exp\left(-\frac{X^2}{2\sigma_{xx}(t)}\right), \qquad (8)$$

where the dispersion $\sigma_{xx}(t)$ of the homodyne quadrature X is

$$\sigma_{xx}(t) = \mu^2\frac{|\mathcal{E}(t)|^2}{2} + \nu^2\frac{|\dot{\mathcal{E}}(t)|^2}{2} + \frac{1}{2}\mu\nu\left(|\mathcal{E}(t)\dot{\mathcal{E}}(t)|^2 - 1\right), \tag{9}$$

with μ and ν being the real parameters.

The optical tomogram $w(X, t \mid \theta)$ of squeezed vacuum state of the superconducting circuit has the form (8) with the dispersion $\sigma_{xx}(t)$ of the homodyne quadrature X dependent on the local oscillator phase as

$$\sigma_{xx}(t) = \frac{\cos^2\theta|\mathcal{E}(t)|^2}{2} + \frac{\sin^2\theta|\dot{\mathcal{E}}(t)|^2}{2} + \frac{\sin 2\theta}{2}\left(|\mathcal{E}(t)\dot{\mathcal{E}}(t)|^2 - 1\right). \tag{10}$$

The squeezed coherent state of the parametric superconducting circuit $\mid \alpha, t\rangle$ is such that $\hat{a}(t) \mid \alpha, t\rangle = \alpha \mid \alpha, t\rangle$ has the symplectic tomogram in the form of normal distribution with dispersion $\sigma_{xx}(t)$ given by (9) and the mean value of the homodyne quadrature X

$$\langle X(t)\rangle = \mu\frac{\mathcal{E}^\star(t)\alpha + \mathcal{E}(t)\alpha^\star}{\sqrt{2}} + \nu\frac{\dot{\mathcal{E}}^\star(t)\alpha + \dot{\mathcal{E}}(t)\alpha^\star}{\sqrt{2}}. \tag{11}$$

Thus, the optical tomogram of the squeezed coherent state of the superconducting circuit reads

$$w_\alpha(X, t \mid \theta) = \frac{1}{\sqrt{2\pi\sigma_{xx}(t)}}\exp\left(-\frac{X - \bar{X}_\alpha^2(t)}{2\sigma_{xx}(t)}\right), \tag{12}$$

where $\sigma_{xx}(t)$ is given by (10), and the mean value of the homodyne quadrature is

$$\bar{X}_\alpha(t) = \sqrt{2}\,\mathrm{Re}\left[\alpha\left(\mathcal{E}^\star(t)\cos\theta + \dot{\mathcal{E}}^\star(t)\sin\theta\right)\right]. \tag{13}$$

One can check that tomogram (12) satisfies the entropic inequality valid for an arbitrary optical tomogram [38]

$$-\int w(X, t \mid \theta)\ln w(X, t \mid \theta)\,dX - \int w(X, t \mid \theta + \pi/2)\ln w(X, t \mid \theta + \pi/2)\,dX \geq \ln(\pi e). \tag{14}$$

For example, the state $\mid 1, t\rangle = \hat{a}^\dagger(t) \mid 0, t\rangle$ with the wave function

$$\psi_1(x, t) = -\frac{i}{\sqrt{2}}\left(\mathcal{E}^\star(t)\hat{p} - \dot{\mathcal{E}}^\star(t)x\right)\psi_0(x, t), \tag{15}$$

which is the second component in the superconducting qubit state with the first component given by (3), has the tomogram

$$w_1(X, t \mid \theta) = \pi^{-1/2}(2\sigma_{xx}(t))^{-3/2}X^2\exp\left(-\frac{X^2}{2\sigma_{xx}(t)}\right), \tag{16}$$

where the dispersion $\sigma_{xx}(t)$ is given by (10). Tomogram (16) satisfies the entropic inequality (14). Optical tomograms of the superconducting circuit states $\mid n, t\rangle = \frac{\hat{a}^\dagger(t)^n}{\sqrt{n!}} \mid 0, t\rangle$ are the probability distributions

$$w_n(X, t \mid \theta) = w_0(X, t \mid \theta)\frac{1}{2^n n!}H_n^2\left(\frac{X}{\sqrt{2\sigma_{xx}(t)}}\right), \tag{17}$$

where $H_n(y)$ are Hermite polynomials. The dependence of the tomogram on the local oscillator phase θ is given by the dependence (10) of the dispersion $\sigma_{xx}(t)$ on this parameter.

If the states $\mid n, t\rangle$ with $n = 0, 1, 2, \ldots, N = 2j+1$ are excited, the parametric superconducting circuit can be interpreted as an artificial atom with N levels or qudit with $j = (N-1)/2$.

3. Information inequalities for single qudit states

In this section, we discuss the entropic and information relations, such as equalities and inequalities, as well as quantum correlations of qudit observables for single qudits. Qudit can be realized either by a spin-j particle or the N-level atom. These systems can also be considered as artificial atoms realized by superconducting circuits. In this consideration, we follow [24, 39–41] where a map of integers $1, 2, \ldots, N = mn$ onto pairs of integers (jk), $j = 1, 2, \ldots, m$ and $k = 1, 2, \ldots, n$ or, in the case of $N = n_1 n_2 n_3$, on triples of integers (jkl), $j = 1, 2, \ldots, n_1$, $k = 1, 2, \ldots, n_2$ and $l = 1, 2, \ldots, n_3$ was used. This map provides a possibility to apply the relations like entropic inequalities (known for bipartite and tripartite quantum systems) to indivisible systems like a superconducting-circuit qudit.

We demonstrate the inequalities for the probability distribution $P_s = (P_1, P_2, \ldots, P_N)$, where the even number $N = n_1 n_2$, $n_1 = 2$, and $n_2 = N/2$. Then we label the integers s, where $s = 1, 2, \ldots, N$, by the pairs of integers (jk), where $j = 1, 2$ and $k = 1, 2, \ldots, N/2$. We obtain the same set of numbers $P_s \equiv P_{s(jk)} \equiv P_{jk}$. The inequality, known as the subadditivity condition for the joint probability distribution P_{jk}, for P_s reads

$$-\sum_{s=1}^{N} P_s \ln P_s \leq -\left(\sum_{s=1}^{N/2} P_s\right) \ln \left(\sum_{s=1}^{N/2} P_s\right) - \left(\sum_{s=N/2}^{N} P_s\right) \ln \left(\sum_{s=N/2}^{N} P_s\right) - \sum_{k=1}^{N/2} (P_k + P_{k+N/2}) \ln (P_k + P_{k+N/2}). \quad (18)$$

4. Hermitian matrix inequalities

Given Hermitian $N \times N$-matrix h, where $N = nm$. Let the eigenvalues of the matrix h have values with a minimal eigenvalue h_0. The Hermitian $N \times N$-matrix $\rho(x)$, where $x > |h_0|$ given in the block form with $n \times n$-blocks

$$\rho(x) = (Nx + \operatorname{Tr} h)^{-1} (h^{jk} + x 1_n \delta^{jk}), \qquad j, k = 1, 2, \ldots, m, \quad (19)$$

is the nonnegative matrix with $\operatorname{Tr} \rho(x) = 1$, where 1_n is the identity $n \times n$-matrix. For the density matrix $h = \rho$, the map $\rho \to \rho(x)$ given by (19) reflects the fact that the noise contribution is taken into account.

One can check that two Hermitian matrices, namely, $m \times m$-matrix $\rho(1, x)$ with matrix elements

$$\rho_{jk}(1, x) = (Nx + \operatorname{Tr} h)^{-1} \left[\operatorname{Tr}(h^{jk}) + nx\delta^{jk}\right] \quad (20)$$

and $n \times n$-matrix $\rho(2, x)$ of the form

$$\rho(2, x) = (Nx + \operatorname{Tr} h)^{-1} \left(mx 1_n + \sum_{k=1}^{m} h^{kk}\right) \quad (21)$$

are nonnegative matrices and $\operatorname{Tr} \rho(1, x) = \operatorname{Tr} \rho(2, x) = 1$. The matrices satisfy the entropic inequality for the mutual information $I(x)$ of the form

$$I(x) = \operatorname{Tr} \rho(x) \ln \rho(x) - \operatorname{Tr} \rho(1, x) \ln \rho(1, x) - \operatorname{Tr} \rho(2, x) \ln \rho(2, x) \geq 0. \quad (22)$$

If $h_0 \geq 0$, one can assume $x = 0$. In this case, inequality (22) for an arbitrary Hermitian $N \times N$-matrix h reads

$$I(x) = \operatorname{Tr} \{h \ln(h / \operatorname{Tr} h)\} - \operatorname{Tr} \left\{\left[\operatorname{Tr} h^{jk}\right] \ln \left[\left(\operatorname{Tr} h^{jk}\right) / \operatorname{Tr} h\right]\right\} - \operatorname{Tr} \left\{\left[\sum_{k=1}^{m} h^{kk}\right] \ln \left[\left(\sum_{k=1}^{m} h^{kk}\right) / \operatorname{Tr} h\right]\right\} \geq 0. \quad (23)$$

Now we present this inequality on an example of the 3×3-matrix

$$h = \begin{pmatrix} h_{11} & h_{12} & h_{13} \\ h_{21} & h_{22} & h_{23} \\ h_{31} & h_{32} & h_{33} \end{pmatrix}.$$

Since $N = 3$, we may assume this matrix as $\tilde{h} = \begin{pmatrix} h & 0 \\ 0 & 0 \end{pmatrix}$. Then taking in the previous formulas $m = n = 2$, we obtain the 4×4-matrix $\rho(x)$ (20) as follows:

$$\rho(x) = (3x + h_{11} + h_{22} + h_{33})^{-1} \begin{pmatrix} h^{11} + x1_2 & h^{12} \\ h^{21} & h^{22} + x\mathcal{P}_1 \end{pmatrix}, \tag{24}$$

where $\mathcal{P}_1 = \begin{pmatrix} 1 & 0 \\ 0 & 0 \end{pmatrix}$ and 2×2-blocks h^{jk} $(j, k = 1, 2)$ read

$$h^{11} = \begin{pmatrix} h_{11} & h_{12} \\ h_{21} & h_{22} \end{pmatrix}, \quad h^{12} = \begin{pmatrix} h_{13} & 0 \\ 0 & 0 \end{pmatrix}, \quad h^{21} = \begin{pmatrix} h_{31} & 0 \\ 0 & 0 \end{pmatrix}, \quad h^{22} = \begin{pmatrix} h_{33} & 0 \\ 0 & 0 \end{pmatrix}.$$

Then the 2×2-matrix $\rho(1, x)$ has the form

$$\rho(1, x) = (3x + h_{11} + h_{22} + h_{33})^{-1} \begin{pmatrix} h_{11} + h_{22} + 2x & h_{13} \\ h_{31} & h_{33} + x \end{pmatrix}, \tag{25}$$

and the 2×2-matrix $\rho(2, x)$ reads

$$\rho(2, x) = (3x + h_{11} + h_{22} + h_{33})^{-1} \begin{pmatrix} h_{11} + h_{33} + 2x & h_{12} \\ h_{21} & h_{22} + x \end{pmatrix}. \tag{26}$$

Now we are in the position to present the entropic inequality which is satisfied by the matrix elements of the 3×3-matrix h for $x > h_0$; it is

$$\begin{aligned} I = \mathrm{Tr}&\left\{\begin{pmatrix} h_{11}+x & h_{12} & h_{13} \\ h_{21} & h_{22}+x & h_{23} \\ h_{31} & h_{32} & h_{33}+x \end{pmatrix} \ln\left[\begin{pmatrix} h_{11}+x & h_{12} & h_{13} \\ h_{21} & h_{22}+x & h_{23} \\ h_{31} & h_{32} & h_{33}+x \end{pmatrix}(3x+h_{11}+h_{22}+h_{33})^{-1}\right]\right\} \\ &-\mathrm{Tr}\left\{\begin{pmatrix} h_{11}+h_{22}+2x & h_{13} \\ h_{31} & h_{33}+x \end{pmatrix} \ln\left[\begin{pmatrix} h_{11}+h_{22}+2x & h_{13} \\ h_{31} & h_{33}+x \end{pmatrix}(3x+h_{11}+h_{22}+h_{33})^{-1}\right]\right\} \\ &-\mathrm{Tr}\left\{\begin{pmatrix} h_{11}+h_{33}+2x & h_{13} \\ h_{21} & h_{22}+x \end{pmatrix} \ln\left[\begin{pmatrix} h_{11}+h_{33}+2x & h_{13} \\ h_{21} & h_{22}+x \end{pmatrix}(3x+h_{11}+h_{22}+h_{33})^{-1}\right]\right\} \geq 0. \end{aligned} \tag{27}$$

If the matrix h is nonnegative Hermitian matrix, one has inequality (27), where $x = 0$; this inequality reads

$$\begin{aligned} I = \mathrm{Tr}&\left\{\begin{pmatrix} h_{11} & h_{12} & h_{13} \\ h_{21} & h_{22} & h_{23} \\ h_{31} & h_{32} & h_{33} \end{pmatrix} \ln\left[\begin{pmatrix} h_{11} & h_{12} & h_{13} \\ h_{21} & h_{22} & h_{23} \\ h_{31} & h_{32} & h_{33} \end{pmatrix}(h_{11}+h_{22}+h_{33})^{-1}\right]\right\} \\ &-\mathrm{Tr}\left\{\begin{pmatrix} h_{11}+h_{22} & h_{13} \\ h_{31} & h_{33} \end{pmatrix} \ln\left[\begin{pmatrix} h_{11}+h_{22} & h_{13} \\ h_{31} & h_{33} \end{pmatrix}(h_{11}+h_{22}+h_{33})^{-1}\right]\right\} \\ &-\mathrm{Tr}\left\{\begin{pmatrix} h_{11}+h_{33} & h_{13} \\ h_{21} & h_{22} \end{pmatrix} \ln\left[\begin{pmatrix} h_{11}+h_{33} & h_{13} \\ h_{21} & h_{22} \end{pmatrix}(h_{11}+h_{22}+h_{33})^{-1}\right]\right\} \geq 0. \end{aligned} \tag{28}$$

If in inequality (28) $\mathrm{Tr}\, h = 1$ and $h \geq 0$, one arrives at the inequality obtained for the qutrit density matrix in [42]. The discussed inequalities describe relations for the density matrix of qutrit corresponding to an artificial three-level atom.

5. Classical observables as "classical states"

Given random variables $j = 1, 2, \ldots, N$ and the probability distribution P_j, which we call "the classical state." One has $\sum_{j=1}^{N} P_j = 1$. Also given a real function f_j of random variables, one can interpret the function f_j as an observable. The means and higher moments are

$$\langle f \rangle = \sum_{j=1}^{N} f_j P_j, \qquad \langle f^k \rangle = \sum_{j=1}^{N} f_j^k P_j. \tag{29}$$

One can construct a new function $F_j = (f_j + x)n$, where $x > |f_0|$. The number f_0 is a minimum value of the observable f_j, and it can be negative. The normalization factor $n = \left[\left(\sum_{j=1}^{N} f_j\right) + Nx\right]^{-1}$ and the function F_j is nonnegative and normalized, $\sum_{j=1}^{N} F_j = 1$. We can consider the function F_j as a probability distribution, which we associate with the observable f_j, and this fact provides the possibility to extend some relations known for the probability distributions and apply these relations to the observables. For example, we can introduce the Shannon entropy [43] for the observable f_j as follows:

$$S_f(x) = -\sum_{j=1}^{N} [(f_j + x)n \ln(f_j + x)n] \geq 0. \tag{30}$$

Also we can associated the Tsallis q-entropy [44] with the observable f_j as

$$T_f(x) = \frac{1}{1-q}\left\{\sum_{j=1}^{N} [(f_j + x)n]^q - 1\right\}. \tag{31}$$

Then we can rewrite the known inequality for relative entropy [2] associated with two probability distributions for two observables $f_j^{(1)}$ and $f_j^{(2)}$; it reads

$$\sum_{j=1}^{N} \left[\left(f_j^{(1)} + x_1\right) n_1 \ln\left(f_j^{(1)} + x_1\right) n_1\right] - \left[\left(f_j^{(1)} + x_1\right) n_1 \ln\left(f_j^{(2)} + x_2\right) n_2\right] \geq 0. \tag{32}$$

Here, x_1 and x_2 are moduli of minimum nonpositive values of observables $f_j^{(1)}$ and $f_j^{(2)}$, respectively. In particular, one can write the inequality for an arbitrary observable f_j and the probability distribution P_j

$$\sum_{j=1}^{N} [P_j \ln P_j - P_j \ln[(f_j + x)n]] \geq 0. \tag{33}$$

Analogously, we can write inequalities for observables in view of Tsallis entropy.

The subadditivity condition is valid for an arbitrary observable f_j as well; for a single system with $N = 4$, we provide the condition in an explicit form. Thus, given an observable f_j, i.e., four numbers f_1, f_2, f_3, and f_4. Assume the number f_4 to be negative, i.e., $f_4 = -|f_4|$ and take the variable $x > |f_4|$. Now we introduce the function $F_j(x)$ taking four values

$$F_1(x) = (f_1 + x)n_4, \quad F_2(x) = (f_2 + x)n_4, \quad F_3(x) = (f_1 + x)n_4, \quad F_4(x) = (f_4 + x)n_4, \tag{34}$$

where $n_4 = (f_1 + f_2 + f_3 + f_4 + 4x)^{-1}$. Then we have the nonnegativity condition for relative entropy associated with the probability distribution P_j and observable f_j of the form

$$\begin{aligned} &P_1 \ln\{P_1[(f_1 + x)n_4]^{-1}\} + P_2 \ln\{P_2[(f_2 + x)n_4]^{-1}\} \\ &+P_3 \ln\{P_3[(f_3 + x)n_4]^{-1}\} + P_4 \ln\{P_4[(f_4 + x)n_4]^{-1}\} \geq 0. \end{aligned} \tag{35}$$

The other known inequality is the subadditivity condition for entropy associated with the observable f_j and two other entropies, which are analogs of "subsystem" states described by two probability distributions

$$\mathcal{P}_1(x) = (f_1 + f_2 + 2x)n_4, \qquad \mathcal{P}_2(x) = (f_3 + f_4 + 2x)n_4$$

and

$$\Pi_1(x) = (f_1 + f_3 + 2x)n_4, \qquad \Pi_2(x) = (f_2 + f_4 + 2x)n_4.$$

The inequality

$$S_1(x) + S_2(x) \geq S(x), \tag{36}$$

where

$$S_1(x) = -\sum_{k=1}^{2} \mathcal{P}_k(x) \ln \mathcal{P}_k(x), \quad S_2(x) = -\sum_{k=1}^{2} \Pi_k(x) \ln \Pi_k(x), \quad S(x) = -\sum_{k=1}^{4} F_k(x) \ln F_k(x),$$

reads

$$\begin{aligned} &-(f_1 + f_2 + 2x) \ln[(f_1 + f_2 + 2x)n_4] - (f_3 + f_4 + 2x) \ln[(f_3 + f_4 + 2x)n_4] \\ &-(f_1 + f_3 + 2x) \ln[(f_1 + f_3 + 2x)n_4] - (f_2 + f_4 + 2x) \ln[(f_2 + f_4 + 2x)n_4] \\ \geq\; &-(f_1 + x) \ln[(f_1 + x)n_4] - (f_2 + x) \ln[(f_2 + x)n_4] \\ &-(f_3 + x) \ln[(f_3 + x)n_4] - (f_4 + x) \ln[(f_4 + x)n_4]. \end{aligned} \tag{37}$$

Thus, the values of an arbitrary classical observable f_j satisfy the inequality, which is an analog of the subadditivity condition for entropies of bipartite classical systems. Formally, inequality (36) is valid for arbitrary real numbers f_j for large numbers x.

6. Quantum inequalities

Hidden correlations described by classical probability distributions and characterized by entropic inequalities take place in quantum systems. For example, the four-level artificial atom realized in the superconducting circuit as a qudit with $j = 3/2$ is described by the density 4×4-matrix

$$\rho = \begin{pmatrix} \rho_{11} & \rho_{12} & \rho_{13} & \rho_{14} \\ \rho_{21} & \rho_{22} & \rho_{23} & \rho_{24} \\ \rho_{31} & \rho_{32} & \rho_{33} & \rho_{34} \\ \rho_{14} & \rho_{24} & \rho_{34} & \rho_{44} \end{pmatrix}.$$

One can check that this Hermitian matrix satisfies the entropic subadditivity condition in spite of the fact that the system under consideration is not bipartite; the inequality reads

$$-\mathrm{Tr}\,(\rho \ln \rho) \leq -\mathrm{Tr}\,[\rho(1) \ln \rho(1)] - \mathrm{Tr}\,[\rho(2) \ln \rho(2)], \tag{38}$$

where $\rho(1) = \begin{pmatrix} \rho_{11} + \rho_{22} & \rho_{13} + \rho_{34} \\ \rho_{31} + \rho_{41} & \rho_{33} + \rho_{44} \end{pmatrix}$ and $\rho(2) = \begin{pmatrix} \rho_{11} + \rho_{33} & \rho_{12} + \rho_{34} \\ \rho_{21} + \rho_{42} & \rho_{22} + \rho_{44} \end{pmatrix}$. An observable for the artificial atom is described by the Hermitian 4×4-matrix

$$f = \begin{pmatrix} f_{11} & f_{12} & f_{13} & f_{14} \\ f_{21} & f_{22} & f_{23} & f_{24} \\ f_{31} & f_{32} & f_{33} & f_{34} \\ f_{14} & f_{24} & f_{34} & f_{44} \end{pmatrix}.$$

The matrix $f(1) = \begin{pmatrix} f_{11}+f_{22} & f_{13}+f_{24} \\ f_{31}+f_{42} & f_{33}+f_{44} \end{pmatrix}$ is the Hermitian 4×4-matrix. An analog of the relative entropy nonnegativity can be given through matrices $\rho(1)$ and $f(1)$ as the following inequality:

$$\mathrm{Tr}\left[\begin{pmatrix} \rho_{11}+\rho_{22} & \rho_{13}+\rho_{34} \\ \rho_{31}+\rho_{41} & \rho_{33}+\rho_{44} \end{pmatrix} \ln \begin{pmatrix} \rho_{11}+\rho_{22} & \rho_{13}+\rho_{34} \\ \rho_{31}+\rho_{41} & \rho_{33}+\rho_{44} \end{pmatrix} - \begin{pmatrix} \rho_{11}+\rho_{22} & \rho_{13}+\rho_{34} \\ \rho_{31}+\rho_{41} & \rho_{33}+\rho_{44} \end{pmatrix}\right]$$
$$\times \ \ln\left\{(f_{11}+f_{22}+f_{33}+f_{44}+2x)^{-1} \begin{pmatrix} f_{11}+f_{22}+x & f_{13}+f_{24} \\ f_{31}+f_{42} & f_{33}+f_{44}+x \end{pmatrix}\right\} \geq 0. \quad (39)$$

This new inequality is valid for large x for an arbitrary state of the superconducting circuit with excited four energy levels and the observable.

A new inequality can also be written for qudit tomograms and tomographic symbols of observables. For example, for the case of four-level artificial atom realized by the superconducting circuit, the tomogram of observable f reads

$$w_f(m,\vec{n}) = (ufu^\dagger)_{mm}, \quad (40)$$

where $m = 1,2,3,4$ and u is the unitary 4×4-matrix depending on the unit vector $\vec{n}$. In this case, one has four real numbers: $w_f(1,\vec{n})$, $w_f(2,\vec{n})$, $w_f(3,\vec{n})$, and $w_f(4,\vec{n})$. Then a new quantum inequality for observable f and the state tomogram with the density matrix ρ can be written in the form of tomographic relative entropy nonnegativity as follows:

$$\sum_{m=1}^{4}\left\{w_\rho(m,\vec{n})\ln w_\rho(m,\vec{n}) - w_\rho(m,\vec{n})\ln\left([w_f(m,\vec{n})+x]\left[\sum_{m'=1}^{4}[w_f(m',\vec{n})+x]\right]^{-1}\right)\right\} \geq 0. \quad (41)$$

Thus, we obtained some new quantum inequalities corresponding to quantum correlations available in superconducting qudits taking into account the properties of observables.

7. Conclusions

To conclude, we point out the main results of our study.

We presented a review of the approach for extending the entropic inequalities known for multipartite systems, both classical and quantum, for example, the nonnegativity condition of relative entropy to obtain new inequalities for indivisible systems, including new inequalities for physical observables.

We discussed the example of superconducting circuit and entropic inequalities for an artificial atom with four energy levels and its parametric analog.

We derive a new inequality for the tomographic symbol of an observable of the four-level atom related to quantum correlations in the system.

The new inequalities obtained can be checked experimentally following the procedure considered in [16], where testing entropic inequalities in superconducting qudits were discussed.

The new entropic and information inequalities reflect the presence of quantum correlations in systems like superconducting single qudits. These correlations can be used as a resource for quantum technologies analogously to the correlations related to the entanglement in multiqubit systems and studied as such a resource. In the future publication, we extend our study to the other systems.

References

[1] Shrödinger E 1935 *Naturwissenschaften* **23** 807

[2] Nielsen M A ang Chuang I L 2000 *Quantum Computation and Quantum Information* (Cambridge University Press, Cambridge, UK)
[3] Holevo A S 1982 *Probabilistic and Statistical Aspects of Quantum Theory* (North Holland, Amsterdam)
[4] Lieb E H and Ruskai M B 1973 *J. Math. Phys.* **14** 1938
[5] Ruskai M B 2002 *J. Math. Phys.* **43** 4358; Erratum *ibid* 2005 **46** 019901
[6] Nielsen N A and Petz D A 2004 A simple proof of the strong subadditivity inequality; arXiv:quant-ph/0408130
[7] Devoret M H and Schoelkopf R J 2013 *Science* **339** 1169
[8] Shalibo Y, Resh R, Fogel O, Shwa D, Bialczak R, Martinis J M and Katz N 2013 *Phys. Rev. Lett.* **110** 100404
[9] Braumüller J, Cramer J, Schlör S, Rotzinger H, Radtke L, Lukashenko A, Yang P, Marthaler M, Guo L, Ustinov A V and Weides M 2015 *Phys. Rev.* B **91** 054523
[10] Pashkin Yu A, Yamamoto T, Astafiev O, Nakamura Y, Averin D V and Tsai J S 2003 *Nature* **421** 823
[11] Dodonov V V, Man'ko V I and Man'ko O V 1989 *J. Sov. Laser Res.* **10** 413
[12] Man'ko O V 1994 *J. Korean Phys. Soc.* **27** 1
[13] Man'ko V I 1991 *J. Sov. Laser Res.* **12** 383
[14] Fedorov A K, Kiktenko E O, Man'ko O V and Man'ko V I 2015 *Phys. Rev.* A **91** 042312
[15] Kiktenko E O, Fedorov A K, Strakhov A A and Man'ko V I 2015 *Phys. Lett.* A **379** 1409
[16] Glushkov E, Glushkova A and Manko V I 2015 *J. Russ. Laser Res.* **36** 448
[17] Takashima K, Hatakenaka N, Kurihara S and Zeilinger A 2008 *J. Phys. A: Math. Gen* **41** 164036
[18] Fujii T, Matsuo S, Hatakenaka S, Kurihara S and Zeilinger A 2011 *Phys. Rev.* B **84** 174521
[19] Takashima K, Matsuo S, Fujii T, Hatakenaka S, Kurihara S and Zeilinger A 2009 *J. Phys. Conf. Ser.* **150** 052260
[20] Dodonov V V and Dodonov A V 2008 *J. Phys. Conf. Ser.* **99** 012006
[21] Fedorov A K, Kiktenko E O, Man'ko O V and Man'ko V I 2015 *Phys. Scr.* **90** 055101
[22] Man'ko M A and Man'ko V I 2014 *Int. J. Quantum Inf.* **12** 156006
[23] Man'ko M A and Man'ko V I 2014 *J. Russ. Laser Res.* **36** 301
[24] Man'ko M A and Man'ko V I 2015 *Entropy* **17** 2876
[25] Man'ko V I and Markovich L A 2015 *J. Russ. Laser Res.* **36** 343
[26] Araki H and Lieb E H 1970 *Commun. Math. Phys.* **18** 160
[27] Figueroa A, López J, Castaños O, López-Peña R, Man'ko M A and Man'ko V I 2015 *J. Phys. A: Math. Theor.* **48** 065301
[28] Doskoch I Ya and Man'ko M A 2015 *J. Russ. Laser Res.* **36** 503
[29] Heisenberg W 1927 *Z. Phys.* **43** 172
[30] Schrödinger E 1930 *Ber. Kgl. Akad. Wiss. Berlin* **24** 296
[31] Robertson H P 1929 *Phys. Rev.* **35** 667
[32] Dodonov V V and Man'ko V I 1987 *Invariants and the Evolution of Nonstationary Quantum Systems. Proceedings of the Lebedev Physical Institute*, Nauka, Moscow, Vol. 183 [translated by Nova Science, New York, 1989]
[33] Johansson J R, Johansson G, Wilson C M and Nori E 2009 *Phys. Rev. Lett.* **103** 147003
[34] Malkin I A and Man'ko V I 1970 *Phys. Lett.* A **32** 243
[35] Guerrero J and López-Ruiz F F 2015 *Phys. Scr.* **90** 074046
[36] Sanborn B, Suslov S S and Vinet L 2011 Dynamical invariants and Berry phase for generalized driven harmonic oscillator; arXiv:math-ph/1108.5144
[37] Mancini S, Man'ko V I and Tombesi P 1996 *Phys. Lett.* A **213** 1
[38] Man'ko M A and Man'ko V I 2011 *Found Phys.* **41** 330
[39] M. A. Man'ko, V. I. Man'ko, G. Marmo, A. Simoni and F. Ventriglia, Introduction to tomography, classical and quantum *Nuovo Cimento, Colloquia and Communications in Physics, Mathematical Structures in Quantum Systems and Applications* (2013), Vol. 36, Ser. 3, p. 163.
[40] Chernega V N and Man'ko O V 2015 *Phys. Scr.* **90** 074052
[41] Man'ko M A and Man'ko V I 2014 *J. Russ. Laser Res.* **35** 509
[42] Chernega V N and Man'ko V I 2007 *J. Russ. Laser Res.* **28** 103
[43] Shannon C E 1948 A mathematical theory of communication *Bell Syst. Tech. J.* **27** 379; 623
[44] Tsallis C 2001 Nonextensive statistical mechanics and thermodynamics: historical background and present status in: Abe S and Okamoto Y (Eds.) *Nonextensive Statistical Mechanics and Its Applications, Lecture Notes in Physics* Springer, Berlin, Vol. 560, p. 3

Periodic Solutions, Eigenvalue Curves, and Degeneracy of the Fractional Mathieu Equation

A Parra-Hinojosa[1] and J C Gutiérrez-Vega[2]

[1] Department of Informatics, Technische Universität München. Boltsmannstr 3. Garching 85748, Germany

[2] Photonics and Mathematical Optics Group, Tecnológico de Monterrey, Monterrey 64849, México

E-mail: `juliocesar@itesm.mx`

Abstract. We investigate the eigenvalue curves, the behavior of the periodic solutions, and the orthogonality properties of the Mathieu equation with an additional fractional derivative term using the method of harmonic balance. The addition of the fractional derivative term breaks the hermiticity of the equation in such a way that its eigenvalues need not be real nor its eigenfunctions orthogonal. We show that for a certain choice of parameters the eigenvalue curves reveal the appearance of degenerate eigenvalues. We offer a detailed discussion of the matrix representation of the differential operator corresponding to the fractional Mathieu equation, as well as some numerical examples of its periodic solutions.

1. Introduction

There is an extensive literature dedicated to the Mathieu equation [1–3]

$$\left[\frac{\mathrm{d}^2}{\mathrm{d}t^2} + a - 2q\cos 2t\right]\Theta(t;q) = 0, \tag{1}$$

where $q \in \mathbb{R}$ is an equation parameter and $a \in \mathbb{C}$ is the eigenvalue (or characteristic number) of the equation. A classical problem associated with this equation is the determination of the eigenvalues $a(q)$ for which the solution $\Theta(t;q)$ is periodic in t.

Different forms of the Mathieu equation have also been studied, which include damping [4–7] and nonlinearities [8, 9], as well as quasi-periodic [10] and nonhomogeneous [11] forms.

In a recent paper, Rand et al. [12] investigated the fractional Mathieu equation

$$\left[\frac{\mathrm{d}^2}{\mathrm{d}t^2} + a - 2q\cos 2t + c\ {}_{-\infty}\mathrm{D}_t^{\alpha}\right]\Theta(t;q) = 0, \tag{2}$$

where $c \in \mathbb{R}$ is a constant and ${}_{-\infty}\mathrm{D}_t^{\alpha}$ denotes the Riemann-Liouville fractional derivative of order $\alpha \geq 0$ and lower limit $b = -\infty$ defined as [13]

$${}_{-\infty}\mathrm{D}_t^{\alpha} f(t) = \frac{1}{\Gamma(n-\alpha)}\frac{\mathrm{d}^n}{\mathrm{d}t^n}\int_{b=-\infty}^{t}\frac{f(\xi)}{(t-\xi)^{1+\alpha-n}}\mathrm{d}\xi, \qquad \alpha \geq 0, \tag{3}$$

with n being the smallest integer exceeding α. Rand et al. studied the case $\alpha \in (0,1)$ and obtained approximated series expressions of the first two eigenvalue curves $a(q)$ of the fractional Mathieu equation for small values of the parameter q.

There are several contributions in this paper that we consider relevant. First, we extend the analysis of the fractional Mathieu equation (2) presented in [12] by studying the behavior of the eigenvalue curves for $\alpha \in (0,2)$ and a much wider range of the parameter q. This more detailed analysis of the eigenvalue curves reveals the existence of degenerate eigenvalues of the fractional Mathieu equation which was not observed in [12], since the authors used a Fourier expansion of the solution with too few terms. A similar result has been observed for a fractional Ince equation [14]. We describe the properties of the fractional Mathieu functions (the solutions of (2)) in terms of periodicity and parity. Additionally, we discuss the characteristics of the degenerate solutions of the equation (corresponding to the degenerate eigenvalues).

Our analysis begins by analyzing the periodic solutions of Eq. (2) and the resulting matrix equations. We then illustrate the phenomenon of degeneracy in the eigenvalue curves. Finally we show some numerical examples of the solutions of the fractional Mathieu equation.

2. Solutions of the fractional Mathieu equation

We begin by recalling that the even and odd periodic solutions of the classical Mathieu equation (1) can be expanded in terms of Fourier series. The corresponding trigonometric expansions fall into four classes, according to their symmetry or antisymmetry, about $t = 0$ and $t = \pi/2$, namely [2]

$$\mathrm{ce}_{2n}(t;q) = \sum_{r=0}^{\infty} A_{2r}^{(2n)}(q) \cos 2rt, \tag{4a}$$

$$\mathrm{ce}_{2n+1}(t;q) = \sum_{r=0}^{\infty} A_{2r+1}^{(2n+1)}(q) \cos (2r+1)t, \tag{4b}$$

$$\mathrm{se}_{2n+1}(t;q) = \sum_{r=0}^{\infty} B_{2r+1}^{(2n+1)}(q) \sin (2r+1)t, \tag{4c}$$

$$\mathrm{se}_{2n+2}(t;q) = \sum_{r=0}^{\infty} B_{2r+2}^{(2n+2)}(q) \sin (2r+2)t, \tag{4d}$$

for $n = 0, 1, 2, \ldots$ To each family there belongs an infinite set of real eigenvalues, labeled a_{2n}, a_{2n+1}, b_{2n+1}, and b_{2n+2}, correspondingly. For $q > 0$ they follow the order $a_0 < b_1 < a_1 < b_2 < a_2 < \cdots$. Additionally, the Mathieu functions satisfy the following orthogonality properties [1]

$$\int_0^{2\pi} \mathrm{ce}_m(t;q)\mathrm{ce}_n(t;q)\mathrm{d}t = \int_0^{2\pi} \mathrm{se}_m(t;q)\mathrm{se}_n(t;q)\mathrm{d}t = \begin{cases} \pi, & \text{if } m = n \\ 0, & \text{if } m \neq n \end{cases}, \tag{5}$$

$$\int_0^{2\pi} \mathrm{ce}_m(t;q)\mathrm{se}_n(t;q)\mathrm{d}t = 0, \quad m, n \text{ positive integers}. \tag{6}$$

Equation (5) provides the normalization condition for the Fourier coefficients of the Mathieu functions, namely

$$2\{A_0^{(2n)}\}^2 + \sum_{r=1}^{\infty}\{A_{2r}^{(2n)}\}^2 = \sum_{r=1}^{\infty}\{A_{2r+1}^{(2n+1)}\}^2 = \sum_{r=1}^{\infty}\{B_{2r+1}^{(2n+1)}\}^2 = \sum_{r=1}^{\infty}\{B_{2r+2}^{(2n+2)}\}^2 = 1. \tag{7}$$

Our goal is to find periodic solutions of the fractional Mathieu equation (2). To do this, let us consider a solution in the form of a Fourier series of the form

$$\Theta(t;q) = A_0 + \sum_{r=1}^{\infty} A_r \cos rt + B_r \sin rt. \tag{8}$$

Note that $\Theta(t;q)$ can be periodic thanks to the choice $b = -\infty$ in Eq. (3). This is not the case for the standard choice of $b = 0$ (i.e. the initial time of the process, $t = 0$), which follows from the following theorem [15]:

Theorem 1 *Suppose that $f(t)$ is $(n-1)$-times continuously differentiable and $f^{(n)}(t)$ is bounded. If $f(t)$ is a non-constant periodic function with period T, the function ${}_0\mathrm{D}_t^\alpha f(t)$, where $0 < \alpha \notin \mathbb{N}$ and n is the first integer greater than α, cannot be a periodic function of period T.*

Corollary 1 *A differential equation of fractional order in the form*

$$ {}_0\mathrm{D}_t^\alpha f(t) + \Psi(f(t), f^1(t), ..., f^{(n)}(t)) = 0, \tag{9}$$

where $0 < \alpha \notin \mathbb{N}$, cannot have any non-constant smooth periodic solution.

To understand why this is the case, note that the Fourier series solution involves sine and cosine functions whose Riemann-Liouville fractional derivatives with lower limit $b = 0$ are non-periodic functions. They are given by [13]

$$ {}_0\mathrm{D}_t^\alpha \cos rt = (rt)^{-\alpha} E_{2,1-\alpha}(-r^2t^2), \tag{10a}$$

$$ {}_0\mathrm{D}_t^\alpha \sin rt = (rt)^{1-\alpha} E_{2,2-\alpha}(-r^2t^2), \tag{10b}$$

where $E_{\alpha,\beta}(z)$ is the two-parameter Mittag-Leffler function [13]

$$E_{\alpha,\beta}(z) = \sum_{k=0}^{\infty} \frac{z^k}{\Gamma(\alpha k + \beta)}, \qquad \begin{cases} \alpha > 0 \\ \beta > 0 \end{cases} . \tag{11}$$

Nevertheless, for large values of t, Eqs. (10) converge to the following periodic functions:

$$ {}_0\mathrm{D}_t^\alpha \cos rt = r^\alpha \cos(rt + \alpha\pi/2), \qquad t \to \infty, \tag{12a}$$

$$ {}_0\mathrm{D}_t^\alpha \sin rt = r^\alpha \sin(rt + \alpha\pi/2), \qquad t \to \infty. \tag{12b}$$

Mathematically, setting $b = -\infty$ in Eq. (3) leads to the same result, i.e.

$$ {}_{-\infty}\mathrm{D}_t^\alpha \cos rt = r^\alpha \cos(rt + \alpha\pi/2), \tag{13a}$$

$$ {}_{-\infty}\mathrm{D}_t^\alpha \sin rt = r^\alpha \sin(rt + \alpha\pi/2), \tag{13b}$$

$\alpha \in \mathbb{R}$. In more physical terms, the solutions are valid for the steady state of the system. Also, for this choice, the Riemann-Liouville and Caputo definitions of the fractional derivative are equivalent [13].

The Fourier solution (8) of the fractional Mathieu equation includes the four solutions (4) of the non-fractional Mathieu equation (1). Following the method of harmonic balance, we insert the series solution (8) into the fractional Mathieu equation (2) and collect terms to obtain the following recurrence relations:

$$aA_0 - qA_2 = 0, \tag{14a}$$

$$[a - 1 + q + c\cos(\alpha\pi/2)]B_1 - c\sin(\alpha\pi/2)A_1 - qB_3 = 0, \tag{14b}$$

$$c\sin(\alpha\pi/2)B_1 + [a - 1 - q + c\cos(\alpha\pi/2)]A_1 - qA_3 = 0, \tag{14c}$$

$$[a - 4 + c2^\alpha \cos(\alpha\pi/2)]B_2 - c2^\alpha \sin(\alpha\pi/2)A_2 - qB_4 = 0, \tag{14d}$$

$$-2qA_0 + c2^\alpha \sin(\alpha\pi/2)B_2 + [a - 4 + c2^\alpha \cos(\alpha\pi/2)]A_2 - qA_4 = 0. \tag{14e}$$

For $r > 3$,

$$-qB_{r-2} + [a - r^2 + cr^\alpha \cos(\alpha\pi/2)]B_r - cr^\alpha \sin(\alpha\pi/2)A_r - qB_{r+2} = 0, \tag{14f}$$

$$-qA_{r-2} + cr^\alpha \sin(\alpha\pi/2)B_r + [a - r^2 + cr^\alpha \cos(\alpha\pi/2)]A_r - qA_{r+2} = 0, \tag{14g}$$

where we used ${}_{-\infty}\mathrm{D}_t^\alpha A_0 = 0$ for $\alpha > 0$.

Note that these recurrence relation are now coupled in A_r and B_r. Additionally we can recognize two independent systems of equations, one involving only even subscripts (A_0, B_2, A_2,...) and the other involving odd subscripts (B_1, A_1, B_3,...). This means that there exist only two kinds of solutions, instead of four, as is the case of the classical Mathieu equation. The first solution is π-periodic, whereas the second is 2π-periodic, namely

$$\Theta_{2n}^{(\alpha)}(t;q) = A_0^{(2n)}(q) + \sum_{r=1}^{\infty} A_{2r}^{(2n)}(q)\cos 2rt + B_{2r}^{(2n)}(q)\sin 2rt, \tag{15a}$$

$$\Theta_{2n+1}^{(\alpha)}(t;q) = \sum_{r=0}^{\infty} A_{2r+1}^{(2n+1)}(q)\cos(2r+1)t + B_{2r+1}^{(2n+1)}(q)\sin(2r+1)t, \tag{15b}$$

where the order of the fractional derivative is indicated by the superscript (α). The eigenvalues of $\Theta_{2n}^{(\alpha)}(t;q)$ can be labeled as $a_{2n}^{(\alpha)}(q)$ and those of $\Theta_{2n+1}^{(\alpha)}(t;q)$ as $a_{2n+1}^{(\alpha)}(q)$. They follow the order $a_0 < a_2 < a_4$... and $a_1 < a_3 < a_5$..., respectively. Solutions (15) can be normalized in an analogous way as the Mathieu functions [Eqs. (7)].

3. Coefficient matrices and eigenvalue curves

Recurrence relations (14) may be written in matrix form in order to obtain the eigenvalue curves. Let M and N denote the coefficient matrices corresponding to the π- and 2π-periodic solutions of the fractional Mathieu equation, such that

$$M\Theta_{2n}^{(\alpha)} = a_{2n}^{(\alpha)}\Theta_{2n}^{(\alpha)}, \quad \text{and} \quad N\Theta_{2n+1}^{(\alpha)} = a_{2n+1}^{(\alpha)}\Theta_{2n+1}^{(\alpha)}, \tag{16}$$

i.e., they are the matrix representation of the fractional operator $[-\mathrm{d}^2/\mathrm{d}t^2 + 2q\cos 2t - c\,{}_{-\infty}\mathrm{D}_t^\alpha]$. By denoting $C \equiv c\cos(\alpha\pi/2)$ and $S \equiv c\sin(\alpha\pi/2)$ we get

$$M = \begin{pmatrix} 0 & 0 & \sqrt{2}q & 0 & 0 & 0 & 0 & \dots \\ 0 & 4-2^\alpha C & 2^\alpha S & q & 0 & 0 & 0 & \dots \\ \sqrt{2}q & -2^\alpha S & 4-2^\alpha C & 0 & q & 0 & 0 & \dots \\ 0 & q & 0 & 16-4^\alpha C & 4^\alpha S & q & 0 & \dots \\ 0 & 0 & q & -4^\alpha S & 16-4^\alpha C & 0 & q & \dots \\ \vdots & \vdots & \vdots & \vdots & \vdots & \vdots & \vdots & \ddots \end{pmatrix}, \tag{17}$$

$$N = \begin{pmatrix} 1-q-C & S & q & 0 & 0 & 0 & \dots \\ -S & 1+q-C & 0 & q & 0 & 0 & \dots \\ q & 0 & 9-3^\alpha C & 3^\alpha S & q & 0 & \dots \\ 0 & q & -3^\alpha S & 9-3^\alpha C & 0 & q & \dots \\ \vdots & \vdots & \vdots & \vdots & \vdots & \vdots & \ddots \end{pmatrix}. \tag{18}$$

The M matrix has the coefficients corresponding to A_0, B_2, A_2, B_4, A_4,... on its columns, while the columns of the N matrix follow the order B_1, A_1, B_3, A_3,... We can write one single

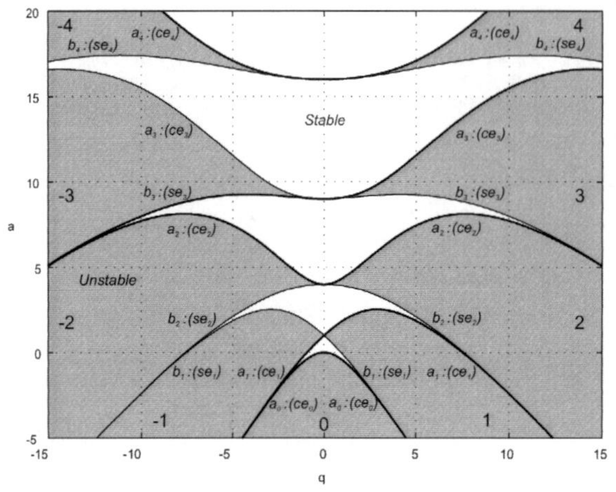

Figure 1. Eigenvalue curves of the standard Mathieu equation ($c = 0$) calculated with Eq. (8) using $r_{max} = 37$. Thick and thin lines denote curves for even ce$(t;q)$ and odd se$(t;q)$ Mathieu functions, respectively. The characteristic curves divide the plane into regions of stability (white) and instability (gray). The unstable regions are numbered sequentially for convenience.

matrix for all the coefficients, ordered as A_0, B_1, A_1, B_2, A_2, B_3, A_3, ... The matrix equation takes the form

$$\begin{pmatrix} 0 & 0 & 0 & 0 & \sqrt{2}q & 0 & 0 & 0 & 0 & 0\,0\,0\,0 & \ldots \\ 0 & 1-q-C & S & 0 & 0 & q & 0 & 0 & 0 & 0\,0\,0\,0 & \ldots \\ 0 & -S & 1+q-C & 0 & 0 & 0 & q & 0 & 0 & 0\,0\,0\,0 & \ldots \\ 0 & 0 & 0 & 4-2^\alpha C & 2^\alpha S & 0 & 0 & q & 0 & 0\,0\,0\,0 & \ldots \\ \sqrt{2}q & 0 & 0 & -2^\alpha S & 4-2^\alpha C & 0 & 0 & 0 & q & 0\,0\,0\,0 & \ldots \\ 0 & q & 0 & 0 & 0 & 9-3^\alpha C & 3^\alpha S & 0 & 0 & q\,0\,0\,0 & \ldots \\ 0 & 0 & q & 0 & 0 & -3^\alpha S & 9-3^\alpha C & 0 & 0 & 0\,q\,0\,0 & \ldots \\ 0 & 0 & 0 & q & 0 & 0 & 0 & 16-4^\alpha C & 4^\alpha S & 0\,0\,q\,0 & \ldots \\ 0 & 0 & 0 & 0 & q & 0 & 0 & -4^\alpha S & 16-4^\alpha C & 0\,0\,0\,q & \ldots \\ \vdots & \vdots & \vdots & \vdots & \vdots & \vdots & \vdots & \vdots & \vdots & \vdots\,\vdots\,\vdots\,\vdots & \ddots \end{pmatrix} \begin{pmatrix} \sqrt{2}A_0 \\ B_1 \\ A_1 \\ B_2 \\ A_2 \\ B_3 \\ A_3 \\ B_4 \\ A_5 \\ \vdots \end{pmatrix} = a \begin{pmatrix} \sqrt{2}A_0 \\ B_1 \\ A_1 \\ B_2 \\ A_2 \\ B_3 \\ A_3 \\ B_4 \\ A_5 \\ \vdots \end{pmatrix}. \quad (19)$$

The coefficient matrix (19) depends on the values of c, α, and q. Fixing c and α, one can draw the curves $a(q)$ by calculating the eigenvalues of (19). The coefficients of the solutions (15) are the entries of the corresponding eigenvectors. Numerically, the matrix has to be truncated up to a certain dimension (the solutions (15) are truncated up to a given r_{max}) and then one can obtain its eigenvalues for a given domain of q.

To make suitable comparisons, in Fig. 1 we plot the known eigenvalue curves $a(q)$ and $b(q)$ for the non-fractional ($c = 0$) standard Mathieu equation [2]. The coefficient matrix in Eq. (19) becomes Hermitian, which results in it having an infinite set of real discrete eigenvalues and orthogonal eigenvectors. Thick and thin lines denote characteristic curves for even $\mathrm{ce}_m(t;q)$ and odd $\mathrm{se}_m(t;q)$ Mathieu functions, respectively. Even-order and odd-order curves are symmetrical and anti-symmetrical about the a-axis, respectively. Except for the trivial case $q = 0$, the curves do not intersect and thus divide the (q, a)-plane into regions where the solutions of the Mathieu equation Eq. (1) are stable (white regions) or unstable (gray regions). We have numbered the unstable regions for ease of reference, as shown in Fig. 1. Note that the j-th unstable region touches the axis $q = 0$ at values $a = |j|^2$.

The appearance of the fractional derivative gives rise to off-diagonal terms of the form $\pm r^\alpha S$ in the coefficient matrix Eq. (19), making it non-Hermitian. Thus, for nonzero c and α one cannot expect its eigenvalues to be all real nor its eigenvectors to be orthogonal. In Fig. 2 we

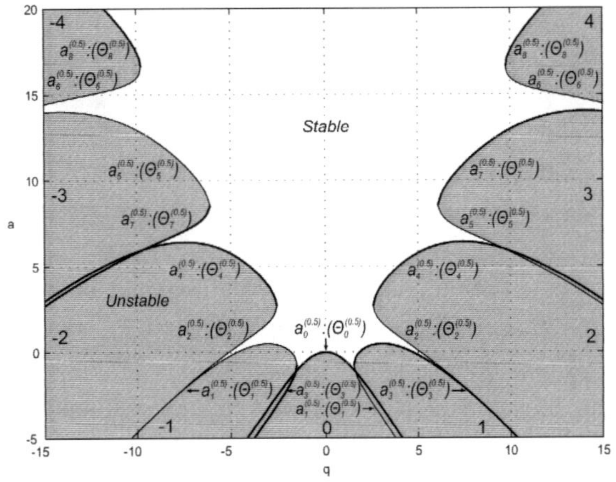

Figure 2. Eigenvalue curves of the fractional Mathieu equation ($c = 2$, $\alpha = 0.5$) calculated with Eq. (8) using $r_{max} = 37$. Cross points of the curves represent degenerate eigenvalues of the fractional Mathieu equation. Unstable regions are numbered sequentially starting from the lowest curve.

plotted the eigenvalue curves $a(q)$ for the fractional Mathieu equation with $c = 2$ and $\alpha = 0.5$. Once again the curves divide the (q, a)-plane into regions where the solutions of the fractional Mathieu equation are stable (white region) or unstable (gray region). The unstable regions are numbered sequentially following the same convention as in Fig. 1. In general for $c > 0$ and $\alpha \in (0, 1)$ we observe the following:

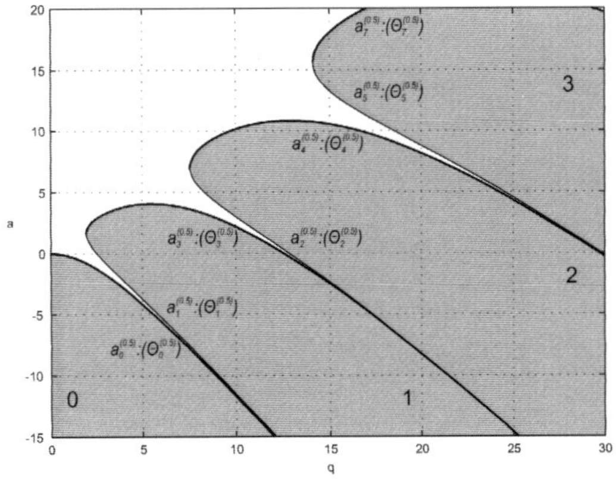

Figure 3. Eigenvalue curves of the fractional Mathieu equation ($c = 2$, $\alpha = 1.3$) calculated with a truncated solution [Eq. (8)] using $r_{max} = 37$. The curves, unlike the case $\alpha \in (0, 1)$, no longer cross.

- Except for the lowest-order curve $a_0(q)$, each eigenvalue curve of the fractional Mathieu equation becomes "detached" from the a-axis. The upper and lower boundaries of the unstable regions correspond to the original curves $a(q)$ and $b(q)$ of the standard Mathieu equation and are depicted using different line-widths in Fig. 2.
- The upper and lower eigenvalue curves of an unstable region connect smoothly, forming a single continuous boundary-curve. For $q > 0$, the *turning-point* of the boundary-curve is

the minimum value of q for which the eigenvalue a is real. For each boundary-curve, the solutions above and below of the turning-points are labeled with different subindexes.

- Mathieu solutions corresponding to different points (q, a) lying on the same boundary-curve have the same periodicity, namely, π and 2π for boundary-curves of even and odd unstable regions, respectively.
- For a given q, there is only a finite number of real eigenvalues, which increases as q increases. Consequently, unlike the classical Mathieu equation, the fractional Mathieu equation has a finite number of real periodic solutions for a given value of q.
- In the "detached" (stable) region, the eigenvalues become complex, and they appear in complex conjugate pairs (i.e. $a_{2n+1}^{(\alpha)} = \bar{a}_{2n+3}^{(\alpha)}$ and $a_{2n+2}^{(\alpha)} = \bar{a}_{2n+4}^{(\alpha)}$, n even).
- There exist real values q^* where the eigenvalue curves intersect, indicating the existence of degenerate eigenvalues or *double points* of the fractional Mathieu equation. In particular, the crossings occur for all pairs of curves $\{a_{2n}^{(\alpha)}, a_{2n+1}^{(\alpha)}\}$ when $q > 0$, and for the curves $\{a_{4n}^{(\alpha)}, a_{4n+2}^{(\alpha)}\}$ and $\{a_{4n+1}^{(\alpha)}, a_{4n+3}^{(\alpha)}\}$ when $q < 0$. Incidentally, it is known that the classical Mathieu equation (1) exhibits double eigenvalues for complex values of q. (See [16, 17] for the case of purely imaginary q.)
- The adjacent curves become asymptotically parallel after the crossing, therefore they only cross once.

For $\alpha \in [1, 2]$, the eigenvalue curves no longer cross and thus there are no degenerate eigenvalues, as shown in Fig. 3. As expected, when $\alpha = 2$, we recover the standard Mathieu equation scaled by a factor $1 + c$, and the curves become "reattached", as shown in Fig. 1.

4. Behavior of the fractional Mathieu functions

The behavior of the solutions of the fractional Mathieu equation 2 is fairly complicated, particularly because we have to understand the dependence of the functions on the variable t and the parameters q and c and the fractional-order α. Let us consider first the effect of varying the order of the fractional derivative term. The behavior of the normalized fractional solution $\Theta_3^{(\alpha)}(t, 2)$ with $c = 1$ and $q = 2$ as a function of t is depicted in Fig. 4 for the range $\alpha \in [0, 2]$. Solid curves correspond to $\alpha = \{0, 0.5, 1, 1.5, 2\}$. When α vanishes, the fractional Mathieu equation reduces to the standard Mathieu equation and $\Theta_3^{(\alpha)}(t; q)$ becomes $\mathrm{ce}_1(t; q)$. Similarly, when $\alpha = 2$, the fractional Mathieu equation reduces to a scaled version of the standard Mathieu equation and $\Theta_3^{(\alpha)}(t; q)$ now becomes $\mathrm{ce}_1(t; q/(1 + c))$. In this way, the Mathieu functions $\mathrm{ce}_1(t; q)$ and $\mathrm{ce}_1(t; q/(1 + c))$ have been smoothly connected by varying the order of the fractional term of the fractional Mathieu equation. Note that whereas the even and odd integer-order Mathieu functions are symmetrical and antisymmetrical about the origin, respectively, the fractional solutions (i.e. $\alpha \neq 0, 2$) do not have a definite parity.

Unlike the Mathieu functions, two solutions of the fractional Mathieu equation belonging to the same family (Eqs. (15a) or (15b)) with different eigenvalues a are not orthogonal for a given q. This can be observed by letting Θ_m and Θ_n (omitting the superscript α) be two solutions of the same family (π- or 2π-periodic) with equal q and different eigenvalues a_m, a_n. By writing $\mathrm{D}^\alpha \equiv {}_{-\infty}\mathrm{D}_t^\alpha$ the solutions satisfy the equations

$$\Theta_m'' + (a_m - 2q\cos 2t)\Theta_m + c\mathrm{D}^\alpha\Theta_m = 0, \tag{20a}$$

$$\Theta_n'' + (a_n - 2q\cos 2t)\Theta_n + c\mathrm{D}^\alpha\Theta_n = 0. \tag{20b}$$

Multiplying the first equation by Θ_n and the second equation by Θ_m and subtracting both, we get

$$\Theta_m''\Theta_n - \Theta_n''\Theta_m = (a_n - a_m)\Theta_m\Theta_n + c[\Theta_m\mathrm{D}^\alpha\Theta_n - \Theta_n\mathrm{D}^\alpha\Theta_m]. \tag{21}$$

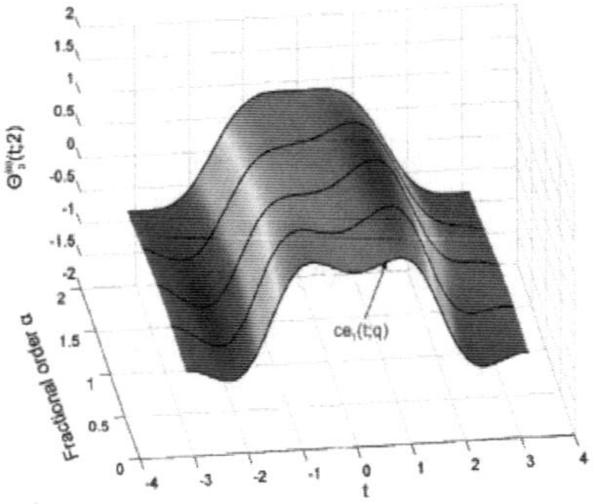

Figure 4. $\Theta_3^{(\alpha)}(t;2)$ solution of the fractional Mathieu equation for $\alpha \in [0,2]$. Solid curves correspond to $\alpha = \{0, 0.5, 1, 1.5, 2\}$.

Integrating both sides between t_1 and t_2, we get, after integrating by parts the LHS

$$[\Theta'_m\Theta_n - \Theta'_n\Theta_m]_{t_1}^{t_2} = (a_n - a_m)\int_{t_1}^{t_2}\Theta_m\Theta_n \mathrm{d}t + c\int_{t_1}^{t_2}[\Theta_m \mathrm{D}^\alpha\Theta_n - \Theta_n \mathrm{D}^\alpha\Theta_m]\mathrm{d}t. \quad (22)$$

Since the Θ solutions are π- or 2π-periodic, we set $t_1 = 0$ and $t_2 = 2\pi$, so that the LHS vanishes, and we obtain

$$(a_n - a_m)\int_0^{2\pi}\Theta_m\Theta_n \mathrm{d}t + c\int_0^{2\pi}[\Theta_m \mathrm{D}^\alpha\Theta_n - \Theta_n \mathrm{D}^\alpha\Theta_m]\mathrm{d}t = 0. \quad (23)$$

In the case $c = 0$ (or $\alpha = 0$), this shows that two functions of the same family with different eigenvalues are orthogonal. We can show by direct substitution that the second member of the LHS does not vanish. Take, for example, two solutions of the 2π-periodic family, say

$$\Theta_m(t;q) = \sum_{r=1,3,\ldots}^{\infty} A_r^{(m)}\cos rt + B_r^{(m)}\sin rt, \quad (24)$$

$$\Theta_n(t;q) = \sum_{r=1,3,\ldots}^{\infty} A_r^{(n)}\cos rt + B_r^{(n)}\sin rt. \quad (25)$$

For this case, we have

$$\int_0^{2\pi}[\Theta_m \mathrm{D}^\alpha\Theta_n - \Theta_n \mathrm{D}^\alpha\Theta_m]\mathrm{d}t = 2\pi\sin(\alpha\pi/2)\sum_{r=1,3,\ldots}^{\infty} r^\alpha[A_r^{(n)}B_r^{(m)} - A_r^{(m)}B_r^{(n)}], \quad (26)$$

$$\neq 0, \quad (27)$$

which means

$$\int_0^{2\pi}\Theta_m\Theta_n \mathrm{d}t \neq 0. \quad (28)$$

Therefore, in general $\Theta_m^{(\alpha)}(t;q)$ and $\Theta_n^{(\alpha)}(t;q)$ are not orthogonal.

5. Behavior of degenerate solutions of the fractional Mathieu equation

Since the fractional Mathieu equation has a periodic solution for each point (q, a) lying on the eigenvalue curves in Fig. 2, the cross-points between adjacent curves define degenerate Mathieu

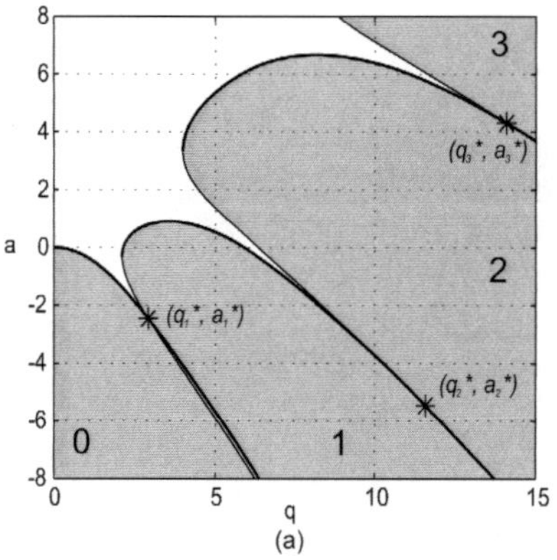

(a)

Crossing points

	q_i^*	a_i^*
i=1	2.9314097325	-2.46655990
i=2	11.5799970244	-5.45929743
i=3	14.1259933909	4.29884188

(b)

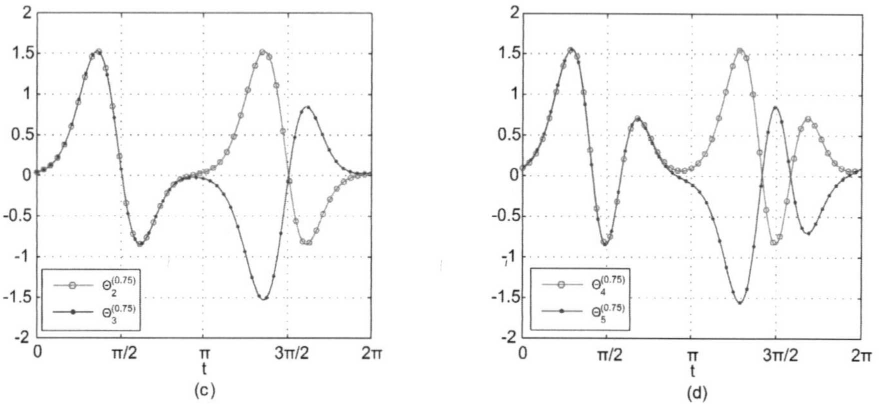

(c) (d)

Figure 5. (a) (b) The first three eigenvalue curves and its corresponding cross-points for the fractional Mathieu equation with $c = 2$, $\alpha = 0.75$. (c) (d) Degenerate fractional Mathieu functions for the pairs (q_2^*, a_2^*) and (q_3^*, a_3^*).

solutions with the same values of q and a. Let us illustrate the behavior of the degenerate solutions. Figure 5(a) shows the first three eigenvalue curves $a(q)$ for $\alpha = 0.75$ and $c = 2$. The crossing points locating the degenerate eigenvalues are marked with asterisks on the curves and their numerical values are given up to ten decimals of precision in Fig. 5(b). Explicitly we have

$$a_0^{(\alpha)}(q_1^*) = a_1^{(\alpha)}(q_1^*), \quad a_2^{(\alpha)}(q_2^*) = a_3^{(\alpha)}(q_2^*), \quad a_4^{(\alpha)}(q_3^*) = a_5^{(\alpha)}(q_3^*). \tag{29}$$

The solutions can be easily plotted once the eigenvectors of matrices (17) and (18) have been calculated.

The behavior of the second and third pairs of degenerate fractional Mathieu functions is depicted in Figs. 5(c) and 5(d). Consider, for instance, the pair (q_2^*, a_2^*). The degenerate functions $\Theta_2^{(0.75)}(t; q_2^*)$ and $\Theta_3^{(0.75)}(t; q_2^*)$ satisfy the same differential equation, namely

$$\left[\frac{\mathrm{d}^2}{\mathrm{d}t^2} + a_2^* - 2q_2^* \cos 2t + 2 \ {}_{-\infty}\mathrm{D}_t^{0.75}\right] \Theta^{(0.75)}(t; q_2^*) = 0. \tag{30}$$

Note that the even-order functions (red-circle lines) are π-periodic whereas odd-order functions (blue-point lines) are 2π-periodic. In general, in any pair of degenerate solutions one of them has period π and the other 2π. In particular, the π-periodic functions (red-circle lines) shown in Figs. 5(c) and 5(d) belong to the same boundary-curve of the second unstable region.

6. Conclusions

This work extends the analysis given in [12] concerning the solutions and the eigenvalue curves of the fractional Mathieu equation. The matrix form of the problem indicates that only two families of solutions exist, unlike the ordinary Mathieu equation. These solutions, valid for the steady state of the system to be studied, are not orthogonal and have no defined parity. We showed the appearance of degenerate eigenvalues for $c > 0$, $\alpha \in (0, 1)$ (seen as cross-points of the eigenvalue curves), which had not been observed in the previous treatment of the fractional Mathieu equation. This gives rise to new questions regarding the stability of the solutions in the overlapping regions on the (q, a)-plane.

Acknowledgements
This research was supported by Consejo Nacional de Ciencia y Tecnologia Mexico (Grant No. 82407) and by Tecnologico de Monterrey.

References

[1] Arscott F M 1964 *Periodic Differential Equations* (London: Pergamon Press)
[2] McLachlan N W 1951 *Theory and Application of Mathieu Functions* (Oxford: Clarendon Press)
[3] Gutiérrez-Vega JC, Rodríguez-Dagnino RM, Meneses-Nava MA, and Chávez-Cerda S 2003 Mathieu functions, a visual approach *Am J Phys* **71** 233–42
[4] Taylor J H, Narendra K S 1969 Stability regions for the damped Mathieu equation *SIAM J Appl Math* **17**, 343–52
[5] Richards J A 1976 Stability diagram approximation for the lossy Mathieu equation *SIAM J Appl Math* **30** 240–47
[6] Ramani D V, Rand R H and Keith W L 2001 Perturbation solution for bifurcation in the quadratically damped Mathieu equation *Proceedings of the 2001 ASME Design Engineering Technical Conferences* (Pittsburgh, PA)
[7] Gunderson H, Rigas H, VanVleck F S 1974 A technique for determining stability regions for the damped Mathieu equation *SIAM J Appl Math* **26** 345–49
[8] El-Dib Y O 2001 Nonlinear Mathieu equation and coupled resonance mechanism *Chaos, Solitons & Fractals* **12** 705–20
[9] Ng L, Rand R H 2002 Bifurcations in a Mathieu equation with cubic nonlinearities *Chaos, Solitons & Fractals* **14** 173–81
[10] Zounes R S, Rand R H 1998 Transition curves for the quasi-periodic Mathieu equation *SIAM J Appl Math* **58** 1094–115
[11] Insperger T, Stépán G 2002 Stability chart for the delayed Mathieu equation *Proceedings: Mathematical, Physical and Engineering Sciences* **458** 1989–98
[12] Rand RH, Sah SM, Suchorsky MK 2010 Fractional Mathieu equation *Commun Nonlinear Sci Numer Simulat* **15** 3254–62
[13] Podlubny I 1990 *Fractional Differential Equations* (San Diego: Academic Press)
[14] Parra Hinojosa A, Gutiérrez Vega J C 2013 Fractional Ince equation with a Riemann–Liouville fractional derivative. Applied Mathematics and Computation **219** 10695–705
[15] Tavazoei M S 2010 A note on fractional-order derivatives of periodic functions *Automatica* **46** 945–48
[16] Blanch G, Clemm D S 1969 The double points of Mathieu's differential equation *Math Comp* **23** 97–108
[17] Shivakumar P N, Xue J 1999 On the double points of a Mathieu equation *J Comput Appl Math* **107** 111–25

Coupling n-level Atoms with ℓ-modes of Quantised Light in a Resonator

O Castaños, S Cordero, E Nahmad-Achar and R López-Peña

Instituto de Ciencias Nucleares, Universidad Nacional Autónoma de México, Apartado Postal 70-543, 04510 México DF, Mexico

E-mail: ocasta@nucleares.unam.mx

Abstract. We study the quantum phase transitions associated to the Hamiltonian of a system of n-level atoms interacting with ℓ modes of electromagnetic radiation in a resonator. The quantum phase diagrams are determined in analytic form by means of a variational procedure where the test function is constructed in terms of a tensorial product of coherent states describing the matter and the radiation field. We demonstrate that the system can be reduced to a set of Dicke models.

1. Introduction

The quantum electrodynamics (QED) in cavities has its origin in the works of Einstein on the statistical theory of light and its interaction with matter. The photoelectric effect and the ultraviolet photoionization are considered the first applications of QED, where the quantisation of the electromagnetic field is fundamental [1]. The quantized coupling between matter and radiation was firstly proposed in the Jaynes-Cummings model (JCM) [2]. However, the JCM model was regarded as a toy model because an excited atom in free space decays in an irreversible form. This situation changes dramatically when the atom is inside a resonator because, depending on its quality factor, the half-life of a photon in the cavity can be up to one second. This means that the atom can now emit and absorb the photon in a reversible manner [3, 4].

A schematic model of a resonator is formed by two mirrors in whose interior is an atom at rest (or passing through the cavity). There are 4 parameters appearing in the physical system: the atom-field coupling, μ, the dipolar atomic radiation of the atom to modes of the electromagnetic field different from the mode of the cavity, γ, the rate of photon decay in the cavity κ (related to its quality factor), and, for atoms passing through, the transit time through the cavity T. The strong coupling regime considers that $\mu \gg \{\gamma, \kappa, T^{-1}\}$. This regime has been used to study entanglement properties between matter and radiation; in particular Haroche and his collaborators have generated Schrödinger cat states using the entanglement properties as a source of information [5]. There are other procedures that use the entanglement properties to generate Schrödinger cat states, for example Wineland and his group used the entanglement between the vibrational states of the molecules with their corresponding hyperfine states [6].

Although the JCM model captures the essence of the interaction dynamics between the matter and the radiation field, there are dissipative processess that appear naturally in these systems of cavities and atoms. Recently effects on the dissipation rates of the uncertainty in the position of the atoms when they are interacting with a reservoir has been considered [7]. The dissipative processess are related to the parameters mentioned above γ, κ, and T; additionally one has to consider the position of the atoms. The position dependence can be made negligible by considering Fabry-Perot optical cavities [3]. By means of these physical systems one may transmit information from one atom in a cavity to another atom in a second cavity [8].

In this work we determine the quantum phase transitions for n-level atoms interacting with ℓ modes of electromagnetic radiation in a cavity [9]. We establish analytic expressions for the energies associated to the ground state for arbitrary values of the dipolar strengths, together with their corresponding eigenstates. Additionally we demonstrate that the phase diagram associated to the transition from the normal to the superradiant regime is reduced to a set of phase sub-diagrams, each of them according as the quantum phase transition found by Hepp and Lieb for the Dicke model [10,11]. For the transitions between superradiant regions one finds always first order phase transitions and the different boundaries correspond to Maxwell sets.

2. Model Hamiltonian

The matter and the electromagnetic field are described by the Hamiltonian

$$\boldsymbol{H} = \sum_{j=1}^{n} \hbar\, \omega_j \boldsymbol{A}_{jj} + \sum_{j<k}^{n} \hbar\Omega_{jk}\, \boldsymbol{\nu}_{jk} \,, \tag{1}$$

where one introduces the convention $\omega_1 \leq \omega_2 \leq \cdots \leq \omega_n$. The operators $\boldsymbol{A}_{jj}$ and $\boldsymbol{\nu}_{jk} \equiv \boldsymbol{a}^\dagger_{jk}\boldsymbol{a}_{jk}$ define the number operator of particles in level j and the number of photons in the mode Ω_{jk}, respectively. The detuning parameter of the transition $j \to k$ is denoted by $\Delta_{jk} = \Omega_{jk} - (\omega_k - \omega_j)$, with $\omega_k > \omega_j$ [9]. A schematic diagram of the system is shown in Fig. 1.

The interaction between the matter and the radiation field, under the long wavelength and dipolar approximations, is written in two parts as follows:

$$\boldsymbol{H}_{FM} = \boldsymbol{O}_R + \boldsymbol{O}_{CR} \,.$$

They are called the rotating and the counter-rotating terms, and are given by the expressions

$$\boldsymbol{O}_R \;=\; -\frac{1}{\sqrt{N_a}} \sum_{j=1}^{n-1} \sum_{k=j+1}^{n} \mu_{jk} (\boldsymbol{A}_{jk}\, \boldsymbol{a}^\dagger_{jk} + \boldsymbol{A}_{kj}\, \boldsymbol{a}_{jk}) \,, \tag{2}$$

$$\boldsymbol{O}_{CR} \;=\; -\frac{1}{\sqrt{N_a}} \sum_{j=1}^{n-1} \sum_{k=j+1}^{n} \mu_{jk} (\boldsymbol{A}_{jk}\, \boldsymbol{a}_{jk} + \boldsymbol{A}_{kj}\, \boldsymbol{a}^\dagger_{jk}) \,, \tag{3}$$

where the coupling strength of the matter-field interaction μ_{jk} is

$$\mu_{jk} = \hbar\, (\omega_j - \omega_k) d_{jk} \sqrt{\frac{2\,\pi\,\rho_m}{\hbar\Omega_{jk}}} \,.$$

The symbol d_{jk} is the matrix element of the dipolar operator $\vec{d} = e\,\vec{r}$, Ω_{jk} is a photonic mode of the cavity, and ρ_m indicates the number of atoms in the quantization volume. The matter operators can be realized in terms of bosonic operators $\boldsymbol{A}_{jk} = \boldsymbol{b}^\dagger_j\, \boldsymbol{b}_k$, and they thus satisfy the commutation relations of a unitary algebra in n dimensions, i.e.,

$$[\boldsymbol{A}_{jk}, \boldsymbol{A}_{mn}] = \boldsymbol{A}_{jn}\, \delta_{mk} - \boldsymbol{A}_{mk}\, \delta_{jn} \,, \tag{4}$$

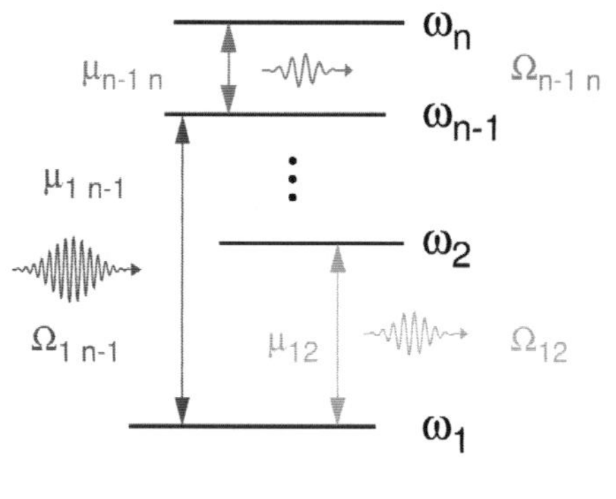

Figure 1. Schematic diagram of an n-level atom interacting with ℓ modes of electromagnetic field, where transitions between arbitrary pair of atomic levels are promoted by one mode of the electromagnetic field.

while the creation and annihilation photon operators $\boldsymbol{a}^\dagger_{jk}$, $\boldsymbol{a}_{jk}$, satisfy the commutation relations of a Heisenberg-Weyl algebra

$$[\boldsymbol{a}_{jk}, \boldsymbol{a}^\dagger_{mn}] = \delta_{jm}\,\delta_{kn}\,, \quad [\boldsymbol{a}_{jk}, \boldsymbol{a}_{mn}] = [\boldsymbol{a}^\dagger_{jk}, \boldsymbol{a}^\dagger_{mn}] = 0\,. \tag{5}$$

In the rotating wave approximation (RWA), the Hamiltonian system has the following constants of motion

$$\boldsymbol{K}_j = \boldsymbol{A}_{jj} + \sum_{k=1}^{j-1} \boldsymbol{\nu}_{kj} - \sum_{k=j+1}^{n} \boldsymbol{\nu}_{jk}\,, \tag{6}$$

because it is straightforward to calculate that $[\boldsymbol{K}_j, \boldsymbol{O}_R] = 0$ for all $j = 1, 2, \cdots, n$. Therefore, for the RWA approximation one has n constants of the motion (notice that they include the total number of particles $N_a = \sum_j Ajj$). From here on, we shall call the Hamiltonian in this approximation the *generalised Tavis-Cummings model* [12].

For the counter-rotating term one finds $[\boldsymbol{K}_j, \boldsymbol{O}_{CR}] \neq 0$. However, it is easy to prove that the parity operators of the mentioned constants of motion are invariants, that is,

$$e^{i\pi \boldsymbol{K}_j}\, \boldsymbol{O}_{CR}\, e^{-i\pi \boldsymbol{K}_j} = \boldsymbol{O}_{CR} \tag{7}$$

because $e^{i\pi \boldsymbol{K}_j}\, \boldsymbol{a}_{jk}\, e^{-i\pi \boldsymbol{K}_j} = -\boldsymbol{a}_{jk}$ and $e^{i\pi \boldsymbol{K}_j}\, \boldsymbol{A}_{jk}\, e^{-i\pi \boldsymbol{K}_j} = -\boldsymbol{A}_{jk}$, for all j, k.

The basis states of the Hilbert space can be defined in terms of Fock states as

$$\{\bigotimes_{j<k} |\nu_{jk}\rangle\} \otimes \{\bigotimes_{m=1}^{n} |n_m\rangle\} \equiv |\nu_{12}\cdots\nu_{1n}\,\nu_{23}\cdots\nu_{2n}\cdots\nu_{(n-1)n}\,;\, n_1,\cdots n_n\rangle\,, \tag{8}$$

with (j, k) indicating the mode of the cavity, and m the atomic quantum level.

In the RWA approximation we have to substitute the quantum labels of the Fock states of the electromagnetic field in terms of $n-1$ constants of motion and one of the labels of the matter Fock states in terms of the number of atoms. In contrast, for the Hamiltonian system without the RWA one has to ask that the basis states have the defined parity of $n-1$ of the n constants of motion, not including the number of atoms. Thus the Hilbert space of the complete Hamiltonian, called the *generalised Dicke model*, is divided into 2^{n-1} subspaces.

3. Quantum phase diagrams

In order to determine the ground state of the Hamiltonian system, one can consider the fidelity susceptibility approach to find the values of the interaction parameters where the ground state changes suddenly [13–16]. However, the size of the matrices increases very much for large values of the coupling parameters of the order of $\mu_{jk} \approx 1$. Instead we use the catastrophe formalism which considers the following procedure [15, 17]:

- Propose a variational state, living in a large enough Hilbert space; we have considered the associated coherent states to the matter and field parts.
- Calculate the expectation value of the Hamiltonian, which defines the energy surface. The energy surface is constructed as an intensive function that depends on variables and parameters, i.e.,
$$\{E(x,c)|x = x_1, x_2, \cdots; c = c_1, c_2, \cdots\},$$
where the x's are order parameters and the c's control parameters. This energy surface defines also a Hamiltonian function in classical mechanics, and can be used to study the corresponding semiclassical dynamics [18]
- The catastrophe term was invented by E. C. Zeeman. Catastrophes are abrupt changes in a state equation as a sudden response to a smooth change in the external conditions [19]. The catastrophe formalism is used to determine the quantum phase diagrams of the system through the expressions
$$\frac{\partial E}{\partial x_k} = 0, \quad x = x^p(c), \quad E_{ij} = \frac{\partial^2 E}{\partial x_i \partial x_j}\Big|_{x^p(c)},$$
which determine the state equation, the critical points, and the stability matrix, respectively.
- When $E^p = E^{p+1}$ for two different critical points denoted by p and $p+1$,
$$\left(\frac{\partial E^p}{\partial c_\alpha} - \frac{\partial E^{p+1}}{\partial c_\alpha}\right)\delta c_\alpha = 0,$$
define the Maxwell sets.

A quantum phase transition occurs when the system jumps from one critical point to another or there are degenerate critical points.

The order of the phase transition can be determined following the Ehrenfest classification. A phase transition takes place between the p and q branches of critical points, is of nth-order, if

$$\lim_{\delta\to 0} \frac{\partial^i E^{(p)}(s)}{\partial s^i}\Bigg|_{s_0-\delta} = \lim_{\delta\to 0} \frac{\partial^i E^{(q)}(s)}{\partial s^i}\Bigg|_{s_0+\delta}, \tag{9}$$

for $i = 0, 1, 2, \ldots, n-1$, but fails for $i = n$ [15].

4. Variational function and the energy surface

Glauber [20] developed the theory of quantum coherence and indicated its importance in optics problems. Sudarshan [21] showed the equivalence of semiclassical and quantum descriptions of beam lights, while Klauder [22] established their connections with the continuous representations theory. Therefore, the variational state for the electromagnetic field is determined by the tensorial product of the coherent states introduced in the sixties

$$|\vec{\alpha}\rangle = \bigotimes_{j<k}^{n} \left\{ e^{-\frac{|\alpha_{jk}|^2}{2}} \sum_{\ell_{jk}} \frac{(\alpha_{jk})^{\ell_{jk}}}{\sqrt{\ell_{jk}!}} |\ell_{jk}\rangle \right\}, \tag{10}$$

while the coherent states of the unitary group in n dimensions U(n) are used for the matter part [23–25]

$$|\vec{\gamma}\rangle = \frac{1}{\sqrt{N_a!}} \left\{ \frac{1}{\sqrt{\sum |\gamma_k|^2}} \sum_k \gamma_k \, \boldsymbol{b}_k^\dagger \right\}^{N_a} |0\rangle \,. \tag{11}$$

with $\boldsymbol{b}_k|0\rangle = 0$ for $k = 1, 2, \cdots, n$. The condensate or the coherent state for the completly symmetric representation of U(n) has a redundant parameter which without loss of generality one takes that $\gamma_1 = 1$ [25]. Therefore the test function is given by $|\vec{\alpha}\rangle \otimes |\vec{\gamma}\rangle$, which depends on $2(n + \ell - 1)$ real parameters.

Then the energy surface, defined as the expectation value of the Hamiltonian with respect to the test function, takes the form

$$E = \sum_{j<k}^{n} \Omega_{jk} \, r_{jk}^2 + \sum_{j}^{n} \omega_j \frac{\rho_j^2}{1+R_2} - 4 \sum_{j<k}^{n} \mu_{jk} \, r_{jk} \frac{\rho_j \, \rho_k \cos\theta_{jk} \cos(\phi_k - \phi_j)}{1+R_2} \,, \tag{12}$$

where we have defined

$$R_2 = \sum_{k=2}^{n} \rho_k^2 \,, \quad \alpha_{jk} = \sqrt{N_a} \, r_{jk} \quad e^{i\theta_{jk}} \, \gamma_k = \rho_k \, e^{i\phi_k} \,,$$

with $j, k = 1, 2, \cdots, n$. Notice that the energy surface is an intensive quantity, i.e., it is independent of the number of atoms.

The critical points are obtained by the following set of derivatives

$$\left(\frac{\partial E}{\partial \rho_2}, \cdots, \frac{\partial E}{\partial \rho_n}, \frac{\partial E}{\partial r_{12}}, \cdots, \frac{\partial E}{\partial r_{n-1n}}, \frac{\partial E}{\partial \theta_{12}}, \cdots, \frac{\partial E}{\partial \theta_{n-1,n}}, \frac{\partial E}{\partial \phi_2}, \cdots, \frac{\partial E}{\partial \phi_n} \right) = 0 \,, \tag{13}$$

which imply the conditions:

$$\theta_{jk}^c = 0, \pi \,, \quad \phi_k^c - \phi_j^c = 0, \pi \,, \quad r_{jk}^c = 2 \, \frac{\mu_{jk} \, \rho_j^c \, \rho_k^c}{\Omega_{jk}(1+R_2)} \,, \tag{14}$$

with $j < k$, and $k = 2, \cdots, n$. We have taken the critical angles equal to zero because in this work we are only considering positive values for the dipolar interactions μ_{ij}. By substituting these conditions into the expression for the energy surface, one gets

$$E = \sum_{j=2}^{n} \frac{\omega_j \, \rho_j^{c\,2}}{1+R_2} - 4 \sum_{j<k}^{n} \frac{\mu_{jk}^2}{\Omega_{jk}} \left(\frac{\rho_j^c \, \rho_k^c}{1+R_2} \right)^2 \,. \tag{15}$$

The energy surface for the RWA approximation is obtained by replacing $\mu_{jk} \to \mu_{jk}^{\rm RWA}/2$ [9], implying a rescaling of the phase diagram. This must be also done for any expectation values of matter-field observables.

The critical values for the variables ρ_j^c are obtained from the set of algebraic equations

$$\frac{\rho_j}{1+R_2} \left(\omega_j - \sum_{k=1}^{n} \frac{\omega_k \, \rho_k^2}{1+R_2} - 4 \sum_{k=1, k\neq j}^{n} \frac{\mu_{jk}^2 \, \rho_k^2}{\Omega_{jk} \, (1+R_2)} + 8 \sum_{j' \leq k'}^{n} \frac{\mu_{j'k'}^2 \, \rho_{j'}^2 \, \rho_{k'}^2}{\Omega_{j'k'} \, (1+R_2)^2} \right) \Bigg|_{\{\rho^c\}} = 0 \,.$$

The critical values of ρ_k corresponding to the minima together with the conditions for their existence are given in Table 1. For $\rho_j^c = 0$, with $j = 2, \cdots, n$ one gets $E_N = 0$, because we took $\omega_1 = 0$. Considering $\rho_s^c = 0$ for some value of s all $(n-1)$ μ_{jk} related to that level can be taken

Regime	Energy	ρ_k^c	Condition
N	0	0	$k = 1, 2, \cdots, n$
S_{1k}	$-\frac{(\omega_k\,\Omega_{1k} - 4\,\mu_{1k}^2)^2}{16\,\Omega_{1k}\,\mu_{1k}^2}$	$\sqrt{\frac{4\,\mu_{1k}^2 - \omega_k\,\Omega_{1k}}{4\,\mu_{1k}^2 + \omega_k\,\Omega_{1k}}}$	$4\,\mu_{1k}^2 - \omega_k\,\Omega_{1k} \geq 0\,, k = 2, \cdots, n$
S_{jk}	$\omega_j - \frac{((\omega_k - \omega_j)\,\Omega_{jk} - 4\,\mu_{jk}^2)^2}{16\,\Omega_{jk}\,\mu_{jk}^2}$	$\sqrt{\frac{4\,\mu_{jk}^2 - (\omega_k - \omega_j)\,\Omega_{jk}}{4\,\mu_{jk}^2 + (\omega_k - \omega_j)\,\Omega_{jk}}}$	$4\,\mu_{jk}^2 - (\omega_k - \omega_j)\,\Omega_{jk} \geq 0$

Table 1. Minima critical values for ρ_k. In the third row we give the expression for η_k^c instead of ρ_k^c (see text), and we take values of $j > 1$ and $k > j$.

equal to zero; thus the expression of the energy surface effectively reduces to that of a system with $n-1$ levels. For each variable $\rho_s^c = 0$ with $s = 2, \cdots, n$ our physical system is reduced to the case of atoms with $n-1$ levels and from them one selects the one with lowest energy, i.e., $E = \text{Min}\{E_2, E_3, \cdots, E_n\}$. For $\rho_j^c \to \infty$ with $\rho_{k\neq j}^c$ finite, one has $E \to \omega_j$ which is greater than zero, and so it is not a minimum value of the energy.

The interaction terms, which give a negative contribution to the energy surface, are different from zero if and only if we assume that the critical points satisfy

$$\varrho_2 \to \infty\,, \quad \varrho_j \to \eta_j\,\varrho_2\,, \tag{16}$$

for $j = 3, \cdots, n$. From the expression of the energy surface one gets

$$E_\infty = \sum_{j=2}^{n} \omega_j \frac{\eta_j^2}{1 + \sum_{k=3}^{n} \eta_k^2} - 4 \sum_{2 \leq j < k}^{n} \frac{\mu_{jk}^2}{\Omega_{jk}} \left(\frac{\eta_j\,\eta_k}{1 + \sum_{\ell=3}^{n} \eta_\ell^2} \right)^2 , \tag{17}$$

where $\eta_2 \equiv 1$ and we have used the relations

$$r_{ij} \to 2\frac{\mu_{ij}}{\Omega_{ij}} \frac{\eta_i\,\eta_j}{1 + \sum_{k=3}^{n} \eta_k^2}\,, \qquad \frac{\varrho_i\,\varrho_j}{1 + \sum_{k=2}^{n} {\varrho_k}^2} \to \frac{\eta_i\,\eta_j}{1 + \sum_{k=3}^{n} \eta_k^2}\,. \tag{18}$$

Comparing (17) with (15) it is clear that we have obtained an equivalent system with $n-1$ levels where the atomic variables ϱ_j are replaced by new variables η_j, which upon finding its critical points leads to an equivalent algebraic system of $n-2$ levels and so on until we reach a two-level system with one radiation mode whose properties, including the complete structure of the phase diagram, have been studied extensively [26–29].

From Table 1 we can see that the minimum energy surface can be divided into two regimes, the normal (N) and the superradiant. This last regime in turn is divided into two sets of monochromatic regions denoted by S_{1k} with $k = 2, \cdots, n$ and S_{jk} with $j > 1$ and $k > j$. In the normal regime there are no photons in the ground state, while for the super-radiant sectors there are mainly photons of the modes Ω_{1k}, with $k = 2, \cdots n$ or of the type Ω_{jk} with $j > 1$ and $k > j$. These regions constitute the quantum phase diagram, i.e., the division of parameter space into sectors where the properties of the ground state change significantly.

For values of the dipolar intensities satisfying

$$4\,\mu_{1k}^2 \leq \omega_k\,\Omega_{1k}\,, \quad 4\,\mu_{jk}^2 \leq (\sqrt{\omega_j} + \sqrt{\omega_k})^2\,\Omega_{jk}\,, \tag{19}$$

the minimum energy value is $E_{min} = 0$.

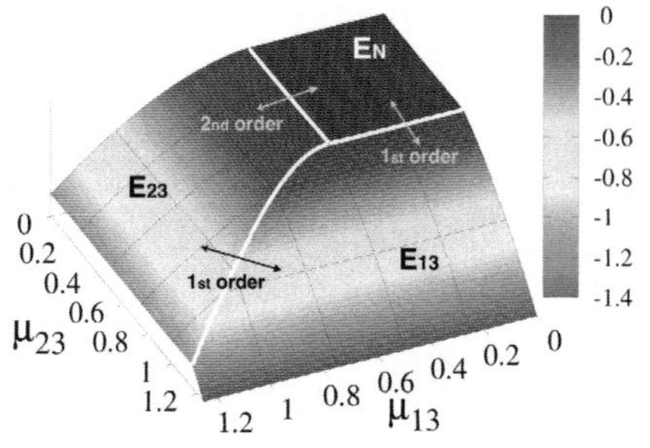

Figure 2. Energy surface as a function of the dipolar intensities for atoms in the Λ-configuration. The separatrix (white line) and the order of the transitions are displayed. The normal region E_N and collective regions E_{13} and E_{23} corresponding to the modes Ω_{13} and Ω_{23}, respectively, are shown. We consider $\Omega_{13} = 1\,, \Omega_{23} = 0.8\,, \omega_2 = 0.2$ and $\omega_3 = 1$.

The points of the quantum phase diagram in the collective region are given by the set of equalities $E_{jk} = E_{j'k'}$ for the modes $\Omega_{jk} \neq \Omega_{j'k'}$:

$$4\,\mu_{jk}^2 - \frac{\Omega_{jk}}{8\,\mu_{j'k'}^2\,\Omega_{j'k'}}\left(\frac{\zeta_j + \zeta_k}{2} \mp \sqrt{\zeta_j\,\zeta_k}\right) = 0\,, \tag{20}$$

$$\zeta_\ell = \left(4\,\mu_{j'k'}^2 - (\omega_{k'} - \omega_{j'})\,\Omega_{j'k'}\right)^2 + 8\,\mu_{j'k'}^2\,\Omega_{j'k'}\,(\omega_\ell - \omega_{j'})\,, \quad \ell = j,\,k\,. \tag{21}$$

As an example we show in Fig. 2 the energy surface for the Λ configuration of 3-level atoms interacting with two modes of the electromagnetic field together with the normal and superradiant regimes. The quantum phase diagram appears in white lines and the order of the transitions is also indicated.

The ground state in the normal regime has all the atoms in their lowest energy level and there are no photons, i.e.,

$$|\psi_N\rangle = \frac{\boldsymbol{b}_1^{\dagger\,N_a}}{\sqrt{N_a!}}|0\rangle_M \otimes |0\rangle_F\,. \tag{22}$$

The collective region, denoted by S_{1k}, with $k = 2, \cdots n$ takes the form

$$|\Psi_{1k}\rangle = \frac{1}{\sqrt{N_a!}}\left(\frac{\boldsymbol{b}_1^\dagger + \varrho_k^c\,\boldsymbol{b}_k^\dagger}{\sqrt{1+\varrho_k^{c\,2}}}\right)^{N_a}|0\rangle_M \otimes e^{-N_a\,r_{1k}^{c\,2}/2}\,e^{\sqrt{N_a}\,r_{1k}^c\,\boldsymbol{a}_{1k}^\dagger}\,|0\rangle_F\,, \tag{23}$$

From this expression one can say that the states are equivalent to those associated to a 2-level system with energy levels (ω_1, ω_k) interacting with one-mode of electromagnetic radiation, Ω_{1k}.

For the monochromatic regions S_{jk}, with $j < k = 3, \cdots, n$ one obtains a similar result

$$|\Psi_{jk}\rangle = \frac{1}{\sqrt{N_a!}}\left(\frac{\boldsymbol{b}_j^\dagger + \eta_k^c\,\boldsymbol{b}_k^\dagger}{\sqrt{1+\eta_k^{c\,2}}}\right)^{N_a}|0\rangle_M \otimes e^{-N_a\,r_{jk}^{c\,2}/2}\,e^{\sqrt{N_a}\,r_{jk}^c\,\boldsymbol{a}_{jk}^\dagger}\,|0\rangle_F\,, \tag{24}$$

where $j > 1$ and the expression

$$r_{jk}^c = \lim_{\varrho_j^c \to \infty} r_{jk}\bigg|_{\varrho_k^c = \eta_k^c\,\varrho_j^c} = \frac{2\,\mu_{jk}\,\eta_k^c}{(1+\eta_k^{c2})\Omega_{jk}} \tag{25}$$

was used. In these cases the 2-level systems are associated to the energy levels (ω_j, ω_k) and mode Ω_{jk}.

Observe that the different expressions for the ground states are separable states, implying that there is no entanglement between the matter and the electromagnetic field parts. It is well known in the literature that this is not true [25], at least for 2-level atoms interacting with one radiation mode. This separability is modified by restoring the symmetry of the Hamiltonian, that is, by taking linear combinations of the previous variational solutions to guarantee that they have good parity properties with respect to the operators K_j defined previously.

From the expressions for the minimum energy value in the different regions and the Ehrenfest classification one finds: second order quantum phase transitions at the crossings $E_N \rightleftharpoons E_{1k}$; first order transitions from the normal to the monochromatic regions $E_N \rightleftharpoons E_{jk}$, and for the transitions$E_{jk} \rightleftharpoons E_{j'k'}$. A first order transition indicates that at least one physical quantity (expectation value of the number of photons, atomic populations, etc.) has a discontinuity.

5. Conclusions

We have determined the phase diagrams for n-level systems interacting with ℓ modes of electromagnetic radiation, with and without the RWA approximation. We have found that there are first and second order phase transitions between the normal and collective regions while the collective regime is divided into monochromatic regions and all the transitions are of first order. This has been supported by the quantum exact calculations of the minimum energy, and expectation values of the number of photons for the three level atomic configuration Ξ interacting with two modes of electromagnetic radiation [9]. We have shown that the Hamiltonian system in the RWA approximations has n constants of motion K_j which in the case without the RWA approximation divides the Hilbert space into 2^{n-1} orthogonal subspaces associated to the parity of $n-1$ constants of motion. For each one of the monochromatic regions from the variational procedure we have found analytic expressions for the eigenstates with minimum energy, which are separable states (Vide Eqs. (23,24)). For this reason the statistical properties of the variational solutions have Poisson distributions for the photon-modes while binomial distributions for the matter parts. The parity symmetry can be restored and we hope to recover properly the entanglement properties between the matter and the modes of the electromagnetic field. This work is in progress.

Acknowledgements

This work was supported by CONACyT (under Project 238494) and DGAPA-UNAM (under Projects IN110114 and IN101614).

References

[1] Kimble H J 1998 *Phys. Scr.* T **76** 127
[2] Jaynes E and Cummings F 1963 *Proc. IEEE* **51** 89
[3] Miller R, Northup T E, Birnbaum K M, Boca A, Boozer A D and Kimble H J 2005 *J. Phys. B: At. Mol. Opt. Phys.* **38** s551
[4] Yoo H I and Eberly J H 1985 *Phys. Rep.* **118** 239
[5] Monroe C *et. al.* 1996 *Science* **272** 1131
[6] Wineland D J, Drullinger R E and Walls F L 1996 *Phys. Rev. Lett.* **40** 1639
[7] Castaños L O and Jáuregui R 2015 *Phys. Rev.* A **83** 062103
[8] Cirac J I, Zoller P, Kimble H J and Mabuchi H 1997 *Phys. Rev. Lett.* **78** 3221
[9] Cordero S, Nahmad-Achar E, López-Peña R and Castaños O 2015 *Phys. Rev A* **92** 053843
[10] Hepp K and Lieb E H 1973 *Ann. Phys.* **76** 360
[11] Dicke R H 1954 *Phys. Rev.* **93** 99
[12] Castaños O, Nahmad-Achar E, López-Peña R 2009 *Phys. Scr.* **80** 055401
[13] Zanardi P and Paunkovic N 2006 *Phys. Rev.* E **74** 031123
[14] Gu SJ 2010 *Int. J. Mod. Phys.* B **24** 4371

[15] Castaños O, López-Peña R, Nahmad-Achar E and Hirsch J G 2012 *J. of Phys.:Conf. Ser.* **387** 012021
[16] Castaños O, López-Peña R, Nahmad-Achar E and Hirsch J G 2012 *J. of Phys.:Conf. Ser.* **403** 012003
[17] Gilmore R 1981 *Catastrophe Theory for Scientist and Engineers* (Wiley, New York)
[18] Pérez-Campos C, González-Alonso J R, Castaños O and López-Peña R 2010 *Ann. Phys.* **325** 325
[19] Poston T and Stewart I 1978 *Catastrophe Theory and its Applications* (Pitman, London)
[20] Glauber R J 1963 *Phys. Rev. Lett.* **10** 84
[21] Sudarshan E C G 1963 *Phys. Rev. Lett.* **10** 277
[22] Klauder J R 1963 *J. Math. Phys.* **4** 1055
[23] Bohr M and Mottelson B R 1980 *Physica Scripta* **22** 468
[24] Ginnocchio J and Kirsom M W 1980 *Phys. Rev. Lett.***44** 1744
[25] López-Peña R, Cordero S, Nahmad-Achar E and Castaños O 2015 *Phys. Scr.* **90** 068016
[26] Castaños O, Nahmad-Achar E, López-Peña R and Hirsch J G 2011 *Phys. Rev.* A **83** 051601
[27] Castaños O, Nahmad-Achar E, López-Peña R and Hirsch J G 2011 *Phys. Rev.* A **84** 013819
[28] Nahmad-Achar E, Castaños O, López-Peña R and Hirsch J G 2013 *Phys. Scr.* **87** 038114
[29] Nahmad-Achar E, Cordero S, Castaños O and López-Peña R 2015 *Phys. Scr.* **90** 074026

Journal of Physics: Conference Series **698** (2016) 012007 doi:10.1088/1742-6596/698/1/012007

First-Order Polynomial Heisenberg Algebras and Coherent States

M Castillo-Celeita and D J Fernández C

Departamento de Física, Cinvestav, A.P. 14-740, 07000 México D.F. México

E-mail: `mfcastillo@fis.cinvestav.mx`, `david@fis.cinvestav.mx`

Abstract. The polynomial Heisenberg algebras (PHA) are deformations of the Heisenberg-Weyl algebra characterizing the underlying symmetry of the supersymmetric partners of the Harmonic oscillator. When looking for the simplest system ruled by PHA, however, we end up with the harmonic oscillator. In this paper we are going to realize the first-order PHA through the harmonic oscillator. The associated coherent states will be also constructed, which turn out to be the well known even and odd coherent states.

1. Introduction

The polynomial deformations of the Heisenberg-Weyl algebra, polynomial Heisenberg algebras (PHA) by short, start to be important in physics [1–6]. For example, the underlying algebraic symmetry for the supersymmetric (SUSY) partners of the harmonic oscillator is precisely described by a PHA. In addition, when looking for the general one-dimensional Schrödinger Hamiltonians ruled by second (third) order PHA, the key turns out to be the determination of a function which must satisfy the Painlevé IV (Painlevé V) equation [5]. As a consequence, through this link it has been possible to design a simple method to generate solutions to these non-linear second-order ordinary differential equations [6].

It is important to note that the spectrum for systems ruled by PHA can be obtained by identifying the physical extremal states, those eigenstates of the Hamiltonian which belong also to the kernel of the annihilation operator. In principle, the number of physical extremal states could be equal to the order of the annihilation operator. However, we have observed in most of the works that the number of extremal states is less than the order of the annihilation operator.

In this paper we would like to work with the simplest systems realizing the PHA. Moreover, we want to explore the possibility that every extremal state, if possible, would have an eigenvalue which is in the spectrum of the Hamiltonian. Once this has been done, we will construct the associated coherent states and we will study their properties.

2. Polynomial Heisenberg algebras

In the standard harmonic oscillator (Heisenberg-Weyl) algebra there are three generators H, a, a^+ satisfying the following commutation relations:

$$[H, a^+] = a^+, \quad [H, a] = -a, \quad [a, a^+] = 1, \tag{1}$$

where there is a linear dependence between the number operator N and the Hamiltonian H:

$$N = a^+ a = H - \tfrac{1}{2}. \qquad (2)$$

On the other hand, the $(m-1)$-th order polynomial Heisenberg algebras are deformations of the oscillator algebra of kind [5]:

$$[H, \mathcal{L}_m^+] = \omega\, \mathcal{L}_m^+, \qquad (3)$$
$$[H, \mathcal{L}_m^-] = -\omega\, \mathcal{L}_m^-, \qquad (4)$$
$$[\mathcal{L}_m^-, \mathcal{L}_m^+] \equiv N_m(H+\omega) - N_m(H) \equiv P_{m-1}(H), \qquad (5)$$

where the analogue of the number operator is now an m-th degree polynomial in H:

$$N_m(H) \equiv \mathcal{L}_m^+\mathcal{L}_m^- = \prod_{i=1}^{m} (H - \mathcal{E}_i)\,. \qquad (6)$$

The most common realization of a PHA is a differential one, where H has the standard one-dimensional Schrödinger form,

$$H = -\frac{1}{2}\frac{d^2}{dx^2} + V(x), \qquad (7)$$

while $\mathcal{L}_m^\pm$ are differential ladder operators of order m-th.

The spectrum of H, Sp(H), can be obtained from the analysis of the Kernel $K_{\mathcal{L}_m^-}$ of $\mathcal{L}_m^-$. In fact, it turns out that

$$\mathcal{L}_m^-\psi = 0 \quad \Rightarrow \quad \mathcal{L}_m^+\mathcal{L}_m^-\psi = \prod_{i=1}^{m} (H - \mathcal{E}_i)\,\psi = 0. \qquad (8)$$

Since $K_{\mathcal{L}_m^-}$ in invariant under H,

$$\mathcal{L}_m^-(H\psi) = (H+\omega)\mathcal{L}_m^-\psi = 0 \quad \forall \quad \psi \in K_{\mathcal{L}_m^-}, \qquad (9)$$

then a basis in $K_{\mathcal{L}_m^-}$ can be chosen such that,

$$H\psi_{\mathcal{E}_i} = \mathcal{E}_i\psi_{\mathcal{E}_i}. \qquad (10)$$

The states $\psi_{\mathcal{E}_i}$ are the so-called extremal states; departing from them we can build several energy ladders with spacing $\Delta E = \omega$.

Depending on how many extremal states $\psi_{\mathcal{E}_i}$ are eigenstates of H (satisfying also the boundary conditions), two possibilities appear:

(a) If s extremal states are eigenstates of H, $\{\psi_{\mathcal{E}_i}, i = 1, \ldots, s\}$, there will be s energy ladders arising from the iterated action of $\mathcal{L}_m^+$ onto $\psi_{\mathcal{E}_i}$.

(b) If for the j-th ladder it turns out that

$$\left(\mathcal{L}_m^+\right)^{n-1}\psi_{\mathcal{E}_j} \neq 0, \qquad \left(\mathcal{L}_m^+\right)^{n}\psi_{\mathcal{E}_j} = 0, \qquad (11)$$

then we will have

$$\mathcal{L}_m^-(\mathcal{L}_m^+)^n\psi_{\mathcal{E}_j} = \mathcal{L}_m^-\mathcal{L}_m^+(\mathcal{L}_m^+)^{n-1}\psi_{\mathcal{E}_j} = \prod_{i=1}^{m} (\mathcal{E}_j + n\omega - \mathcal{E}_i)\,(\mathcal{L}_m^+)^{n-1}\psi_{\mathcal{E}_j} = 0. \qquad (12)$$

Therefore $\mathcal{E}_k = \mathcal{E}_j + n\omega$ for some $k \in \{s+1, \ldots, m\}$, $j \in \{1, \ldots, s\}$; thus, Sp(H) will consist of $s-1$ infinite ladders and a finite one of lenght $(n-1)\omega$ which starts from $\mathcal{E}_j$ and ends at $\mathcal{E}_j + (n-1)\omega$. We can conclude that Sp(H) can have up to m infinite ladders with spacing $\Delta E = \omega$ between steps.

3. Harmonic oscillator and polynomial Heisenberg algebras

It is possible to realize the PHA through the harmonic oscillator Hamiltonian H and deformed versions a_g, a_g^+ of the standard annihilation and creation operators a, a^+. One possibility is to take $a_g = P_{k-1}(H)a$, $a_g^+ = a^+P_{k-1}(H)$, where $P_{k-1}(x)$ is a polynomial of degree $k-1$ with real roots; then, the set of operators $\{H, a_g, a_g^+\}$ satisfy the commutation relations of Eqs. (3-5). However, in these deformations it is typical that some extremal states have formal eigenvalues which do not belong to $\mathrm{Sp}(H)$.

We are looking for deformations of the annihilation and creation operators such that all extremal states become physical and, consequently, from them we will generate infinite ladders of eigenfunctions and eigenvalues of H. For that reason, let us take now a different deformation,

$$a_g = a^2, \qquad a_g^+ = (a^+)^2. \tag{13}$$

The operator set $\{H, a_g, a_g^+\}$ gives place to a first-order PHA with $\omega = 2$ since,

$$[H, a_g] = -2a_g, \tag{14}$$

$$[H, a_g^+] = 2a_g^+, \tag{15}$$

$$[a_g, a_g^+] = N(H+2) - N(H), \tag{16}$$

where

$$N(H) = \left(H - \tfrac{1}{2}\right)\left(H - \tfrac{3}{2}\right). \tag{17}$$

There are two extremal state energies,

$$\mathcal{E}_1 = E_0 = \tfrac{1}{2}, \quad \mathcal{E}_2 = E_1 = \tfrac{3}{2}. \tag{18}$$

Thus, the eigenvalues associated to the j-th ladder are

$$\mathcal{E}_n^j = \mathcal{E}_j + 2n \quad n = 0, 1, \ldots, \quad j = 1, 2. \tag{19}$$

The corresponding eigenstates turn out to be

$$|\psi_n^j\rangle = |2n + j - 1\rangle = \sqrt{\frac{(j-1)!}{(2n+j-1)!}}(a_g^+)^n|j-1\rangle. \tag{20}$$

The spectrum of H becomes,

$$\mathrm{Sp}(H) = \{\mathcal{E}_0^1, \mathcal{E}_1^1, \ldots\} \cup \{\mathcal{E}_0^2, \mathcal{E}_1^2, \ldots\}, \tag{21}$$

which coincides with the harmonic oscillator spectrum, but seen from a different point of view: the Hilbert space $\mathcal{H}$ is now decomposed as the direct sum of two orthogonal supplementary subspaces, namely, $\mathcal{H} = \mathcal{H}_0 \oplus \mathcal{H}_1$ (a representation of both ladders can be seen in Figure 1). Note that states in each subspace have definite parity: the ones in $\mathcal{H}_0$ have positive parity and they are called even states, while those in $\mathcal{H}_1$ have negative parity, being called odd states.

4. Coherent states

Once we have identified the algebraic structure underlying the system under study, let us derive the corresponding coherent states. We will look for them as eigenstates of the deformed annihilation operator:

$$a_g|\alpha\rangle_j = \alpha|\alpha\rangle_j, \quad j = 0, 1, \tag{22}$$

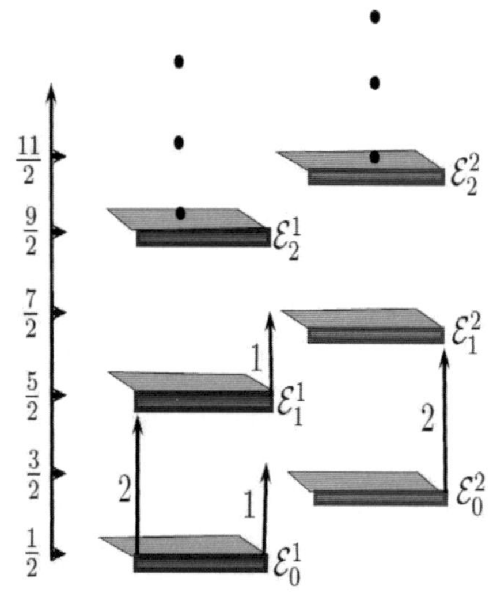

Figure 1: The two independent energy ladders (with level spacing $\Delta E = 2$) associated to the first-order polynomial Heisenberg algebra of Eqs. (14-16). Globally they give place to the harmonic oscillator spectrum with the standard level spacing $\Delta E = 1$.

where

$$|\alpha\rangle_j = \sum_{n=0}^{\infty} C_n |2n+j\rangle. \tag{23}$$

We thus get the following recurrence relationship:

$$C_{n+1} = \frac{\alpha C_n}{\sqrt{(2n+j+2)(2n+j+1)}}, \tag{24}$$

which after iteration leads to:

$$C_n = \sqrt{\frac{j!}{(2n+j)!}}\alpha^n C_0. \tag{25}$$

Using C_0 as a normalization constant, we finally obtain the coherent states we were looking for:

$$|\alpha\rangle_j = \frac{1}{\sqrt{\sum_{n=0}^{\infty} \frac{|\alpha|^{2n}}{(2n+j)!}}} \sum_{n=0}^{\infty} \frac{\alpha^n}{\sqrt{(2n+j)!}} |2n+j\rangle. \tag{26}$$

Some quantities useful to analyze the Heisenberg uncertainty relation for these coherent states are the following:

$$\langle x\rangle_j = \langle p\rangle_j = 0, \tag{27}$$

$$\langle x^2\rangle_j = |a|\alpha\rangle_j|^2 + \tfrac{1}{2} + \mathrm{Re}(\alpha), \tag{28}$$

$$\langle p^2\rangle_j = |a|\alpha\rangle_j|^2 + \tfrac{1}{2} - \mathrm{Re}(\alpha), \tag{29}$$

where

$$|a|\alpha\rangle_j|^2 = \begin{cases} \frac{1}{\sum\limits_{r=0}^{\infty}\frac{|\alpha|^{2r}}{(2r)!}} \sum\limits_{n=0}^{\infty} \frac{|\alpha|^{2n+2}}{(2n+1)!} = |\alpha|\tanh|\alpha| & \text{for } j=0, \\ \frac{1}{\sum\limits_{r=0}^{\infty}\frac{|\alpha|^{2r}}{(2r+1)!}} \sum\limits_{n=0}^{\infty} \frac{|\alpha|^{2n}}{(2n)!} = |\alpha|\coth|\alpha| & \text{for } j=1. \end{cases} \tag{30}$$

Therefore:

$$(\Delta x)_j^2(\Delta p)_j^2 = \begin{cases} \left(|\alpha|\tanh|\alpha| + \frac{1}{2}\right)^2 - [\mathrm{Re}(\alpha)]^2 \geq \frac{1}{4} & \text{for } j=0, \\ \left(|\alpha|\coth|\alpha| + \frac{1}{2}\right)^2 - [\mathrm{Re}(\alpha)]^2 \geq \frac{9}{4} & \text{for } j=1. \end{cases} \tag{31}$$

Plots of these uncertainty relations are shown in Figure 2 for $j = 0$ and Figure 3 for $j = 1$.

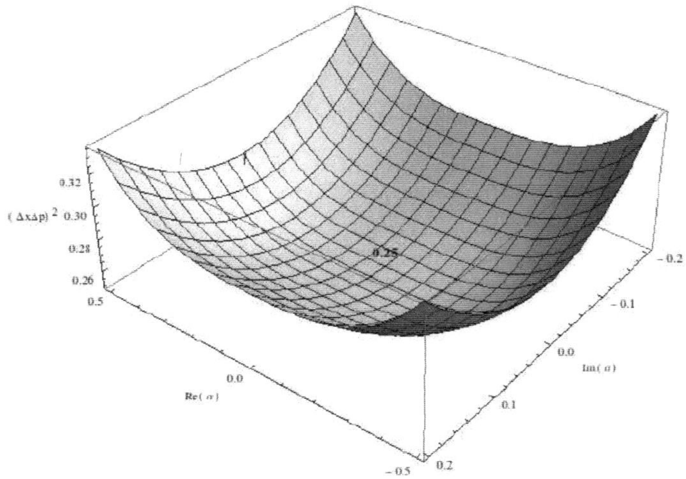

Figure 2: Heisenberg uncertainty relation $(\Delta x)_0^2(\Delta p)_0^2$ of Eq. (31) for $j = 0$.

On the other hand, the mean energy value for a system being in a coherent state becomes:

$$\langle H\rangle_j = |a|\alpha\rangle_j|^2 + \frac{1}{2} = \begin{cases} |\alpha|\tanh|\alpha| + \frac{1}{2} & \text{for } j=0, \\ |\alpha|\coth|\alpha| + \frac{1}{2} & \text{for } j=1. \end{cases} \tag{32}$$

It is also important to guarantee the completeness relation in the subspace $\mathcal{H}_j$ (remember that $\mathcal{H} = \mathcal{H}_0 \oplus \mathcal{H}_1$), namely:

$$\int |\alpha\rangle_j\,{}_j\langle\alpha| d\mu_j(\alpha) = I_j, \quad j = 0, 1, \tag{33}$$

where

$$d\mu_j(\alpha) = \frac{1}{\pi|\alpha|}\left(\sum_{r=0}^{\infty} \frac{|\alpha|^{2r}}{(2r+j)!}\right) f_j(|\alpha|^2) d|\alpha| d\varphi. \tag{34}$$

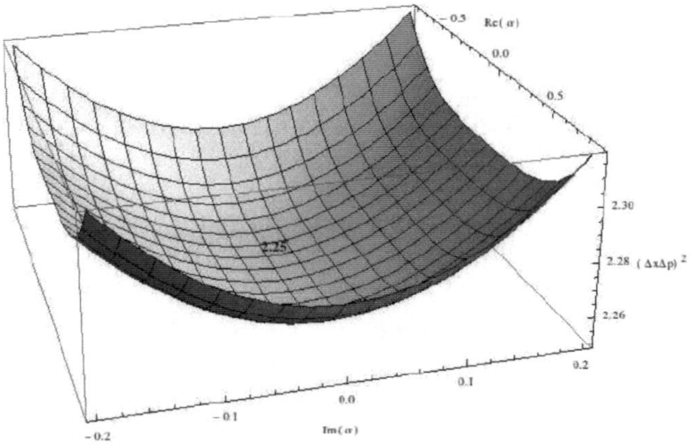

Figure 3: Heisenberg uncertainty relation $(\Delta x)_1^2(\Delta p)_1^2$ of Eq. (31) for $j = 1$.

It turns out that the function $f_j(x)$ must satisfy:

$$\int_0^\infty x^{n-1} f_j(x)dx = \Gamma(2n + j + 1). \tag{35}$$

Thus, provided that Eq. (35) is satisfied, it is true that any state vector can be decomposed in terms of our coherent states.

Finally, the time evolution of a coherent state is quite simple:

$$U(t)|\alpha\rangle_j = e^{-i(j+\frac{1}{2})t}|\alpha(t)\rangle_j, \quad \alpha(t) = \alpha\, e^{-ikt}. \tag{36}$$

Let us consider next the following non-normalized coherent states:

$$|z\rangle = \sum_{n=0}^{\infty} \frac{z^n}{\sqrt{n!}}|n\rangle, \tag{37}$$

$$|z\rangle_j = \sum_{n=0}^{\infty} \frac{z^{2n+j}}{\sqrt{(2n+j)!}}|2n+j\rangle, \quad \alpha = z^2. \tag{38}$$

Up to normalization, the states of Eq. (37) are the standard coherent states and those of Eq. (38) are the even ($j = 0$) and odd ($j = 1$) coherent states [7–14] (see also the discussion at the end of Section 3). The completeness relationship of Eq. (33) guarantees that $|z\rangle$ and $|e^{i\pi}z\rangle$ can be decomposed in terms of $|z\rangle_0$ and $|z\rangle_1$ as:

$$|z\rangle = |z\rangle_0 + |z\rangle_1, \quad |e^{i\pi}z\rangle = |z\rangle_0 - |z\rangle_1. \tag{39}$$

On the other hand, from these equations we can solve $|z\rangle_0$ and $|z\rangle_1$ in terms of $|z\rangle$ and $|e^{i\pi}z\rangle$. After normalization we get that:

$$|z\rangle_0 = \frac{e^{-|z|^2/2}}{\sqrt{2(1+e^{-2|z|^2})}}[|z\rangle + |e^{i\pi}z\rangle], \tag{40}$$

$$|z\rangle_1 = \frac{e^{-|z|^2/2}}{\sqrt{2(1-e^{-2|z|^2})}}[|z\rangle - |e^{i\pi}z\rangle], \tag{41}$$

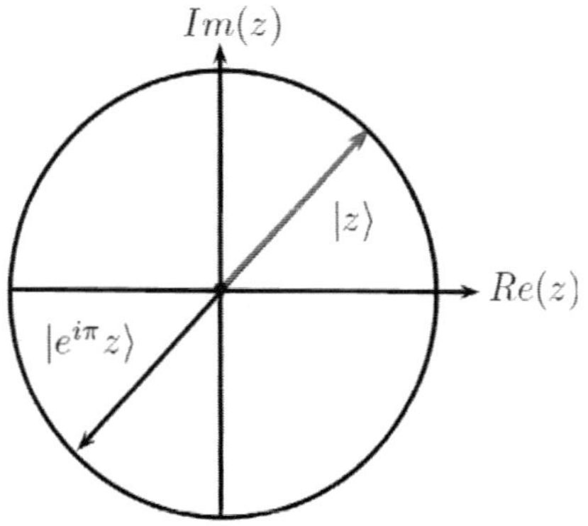

Figure 4: Diagram indicating the position in the complex plane of the standard coherent states which give place to the even and odd coherent states $|z\rangle_0$ and $|z\rangle_1$.

which means that the even and odd coherent states are linear combinations of two standard coherent states with *opposite directions* in the z complex plane, as it is represented in Figure 4. The expressions (40-41) can be used now to identify in a simple way the wave packets for $U(t)|z\rangle_0$ and $U(t)|z\rangle_1$. In fact, the wavefunction for a normalized standard coherent state is [15]:

$$\psi_z(x) = \langle x|z\rangle = \left(\frac{1}{\pi}\right)^{1/4} e^{-\frac{1}{2}(x-\langle x\rangle)^2 + ix\langle p\rangle}, \tag{42}$$

with $\langle x\rangle = \sqrt{2}\,\mathrm{Re}(z)$ and $\langle p\rangle = \sqrt{2}\,\mathrm{Im}(z)$. Thus, the wavefunctions associated to $|z\rangle_0$ and $|z\rangle_1$ in Eqs. (40-41) are:

$$\psi_z^0(x) = \langle x|z\rangle_0 = N_0\left[e^{-\frac{1}{2}(x-\langle x\rangle)^2+ix\langle p\rangle} + e^{-\frac{1}{2}(x+\langle x\rangle)^2-ix\langle p\rangle}\right], \tag{43}$$

$$\psi_z^1(x) = \langle x|z\rangle_1 = N_1\left[e^{-\frac{1}{2}(x-\langle x\rangle)^2+ix\langle p\rangle} - e^{-\frac{1}{2}(x+\langle x\rangle)^2-ix\langle p\rangle}\right], \tag{44}$$

where

$$N_0 = \left(\frac{1}{\pi}\right)^{1/4}\frac{1}{\sqrt{2(1+e^{-\langle x\rangle^2-\langle p\rangle^2})}}, \quad N_1 = \left(\frac{1}{\pi}\right)^{1/4}\frac{1}{\sqrt{2(1-e^{-\langle x\rangle^2-\langle p\rangle^2})}}.$$

Finally, up to global phase factors the wavefunctions $\psi_z^j(x,t) = \langle x|U(t)|z\rangle_j$, $j = 0,1$ arise from the right hand side of Eqs. (43-44) substituting $\langle x\rangle$ by $\langle x\rangle\cos t + \langle p\rangle\sin t$ and $\langle p\rangle$ by $\langle p\rangle\cos t - \langle x\rangle\sin t$. The corresponding probability densities are shown in Figure 5 for $j = 0$ and in Figure 6 for $j = 1$.
Note that for certain times the probability density is maximum at $x = 0$ for the even coherent states, while this never happens for the odd ones. It is interesting to observe also that the coherent states $U(t)|z\rangle_0$ and $U(t)|z\rangle_1$ are cyclic, with a period $\tau = \pi$ which is half the period of the oscillator (2π). This property reflects clearly the very quantum nature of the even and odd coherent states compared with the somehow classical behavior of the standard ones, which always have the period of the oscillator.

5. Conclusions

In this paper we have explored an interesting realization of the first-order polynomial Heisenberg algebra. The generators of such an algebra are the harmonic oscillator Hamiltonian H and the

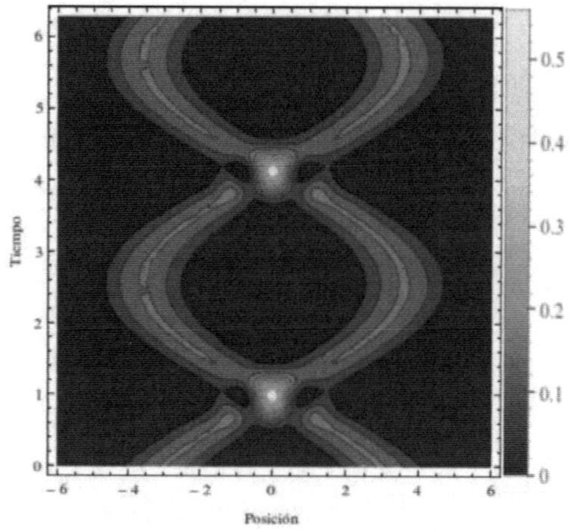

Figure 5: Probability density $|\psi_z^0(x,t)|^2$ for the even coherent states.

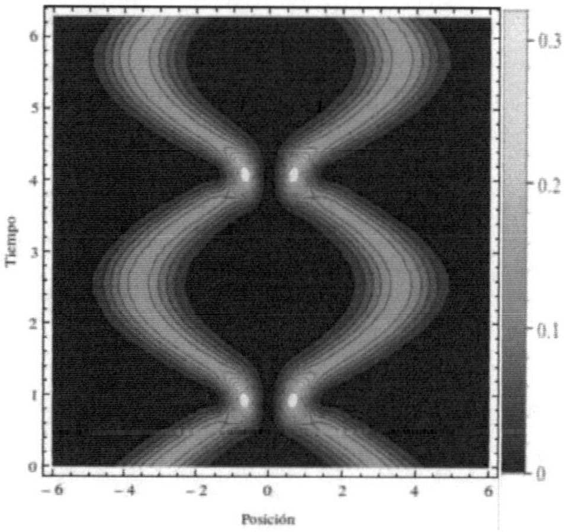

Figure 6: Probability density $|\psi_z^1(x,t)|^2$ for the odd coherent states.

deformed annihilation and creation operators $a_g = a^2$, $a_g^+ = (a^+)^2$. We have shown that the two extremal states involved in our treatment are physical eigenstates of the Hamiltonian. We have seen also that the two ladders generated from these extremal states are of infinite length, and the associated coherent states become the so-called even and odd. We have realized that the period of the even and odd coherent states is a fraction $(1/2)$ of the original period (2π) for the oscillator. This property indicates the very quantum nature of such coherent states. This statement should be reinforced by calculating other quantities indicating how quantum a given state is. We hope to perform this analysis in the near future.

Acknowledgments

The authors acknowledge the support of Conacyt, project 152574.

References

[1] Fernández D J, Hussin V and Nieto L M 1994 Coherent states for isospectral oscillator Hamiltonians *J Phys A: Math Gen* **27** 3547

[2] Fernández D J, Nieto L M and Rosas-Ortiz O 1995 Distorted Heisenberg algebra and coherent states for isospectral oscillator hamiltonians *J Phys A: Math Gen* **28** 2693

[3] Fernández D J and Hussin V 1999 Higher-order SUSY, linearized nonlinear Heisenberg algebra and coherent states *J Phys A: Math Gen* **32** 3603

[4] Fernández D J and Hussin V 2002 Polynomial Heisenberg algebras and higher order supersymmetry *Proceedings of the XXIII International Colloquium on Group Theoretical Methods in Physics Vol, 1*, eds A N Sissakian *et al* (Dubna: JINR) p 121

[5] Carballo J M, Fernández D J, Negro J and Nieto L M 2004 Polynomial Heisenberg algebras *J Phys A: Math Gen* **37** 10349

[6] Bermudez D and Fernández D J 2014 Supersymmetric quantum mechanics and Painlevé equations *AIP Conf Proc* **1575** 50

[7] Dodonov V V 2002 Nonclassical states in quantum optics: a squeezed review of the first 75 years *J Opt B: Quantum Semiclass Opt* **4** R1

[8] Dodonov V V, Malkin I A and Man'ko V I 1974 Even and odd coherent states and excitations of singular oscillator *Physica* **72** 597

[9] Xia Y and Guo G 1989 Nonclassical properties of even and odd coherent states *Phys Lett A* **136** 281

[10] Roy B and Roy P 2000 New nonlinear coherent states and some of their nonclassical properties *J Opt B: Quantum Semiclass Opt* **2** 65

[11] Gerry C C and Knight P L 1997 Quantum superpositions and Schrödinger cat states in quantum optics *Am J Phys* **65** 964

[12] Kira M, Koch S W, Smith R P, Hunter A E and Cundiff S T 2011 Quantum spectroscopy with Schrödinger-cat states *Nature Phys* **7** 799

[13] Castaños O and López-Saldívar J A 2012 Dynamics of Schrödinger cat states *J Phys: Conf Ser* **380** 012017

[14] Riahi N 2013 The position-momentum correlation for quantum and classical probability distributions *Eur J Phys* **34** 461

[15] Cohen-Tannoudji C, Diu B and Laloë F 1977 *Quantum Mechanics Vol. 1* (Paris: Wiley)

Approximate yet Confident Solution for a Parametric Oscillator in a Kerr Medium

R Román-Ancheyta[1], M Berrondo[2] and J Récamier[1]

[1]Instituto de Ciencias Físicas, Universidad Nacional Autónoma de México, C.P. 62210 Cuernavaca, Morelos, México

[2]Dept. Physics and Astronomy, Brigham Young University, Provo, UT 84602, USA

E-mail: ricardo@fis.unam.mx,berrondo@byu.edu,pepe@fis.unam.mx

Abstract. We study the temporal evolution of a coherent state under the action of a parametric oscillator immersed in a nonlinear Kerr-like medium. Applying a self consistent method we obtain an approximate time evolution operator. This operator behaves like a squeezing operator due to the temporal dependence of the oscillator's frequency. We analyze Mandel's parameter, the presence of squeezing in the field quadratures and the generation of photons from the vacuum state.

1. Introduction

Coherent states were introduced by Schrödinger in 1926 since the early stages of quantum mechanics [1] in connection with the classical states of the quantum harmonic oscillator. These quantum states are characterized by the fact that the trajectory of the center of the coherent wave packet evolves in time in the same way as a classical harmonic oscillator and its dispersion takes the minimum value allowed by Heisenberg's principle. Much later, in 1963, Glauber introduced the *field* coherent states, that is, coherent states for the electromagnetic field, as right eigenstates of the annihilation operator. These states have played an important role in quantum optics [2]. The field coherent states can be obtained from any one of the three mathematical definitions: (i) as the right hand eigenstates of the boson annihilation operator $\hat{a}|\alpha\rangle = \alpha|\alpha\rangle$ with α a complex number, (ii) as those states obtained by application of the displacement operator upon the vacuum state of the harmonic oscillator $D(\alpha)|0\rangle = |\alpha\rangle$ with $D(\alpha) = \exp(\alpha\hat{a}^\dagger - \alpha^*\hat{a})$, and (iii) as the quantum states with a minimum uncertainty product $(\Delta p)(\Delta q) = 1/2$ with $\Delta q = \Delta p$. The coherent states obtained from any one of these definitions are identical when one makes use of the harmonic oscillator algebra. Subject to a linear interaction a coherent state evolves into a new coherent state, that is, they show temporal stability [3, 4].

In the presence of a nonlinear interaction, field coherent states evolve into non classical states. This can be achieved experimentally by passing a coherent state through a Kerr medium resulting in the appearance of distinguishable macroscopic superpositions of coherent states, the so called cat states [5, 6].

The parametric harmonic oscillator, namely a harmonic oscillator with a time dependent frequency, has been studied from several points of view: using the method of adiabatic invariants [7–9], super symmetric quantum mechanics [10], algebraic methods [11], and different approximation methods [12] among others. A particularly relevant realization of the parametric oscillator is cavity quantum electrodynamics (CQED) where the frequency of a given field mode in the cavity can change in time due to the motion of the cavity walls or to changes in the dielectric function of the medium [13]. For instance, Wineland et. al. [14] analyzed both theoretically and experimentally the loss of coherence caused by fluctuations in the trap parameters and in the amplitude and frequency of the laser beams, heating due to collisions with background gas, internal state decoherence due to radiative decay, and coupling to spectator levels.

In this work we consider a nonlinear system corresponding to a single mode field propagating in a Kerr-like medium immersed in a cavity with a time dependent frequency. In Section 2 we write the Hamiltonian and construct its time evolution operator. In section 3 we follow the evolution of coherent states under the nonlinear Hamiltonian and analyze some of their statistical properties like the Mandel parameter, the average value of the number operator and the dispersion of the quadratures.

2. Parametric oscillator in a Kerr medium

Consider a parametric harmonic oscillator immersed in a Kerr-like medium. Its Hamiltonian is given by:

$$H(t) = \frac{1}{2}[p^2 + \Omega^2(t)q^2] + H_{Kerr} \tag{1}$$

where $\Omega(t)$ is an explicit time dependent frequency and H_{Kerr} has to do with the Kerr-like medium. We can define the usual annihilation, creation and number operators as:

$$\hat{a} = \frac{1}{\sqrt{2\Omega_0}}(\Omega_0 q + \mathrm{i}p), \quad \hat{a}^\dagger = \frac{1}{\sqrt{2\Omega_0}}(\Omega_0 q - \mathrm{i}p), \quad \hat{n} = \hat{a}^\dagger \hat{a}. \tag{2}$$

where we have set $\hbar = 1$ and we can write the Kerr medium [15] as $H_{Kerr} = \chi\hat{n}^2$, where χ is proportional to a third-order nonlinear susceptibility $\chi^{(3)}$. To be specific, in what follows we will choose $\Omega(t) = \Omega_0[1 + 2\kappa\cos(2\Omega_0 t)]$ [16] with κ and χ small parameters. The Hamiltonian can be written in terms of $\hat{a}^\dagger$, $\hat{a}$ and $\hat{n}$ as [17]:

$$H(t) \quad = \Omega_0(\hat{n} + 1/2) + \chi\hat{n}^2 + g(t)(\hat{a}^2 + \hat{a}^{\dagger 2} + 2\hat{n} + 1) \tag{3}$$

and $g(t) = \Omega_0\kappa\cos(2\Omega_0 t)(1 + \kappa\cos(2\Omega_0 t))$.

The time evolution operator corresponding to the non linear time independent part of the Hamiltonian is given by:

$$U_0 = \exp\left(-\mathrm{i}\Omega_0 t(\hat{n} + 1/2) - \mathrm{i}t\chi\hat{n}^2\right) \tag{4}$$

and we can write the time dependent Hamiltonian in the interaction picture as

$$H_I(t) = g(t)\left(\mathrm{e}^{-2\mathrm{i}\Omega(\hat{n})t}\hat{a}^2 + \hat{a}^{\dagger 2}\mathrm{e}^{2\mathrm{i}\Omega(\hat{n})t} + 2\hat{n} + 1\right) \tag{5}$$

where $\Omega(\hat{n}) = \Omega_0 + 2\chi(1 + \hat{n})$ is a function of the number operator.

The set $\{\hat{a}^{\dagger 2}, \hat{n}, \hat{a}^2, 1\}$ forms the basis of a Lie algebra closed under commutation. However, the Hamiltonian $H_I(t)$ includes terms where the number operator appears in the exponent. Invoking the mean field approximation [18] we make the replacement $\exp[\pm 2\mathrm{i}\Omega(\hat{n})t]$ by $\langle\exp[\pm 2\mathrm{i}\Omega(\hat{n})t]\rangle$ obtaining:

$$H_I(t) = g(t)\left(\hat{a}^2\langle\mathrm{e}^{-2\mathrm{i}\Omega(\hat{n})t}\rangle + \hat{a}^{\dagger 2}\langle\mathrm{e}^{2\mathrm{i}\Omega(\hat{n})t}\rangle + 2\hat{n} + 1\right) \tag{6}$$

where the expectation value is taken with respect to the coherent state in the interaction picture representation $|\alpha_0\rangle_I = U_I|\alpha_0\rangle$. The resulting approximate Hamiltonian is similar to that of a degenerate parametric amplifier where a non linear medium is pumped by a strong laser inducing the emission and absorption of photon pairs [19].

With this simplification, the Hamiltonian in the interaction picture becomes an element of the Lie algebra with time dependent coefficients and the corresponding time evolution operator may be written in the product form [20]

$$H_I(t) = \sum_{n=1}^{4} f_n(t)X_n, \quad U_I(t) = \prod_{n=1}^{4} \mathrm{e}^{\alpha_n(t)X_n}. \tag{7}$$

with initial conditions $\alpha_n(t_0) = 0$, and we have chosen the ordering $X_1 = \hat{a}^{\dagger 2}$, $X_2 = \hat{n}$, $X_3 = \hat{a}^2$ and $X_4 = 1$.

The average takes the form:

$$\langle \mathrm{e}^{\pm 2\mathrm{i}\Omega(\hat{n})t} \rangle = \mathrm{e}^{\pm 2\mathrm{i}(\Omega_0+2\chi)t} \langle \alpha_0 | U_I^\dagger \mathrm{e}^{\pm 4\mathrm{i}\chi t\hat{n}} U_I | \alpha_0 \rangle. \tag{8}$$

Using

$$\begin{aligned} U_I^\dagger \hat{n} U_I &= [2\alpha_1 \mathrm{e}^{-2\alpha_2}]\hat{a}^{\dagger 2} + [1 - 8\alpha_1\alpha_3 \mathrm{e}^{-2\alpha_2}]\hat{n} + [2\alpha_3(4\alpha_1\alpha_3\mathrm{e}^{-2\alpha_2} - 1)]\hat{a}^2 - [4\alpha_1\alpha_3\mathrm{e}^{-2\alpha_2}] \\ &\equiv a_1\hat{a}^{\dagger 2} + a_2\hat{n} + a_3\hat{a}^2 + a_4 \end{aligned} \tag{9}$$

we obtain the following expression for the average:

$$\langle \mathrm{e}^{\pm 2\mathrm{i}\Omega(\hat{n})t} \rangle = \mathrm{e}^{\pm 2\mathrm{i}(\Omega_0+2\chi)t} \langle \alpha_0 | \mathrm{e}^{\pm 4\mathrm{i}\chi t(a_1\hat{a}^{\dagger 2} + a_2\hat{n} + a_3\hat{a}^2 + a_4)} | \alpha_0 \rangle. \tag{10}$$

Here the functions a_i are complex functions of time, and we now face the problem of writing the exponential in a product form.

If $\kappa \ll 1$ then $a_2 \to 1$ and $a_i \to 0$ for $i = 1, 3, 4$. In that case the average is given by:

$$\mathrm{e}^{\pm 2\mathrm{i}(\Omega_0+2\chi)t} \langle \alpha_0 | \mathrm{e}^{\pm 4\mathrm{i}\chi t\hat{n}} | \alpha_0 \rangle = \mathrm{e}^{\pm 2\mathrm{i}(\Omega_0+2\chi)t} \exp[|\alpha_0|^2(\mathrm{e}^{\pm 4\mathrm{i}\chi t} - 1)]. \tag{11}$$

If we take the average between initial coherent states $|\alpha_0\rangle$ instead of the evolved coherent states $|\alpha_0\rangle_I$ we get the same result as that given by Eq. 11 (see [17]). For the general case we make use of the fact that the set $\{\hat{a}^{\dagger 2}, \hat{a}^2, \hat{n}, 1\}$ is closed under commutation and we write [21, 22]:

$$\exp\left[\lambda \sum_{n=1}^{N} \alpha_n \hat{O}_n\right] = \exp[\phi_1(\lambda)\hat{O}_1] \cdots \exp[\phi_N(\lambda)\hat{O}_N] \tag{12}$$

where the set of functions $\{\phi_i(\lambda)\}$ must be determined under the constraint $\phi_i(\lambda = 0) = 0$. Consider first the case with a positive sign in the exponent. Then we write:

$$\langle \mathrm{e}^{2\mathrm{i}\Omega(\hat{n})t} \rangle = \mathrm{e}^{2\mathrm{i}(\Omega_0+2\chi)t} \langle \alpha_0 | \mathrm{e}^{4\mathrm{i}\chi t(a_1\hat{a}^{\dagger 2} + a_2\hat{n} + a_3\hat{a}^2 + a_4)} | \alpha_0 \rangle \tag{13}$$

using Eq. 12 we write

$$\mathrm{e}^{\lambda[\beta_1\hat{a}^{\dagger 2} + \beta_2\hat{n} + \beta_3\hat{a}^2 + \beta_4]} = \mathrm{e}^{\phi_1(\lambda)\hat{a}^{\dagger 2}} \mathrm{e}^{\phi_2(\lambda)\hat{n}} \mathrm{e}^{\phi_3(\lambda)\hat{a}^2} \mathrm{e}^{\phi_4(\lambda)} \tag{14}$$

where we have renamed $\beta_i = 4\mathrm{i}\chi t a_i$. From Eq. 14 we obtain the following set of ordinary differential equations:

$$\frac{d\phi_1}{d\lambda} = \beta_1 + 2\beta_2\phi_1 + 4\beta_3\phi_1^2, \qquad \frac{d\phi_2}{d\lambda} = \beta_2 + 4\beta_3\phi_1$$

$$\frac{d\phi_3}{d\lambda} = \beta_3 e^{2\phi_2}, \qquad \frac{d\phi_4}{d\lambda} = \beta_4 + 2\beta_3\phi_1.$$

This set of equations can be solved analytically (see Appendix A) and as a result we get:

$$l\phi_1(\lambda) = \frac{\beta_1 \sinh(2\Sigma\lambda)}{2\Sigma\cosh(2\Sigma\lambda) - \beta_2\sinh(2\Sigma\lambda)},$$
$$\phi_2(\lambda) = -\ln[\cosh(2\Sigma\lambda) - \frac{\beta_2}{2\Sigma}\sinh(2\Sigma\lambda)]$$
$$\phi_3(\lambda) = \frac{\beta_3 \sinh(2\Sigma\lambda)}{2\Sigma\cosh(2\Sigma\lambda) - \beta_2\sinh(2\Sigma\lambda)},$$
$$\phi_4(\lambda) = (\beta_4 - \beta_2/2)\lambda - \frac{1}{2}\ln[\cosh(2\Sigma\lambda) - \frac{\beta_2}{2\Sigma}\sinh(2\Sigma\lambda)]$$

where $\Sigma = \sqrt{\beta_2^2/4 - \beta_1\beta_3}$. Once we have found the functions $\phi_i(\lambda)$ we can take the average given in Eq. 10, getting:

$$\langle e^{2i\Omega(\hat{n})t}\rangle = e^{2i(\Omega_0+2\chi)t}e^{\phi_4+\phi_3\alpha_0^2+\phi_1\alpha_0^{*2}}e^{|\alpha_0|^2(e^{\phi_2}-1)} \tag{15}$$

and

$$\langle e^{-2i\Omega(\hat{n})t}\rangle = e^{-2i(\Omega_0+2\chi)t}e^{\phi_4^*+\phi_3^*\alpha_0^{*2}+\phi_1^*\alpha_0^2}e^{|\alpha_0|^2(e^{\phi_2^*}-1)} \tag{16}$$

where the functions $\phi_i(\lambda)$ must be evaluated at $\lambda = 1$.

The complex, time dependent functions $\alpha_n(t)$ needed to construct the time evolution operator in the interaction picture are likewise obtained from the following set of coupled, nonlinear, ordinary differential equations obtained after substitution of Eq. 7 into Schrödinger's equation

$$\frac{d\alpha_1}{dt} = -i(f_1 + 2\alpha_1 f_2 + 4\alpha_1^2 f_3), \qquad \frac{d\alpha_2}{dt} = -i(f_2 + 4\alpha_1 f_3)$$
$$\frac{d\alpha_3}{dt} = -if_3 e^{2\alpha_2}, \qquad \frac{d\alpha_4}{dt} = -i(f_4 + 2\alpha_1 f_3)$$

where the functions f_n correspond to the coefficients in the mean field Hamiltonian, namely:

$$f_1 = g(t)\langle e^{2i\Omega(\hat{n})t}\rangle, \quad f_2 = 2g(t), \quad f_3 = g(t)\langle e^{-2i\Omega(\hat{n})t}\rangle, \quad f_4 = g(t).$$

At the initial time t_0, the complex functions $\alpha_n(t_0) = 0$ so that we can construct the functions $f_n(t_0)$ corresponding to the mean field Hamiltonian. With these we integrate for a time Δt and obtain new values for the functions $\alpha_n(t = \Delta t)$. At each step we produce a new value for the functions $f_n(t)$ which is then used to compute the new $\alpha_n(t)$. In this way we couple the differential equations with the expectation value of the exponential of the number operator in a self consistent way [18].

If we take the average value of the number operator (see Eq. 9) between number states $|n_i\rangle$ we obtain:

$$\langle n_i|\hat{n}(t)|n_i\rangle = [1 - 8\alpha_1\alpha_3 e^{-2\alpha_2}]n_i - [4\alpha_1\alpha_3 e^{-2\alpha_2}]. \tag{17}$$

If instead, we take the average value using coherent states $|\alpha\rangle$, then the corresponding expectation value of the number operator is

$$\begin{aligned}\langle\alpha|\hat{n}(t)|\alpha\rangle &= [1 - 8\alpha_1\alpha_3 e^{-2\alpha_2}]|\alpha|^2 - [4\alpha_1\alpha_3 e^{-2\alpha_2}] + \\ &\quad [2\alpha_3(4\alpha_1\alpha_3 e^{-2\alpha_2} - 1)]\alpha^2 + [2\alpha_1 e^{-2\alpha_2}]\alpha^{*2}.\end{aligned} \tag{18}$$

Notice that the behavior of $\langle\hat{n}(t)\rangle$ is different depending upon the kind of averaging that is chosen. In what follows we will use the averaging with respect to coherent states.

3. Statistical properties

3.1. Mandel parameter

The Mandel parameter Q gives information about the nature of the photon statistics of any state. It is defined as:

$$Q = \frac{\langle \hat{n}^2 \rangle - \langle \hat{n} \rangle^2}{\langle \hat{n} \rangle}. \tag{19}$$

For a state with Q in the range $0 \leq Q < 1$ the statistics are sub-Poissonian and for $Q > 1$, super-Poissonian. A field coherent state has $Q = 1$. States with sub-Poissonian distribution are non classical states [23]. When $\kappa = 0$ the time evolution operator is that given by Eq. 4, that is: $U = U_0$ and $\langle \alpha | U_0^\dagger \hat{n} U_0 | \alpha \rangle = \langle \alpha | \hat{n} | \alpha \rangle = |\alpha|^2$ and $\langle \alpha | U_0^\dagger \hat{n}^2 U_0 | \alpha \rangle = |\alpha|^4 + |\alpha|^2$ so that the Mandel parameter is equal to one regardless of the value of χ.

In figure 1 we plot the temporal evolution of the Mandel parameter for fixed α with average photon number $\langle \hat{n} \rangle = 18$ and potential parameters $\kappa = 0.05$, $\chi = 0.0$ and $\Omega_0 = 1$ corresponding to a parametric oscillator. In this case the Mandel parameter starts at one as corresponds to a usual coherent state. When $\alpha = (3, 3)$ (green line) the Mandel parameter takes values smaller than one at the beginning of the evolution corresponding to a non classical state and remains sub-Poissonian for some time until it becomes larger than one and maintains an increasing behavior as a function of time. The results for real $\alpha = (\sqrt{18}, 0)$ are shown in purple. As in the previous case the Mandel parameter starts at one as corresponds to a usual coherent state but in contrast with that case here it is always an increasing function of time taking values corresponding to a classical state. The difference is due to the fact that for the evaluation of the Mandel parameter we have taken the averages from Eq. 18 where the importance of the nature of α (it being real or complex) is evident. The undulations are due to the temporal dependence of the oscillator's frequency. Also shown in the figure are the cases with $\kappa = 0.7$ and $\chi = 0.25$ corresponding to a nonlinear parametric oscillator. We present, in blue, the case for $\alpha = (\sqrt{18}, 0)$ and in red the case for $\alpha = (3, 3)$. The general conduct is similar to that of the previous cases, however the undulations we mentioned above are washed out and we see instead the appearance of strong oscillations between intervals where the Mandel parameter remains practically constant. It is precisely when these oscillations appear that the Mandel parameter changes. Notice that for non negligible values of χ, the nonlinear term H_{Kerr} in the Hamiltonian becomes important in the evolution of the system and is responsible for a slower change in the average value of the number operator.

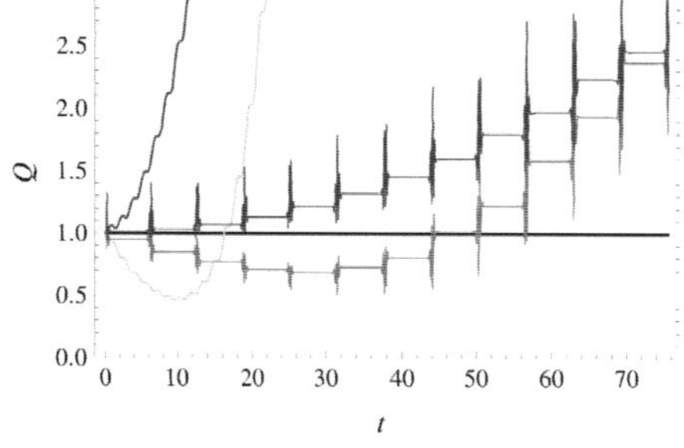

Figure 1. Temporal evolution of the Mandel parameter for $\alpha = \sqrt{18}$ (purple and blue), $\alpha = (3, 3)$ (green and red), corresponding to an initial average number $\langle \hat{n} \rangle = 18$ and parameters $\kappa = 0.05$, $\chi = 0$ (purple and green), $\kappa = 0.7$, $\chi = 0.25$ (blue and red). In all cases $\Omega_0 = 1$ and the initial condition used was $\alpha_0 = \sqrt{18}$.

3.2. Average value of the number operator for the vacuum

If we evaluate the average number of photons with respect to the vacuum state we find an exponential grow when $\chi = 0$ and some undulations due to the temporal dependence of the frequency (see figure 2, purple); this effect is related to the dynamical Casimir effect [24] namely the generation of real photons starting from the vacuum. When the nonlinear term $\chi \neq 0$ the average number of photons with respect to the vacuum state decreases rapidly (see figure 2, black, green and blue) with increasing oscillations as the value of χ increases. This conduct was reported in [25] where they studied the interaction between two two-level atoms in a cavity with time dependent frequency. The similarity with their results is due to the fact that the atoms in the cavity emulate an intensity dependent refraction index.

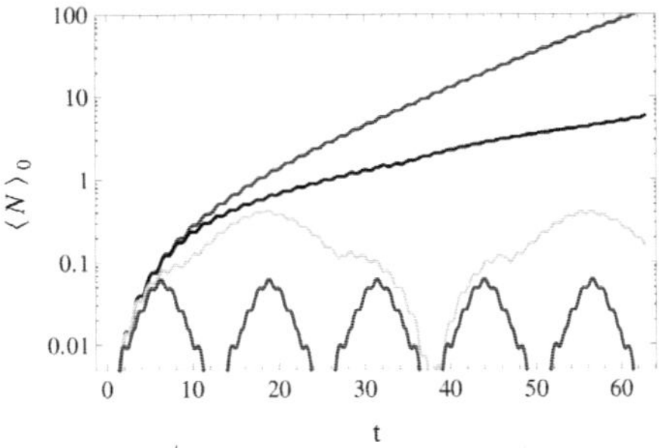

Figure 2. Temporal evolution of the average value of the number operator with respect to the vacuum state. Parameters used: $\chi = 0$ (purple), $\chi = 0.025$ (black), $\chi = 0.085$ (green), $\chi = 0.25$ (blue). In all cases $\kappa = 0.05$ and $\Omega_0 = 1$.

3.3. Squeezing and dispersions

In order to analyze the temporal behavior of different observables let us consider the temporal evolution of the creation-annihilation operators. The creation operator in the Heisenberg picture is given by

$$\hat{a}^\dagger(t) = U_I^\dagger U_0^\dagger \hat{a}^\dagger U_0 U_I \tag{20}$$

Applying the first transformation we get:

$$U_0^\dagger \hat{a}^\dagger U_0 = \mathrm{e}^{\mathrm{i}\chi t\hat{n}^2} \mathrm{e}^{\mathrm{i}\Omega_0 t\hat{n}} \hat{a}^\dagger \mathrm{e}^{-\mathrm{i}\Omega_0 t\hat{n}} \mathrm{e}^{-\mathrm{i}\chi t\hat{n}^2} = \hat{a}^\dagger \mathrm{e}^{\mathrm{i}\chi t(2\hat{n}+1)} \mathrm{e}^{\mathrm{i}\Omega_0 t} \tag{21}$$

The number operator appears in the exponential again. To be consistent with the approximation used to get the interaction picture Hamiltonian (Eq. 6) we replace the exponential by its average value, and apply the second transformation (only to the operator $\hat{a}^\dagger$) obtaining:

$$\hat{a}^\dagger(t) = U^\dagger \hat{a}^\dagger U = \mathrm{e}^{\mathrm{i}(\Omega_0+\chi)t} \langle \mathrm{e}^{2\mathrm{i}\chi t\hat{n}} \rangle U_I^\dagger \hat{a}^\dagger U_I = \mathrm{e}^{\mathrm{i}(\Omega_0+\chi)t} \langle \mathrm{e}^{2\mathrm{i}\chi t\hat{n}} \rangle \mathrm{e}^{-\alpha_2} \left[\hat{a}^\dagger - 2\alpha_3 \hat{a}\right]. \tag{22}$$

Proceeding in a similar form for the annihilation operator we get:

$$\hat{a}(t) = \left[\hat{a}(\mathrm{e}^{\alpha_2} - 4\alpha_1\alpha_3 \mathrm{e}^{-\alpha_2}) + 2\alpha_1 \mathrm{e}^{-\alpha_2} \hat{a}^\dagger\right] \mathrm{e}^{-\mathrm{i}(\Omega_0+\chi)t} \langle \mathrm{e}^{-2\mathrm{i}\chi t\hat{n}} \rangle. \tag{23}$$

The above expressions can be written as:

$$\begin{pmatrix} \hat{a}^\dagger(t) \\ \hat{a}(t) \end{pmatrix} = \begin{pmatrix} a_{11}(t) & a_{12}(t) \\ a_{21}(t) & a_{22}(t) \end{pmatrix} \begin{pmatrix} \hat{a}^\dagger \\ \hat{a} \end{pmatrix}. \tag{24}$$

The complex, time dependent coefficients $a_{ij}(t)$ fulfill the conditions $a_{11}(t) = a_{22}^*(t)$ and $a_{12}(t) = a_{21}^*(t)$ (these have been checked numerically).

The quadratures q and p at time t are given by:

$$q(t) = \frac{1}{\sqrt{2}}\left[\hat{a}(t) + \hat{a}^\dagger(t)\right], \quad p(t) = \frac{\mathrm{i}}{\sqrt{2}}\left[\hat{a}^\dagger(t) - \hat{a}(t)\right] \tag{25}$$

Using the expressions given above we obtain

$$\langle\alpha|q(t)|\alpha\rangle = \frac{1}{\sqrt{2}}\left[\alpha^*(a_{11}(t) + a_{21}(t)) + \alpha(a_{12}(t) + a_{22}(t))\right] \tag{26}$$

and

$$\langle\alpha|p(t)|\alpha\rangle = \frac{\mathrm{i}}{\sqrt{2}}\left[\alpha^*(a_{11}(t) - a_{21}(t)) + \alpha(a_{12}(t) - a_{22}(t))\right]. \tag{27}$$

And for $q(t)^2$, $p(t)^2$ we get:

$$\begin{aligned}\langle\alpha|q(t)^2|\alpha\rangle = \frac{1}{2}[(a_{11}(t) + a_{21}(t))^2\alpha^{*2} + (a_{12}(t) + a_{22}(t))^2\alpha^2 + \\ 2(a_{11}(t) + a_{21}(t))(a_{12}(t) + a_{22}(t))|\alpha|^2 + (a_{11}(t) + a_{21}(t))a_{12}(t) \\ +(a_{12}(t) + a_{22}(t))a_{21}(t) + 1]\end{aligned} \tag{28}$$

$$\begin{aligned}\langle\alpha|p(t)^2|\alpha\rangle = -\frac{1}{2}[(a_{11}(t) - a_{21}(t))^2\alpha^{*2} + (a_{12}(t) - a_{22}(t))^2\alpha^2 + \\ 2(a_{11}(t) - a_{21}(t))(a_{12}(t) - a_{22}(t))|\alpha|^2 + (a_{11}(t) - a_{21}(t))a_{12}(t) \\ +(a_{22}(t) - a_{12}(t))a_{21}(t) - 1]\end{aligned} \tag{29}$$

with these expressions we can obtain the dispersions $\Delta q(t)$, $\Delta p(t)$.

In figure 3 we show the dispersions $\Delta q(t)$ (purple) and Δp (green) for Hamiltonian parameters $\kappa = 0.05$, $\chi = 0.0$, $\Omega_0 = 1$ corresponding to a parametric oscillator. Notice that at the initial time the state is a minimum uncertainty state with $\Delta q(t=0) = \Delta p(t=0) = 1/\sqrt{2}$ and as time evolves the dispersions oscillate and are such that when one increases the other decreases and there are regions where either Δq or Δp take values corresponding to a squeezed state. The amplitude of the oscillations in the dispersions is an increasing function of time. However, the presence of squeezing is maintained. In black we show the product $\Delta q(t)\Delta p(t)$. It can be seen that it oscillates and at those times when the squeezing is present the product corresponds to that of a minimum uncertainty state. For most of the time, the product $\Delta q\Delta p$ is larger and we can say that these states are not minimum uncertainty states but for a limited set of instants of time. A coherent state corresponding to a parametric oscillator does not evolve into a coherent state.

In figure 4 we show the dispersion $\Delta q(t)$ (red) and the product $\Delta q\Delta p$ (black) for Hamiltonian parameters $\kappa = 0.7$, $\chi = 0.25$, $\Omega_0 = 1$, $\alpha = 3 + \mathrm{i}3$ and $\alpha_0 = \sqrt{18}$ corresponding to a nonlinear parametric oscillator. At the initial time $t = 0$ the state is a minimum uncertainty state. At the initial stages of the evolution there is squeezing present, this however does not last long and the dispersion returns to that corresponding to a coherent state until after about $t = 4\pi$ it oscillates for a short time attaining values larger and smaller than $1/\sqrt{2}$ and returning to $1/\sqrt{2}$ until about $t = 8\pi$. This conduct is repeated periodically. In black we show the product $\Delta q(t)\Delta p(t)$ for the same set of potential parameters. It can be seen that the product corresponds practically to a minimum uncertainty state with deviations at those times where squeezing is present.

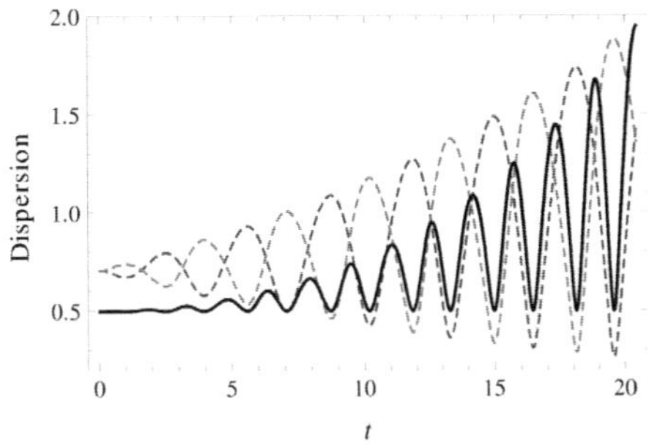

Figure 3. Dispersions Δq (purple) and Δp (green) and their product $\Delta q \Delta p$ (black) with parameters $\chi = 0$, $\kappa = 0.05$, $\alpha = 3 + \mathrm{i}3$, $\alpha_0 = \sqrt{18}$ and $\Omega_0 = 1$.

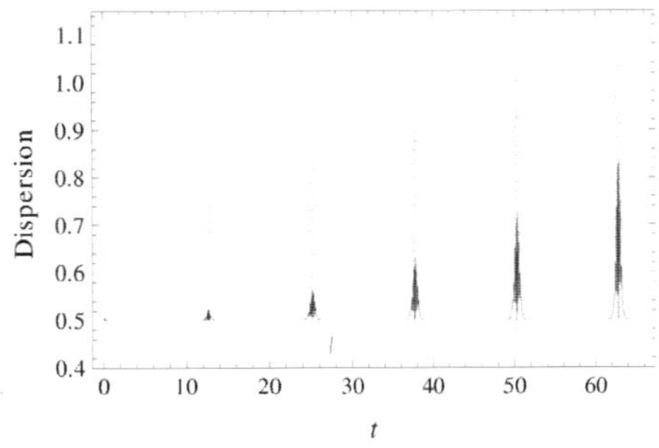

Figure 4. Dispersions Δq (red) and the product $\Delta q \Delta p$ (black) with parameters $\chi = 0.25$, $\kappa = 0.7$, $\alpha = 3 + \mathrm{i}3$, $\alpha_0 = \sqrt{18}$ and $\Omega_0 = 1$.

4. Conclusions

In this work we have built an approximate time evolution operator for a system composed of a parametric oscillator in a nonlinear Kerr-like medium. The Hamiltonian is transformed into the interaction picture and as a result we obtained a time dependent Hamiltonian that contains the number operator in an exponential. In order to solve this problem we approximate the exponential by its average value taken between time dependent coherent states. With this simplification we can write the Hamiltonian in the interaction picture as an element of a finite Lie algebra. The time dependent coefficients of this linear combination are obtained in a self consistent form. The time evolution operator can then be expressed as a product of exponentials. We calculated several statistical properties like the Mandel parameter the dispersions in the quadratures and the generation of photons from the vacuum state. We found squeezing in the quadratures due to the presence of quadratic creation and annihilation operators in the Hamiltonian. The Mandel parameter yields sub-Poissonian statistics at the initial stages of the evolution when α is complex and super-Poissonian statistics when α is real. When α is complex the Mandel parameter is sub-Poissonian at the start and becomes super-Poissonian after some time. For real α the Mandel parameter is always super-Poissonian.

Acknowledgments
We thank Reyes García for the maintenance of our computers and acknowledge partial support from CONACyT through project 166961 and DGAPA-UNAM project IN108413. One of us (MB) would like to thank the Instituto de Ciencias Físicas for its hospitality.

Journal of Physics: Conference Series **698** (2016) 012008 doi:10.1088/1742-6596/698/1/012008

Appendix A

From the set of coupled, nonlinear, ordinary differential equations we got for the functions $\phi_i(t)$ we see that equation 15 has the form of a Riccati equation:

$$y'(\lambda) = q_0(\lambda) + q_1(\lambda)y(\lambda) + q_2(\lambda)y^2(\lambda) \tag{A.1}$$

with:

$$q_0(\lambda) = \beta_1(t), \quad q_1(\lambda) = 2\beta_2(t), \quad q_2(\lambda) = 4\beta_3(t) \tag{A.2}$$

so that,

$$q_2'(\lambda) = 0, \rightarrow \frac{q_2'(\lambda)}{q_2(\lambda)} = 0, \quad q_2(\lambda) \neq 0. \tag{A.3}$$

Introducing the variable $z(\lambda)$ as:

$$\phi_1(\lambda)q_2(\lambda) = -\frac{z'(\lambda)}{z(\lambda)} \tag{A.4}$$

we obtain the following linear, second order equation with constant coefficients

$$z''(\lambda) - 2\beta_2(t)z'(\lambda) + 4\beta_1(t)\beta_2(t)z(\lambda) = 0 \tag{A.5}$$

proposing a solution of the form $z(\lambda) = \exp(\theta\lambda)$ we get the characteristic polynomial

$$\theta^2 - 2\beta_2(t)\theta + 4\beta_1(t)\beta_3(t) = 0 \tag{A.6}$$

with roots:

$$\theta = \beta_2(t) \pm 2\Sigma(t), \quad \Sigma^2(t) = \frac{\beta_2^2(t)}{4} - \beta_1(t)\beta_3(t) \tag{A.7}$$

so the function $z(\lambda)$ can be written as:

$$z(\lambda) = A\exp[(\beta_2(t) + 2\Sigma(t))\lambda] + B\exp[(\beta_2(t) - 2\Sigma(t))\lambda] \tag{A.8}$$

The constants A, B are fixed by the initial condition $\phi_i(t_0) = 0$ and we obtain finally:

$$\phi_1(\lambda) = \frac{\beta_1 \sinh(2\Sigma\lambda)}{2\Sigma\cosh(2\Sigma\lambda) - \beta_2\sinh(2\Sigma\lambda)} \tag{A.9}$$

substitution into the differential equation for $\phi_2(\lambda)$ yields

$$\frac{d\phi_2(\lambda)}{d\lambda} = \frac{\beta_2(t)\cosh(2\Sigma(t)\lambda) - 2\Sigma(t)\sinh(2\Sigma(t)\lambda)}{\cosh(2\Sigma(t)\lambda) - \frac{\beta_2(t)}{2\Sigma(t)}\sinh(2\Sigma(t)\lambda)} \tag{A.10}$$

integrating we get

$$\phi_2(\lambda) = -\ln\left[\cosh(2\Sigma(t)\lambda) - (\beta_2(t)/2\Sigma(t))\sinh(2\Sigma(t)\lambda)\right] \tag{A.11}$$

substitution in the differential equation for $\phi_3(\lambda)$ yields

$$\frac{d\phi_3(\lambda)}{d\lambda} = \frac{4\Sigma^2(t)\beta_3(t)}{[2\Sigma(t)\cosh(2\Sigma(t)\lambda) - \beta_2(t)\sinh(2\Sigma(t)\lambda)]^2} \tag{A.12}$$

integrating we get:

$$\phi_3(\lambda) = \frac{\beta_3(t)\sinh(2\Sigma(t)\lambda)}{2\Sigma(t)\cosh(2\Sigma(t)\lambda) - \beta_2(t)\sinh(2\Sigma(t)\lambda)} \tag{A.13}$$

Finally from the equation for $\phi_4(\lambda)$ we get:

$$\frac{d\phi_4(\lambda)}{d\lambda} = \beta_4(t) + \frac{2\beta_1(t)\beta_3(t)\sinh(2\Sigma(t)\lambda)}{2\Sigma(t)\cosh(2\Sigma(t)\lambda) - \beta_2(t)\sinh(2\Sigma(t)\lambda)} \tag{A.14}$$

integrating we obtain:

$$\phi_4(\lambda) = (\beta_4(t) - \beta_2(t)/2)\lambda - \frac{1}{2}\ln\left[\cosh(2\Sigma(t)\lambda) - \beta_2(t)/(2\Sigma(t))\sinh(2\Sigma(t)\lambda)\right] \tag{A.15}$$

References

[1] Schrödinger E 1926 *Naturwissenschaften* **14** 664
[2] Glauber R J 1963 *Phys. Rev. Lett.* **10** 84-86; Glauber R 1963 *J Phys. Rev.* **131** 2766-2788
[3] Klauder J R 1960 *Ann. Phys.* **11**, 123
[4] Récamier J, García de León P, Jáuregui R, Frank A and Castaños O 2002 *International Journal of Quantum Chemistry* **89** 494-502
[5] Yurke and Stoler 1986 *Phys. Rev. Lett.* **57** 13
[6] Haroche S and Raimond J M 2006 *Exploring the quantum* (Oxford University Press)
[7] Lewis H R Jr. 1967 *Phys. Rev. Lett.* **18** 510; Lewis H R Jr. 1968 *J. Math. Phys.* **9** 1976; Lewis H R Jr. and Riesenfeld W B 1969 *J. Math. Phys.* **10** 1458
[8] Dodonov V V and Man'ko V I 1990 in *Invariants and the evolution of non stationary quantum systems*, ed M A Markov (Lebedev Institute, Moscow); Man'ko V I 1992 in *Symmetries in Physics*, eds A Frank and K B Wolf (Springer Verlag); Dodonov V and Man'ko V I Eds. 2003 *Theory of Nonclassical states of light* (London: Taylor and Francis).
[9] Guerrero J and López-Ruiz F 2013 *Physica Scripta* **87** 038105
[10] Castaños O, Schuch D and Rosas-Ortiz O 2013 *J. Phys. A: Math. Theor.* **46**, 075304
[11] Gazdy B and Micha D 1985 *J. Chem. Phys.* **82**, 4926; Récamier J and Jáuregui R 1997 *Int. J. Quantum Chem.* **62** 125-135
[12] Kiss T, Janszky J and Adam P 1994 *Phys. Rev.* A **49**(6) 4935
[13] Dodonov V V, Klimov A B and Nikonov D E 1993 *Phys. Rev.* A **47**, 4422
[14] Wineland D J, Monroe C, Itano W M, Leibfried D, King B E and Meekhof D M 1998 *J. Res. Natl. Inst. Stand. Tech* **103** 259
[15] Walls D F and Milburn G 1995 *Quantum Optics* (Springer, Berlin)
[16] Dodonov V V, Marchiolli M A, Korennoy Ya A, Man'ko V I and Moukhin Y A 1998 *Phys. Rev.* A **58**(5) 4087-4094
[17] Román-Ancheyta R, Berrondo M and Récamier J 2015 *Journal of the Optical Society of America* B **32** (8) 1651-1655
[18] Berrondo M and Récamier J 2011 *Chem. Phys. Lett.* **503** 180-184
[19] Gerry C C and Knight P L 2005 *Introductory Quantum Optics* Ch. 7 (Cambridge: Cambridge University Press)
[20] Wei J and Norman E 1964 *Proc. Am. Math. Soc.* **15** 327.
[21] Puri R R 2001 *Mathematical Methods of Quantum Optics* (Springer Series in Optical Sciences).
[22] Gilmore R 2002 *Lie Groups,Lie Algebras and some of their applications* (Dover Publications, Inc.)
[23] Gerry C C and Knight P L 2005 *Introductory Quantum Optics* (Cambridge University Press, New York.)
[24] Dodonov V V 2010 *Phys. Scr.* **82** 038105
[25] Dodonov A V and Dodonov V V 2012 *Phys. Rev.* A **85** 055805

Journal of Physics: Conference Series **698** (2016) 012009 doi:10.1088/1742-6596/698/1/012009

Can We Describe Biological Systems with Quantum Mechanics?

C G Granados-Ramírez[1], C G Benítez-Cardoza[2], and M D Carbajal-Tinoco[3]

[1] Grupo de Investigación Bioquímica INS & LIBBIQ-UN Universidad Nacional de Colombia, Cra. 45 No. 26-85 U. Camilo Torres, Bogotá 111321, Colombia

[2] Laboratorio de Investigación Bioquímica, Sección de Estudios de Posgrado, ENMyH-IPN, Guillermo Massieu No. 239, La Escalera Ticomán, 07320 México D.F., Mexico

[3] Departamento de Física, Centro de Investigación y de Estudios Avanzados del IPN, Apartado Postal 14-740, 07000 México D.F., Mexico

E-mail: mdct@fis.cinvestav.mx

Abstract. Quantum Mechanics is the favourite theory to predict the structure of any group of atoms, including biological molecules. Due to numerous difficulties, however, it is necessary to introduce a series of approximations to overcome such impediments. We present a coarse-grained model of circular dichroism (CD) that is based on the theory of optical activity, developed by DeVoe, in order to predict CD spectra. In first stage, we determine the polarisability of individual monomers (residues, in the case of peptides) from experiments of molar absorptivity. The complex polarisabilities are used together with peptide structures obtained by density functional theory and other methods to determine their corresponding CD spectra, which are in reasonable agreement with their experimental counterparts.

1. Introduction

The description and full understanding of biological systems is one of the major scientific tasks for this century. For example, malaria killed about half a million people around the world in 2015 and that is good news because it used to kill twice as many persons, 20 years earlier. In 2015, Youyou Tu received the Nobel prize in medicine for using a plant extract, artemisinin, in the treatment of malaria. In spite of important efforts, there is still no vaccine for malaria and the complete cure remains an open problem as in the case of many other diseases. The comprehension of biological systems can be a challenge because they are composed of a variety of molecules such as lipids, proteins, and nucleic acids like DNA and RNA. These molecules have specific functions that include interaction with other molecules and, eventually, they can modify their shape to perform a certain activity.

The typical biological systems are characterised by experimentally accessible distances (from nanometers to microns) and relatively low energies (of the order of the thermal energy, k_BT). Therefore, the preferred approach to describe such systems should be Quantum Mechanics (QM). For instance, we can propose the solution of the time-independent non-relativistic Schrödinger equations for a system consisting of M nuclei and N electrons that mainly interact through a Coulomb potential. A reduced description, that captures the intended behaviour of the system, requires more than 10^6 atoms (including water molecules). Thus, it is basically impossible to solve the resulting highly-coupled Schrödinger equations, even at the level of the Born-Oppenheimer approximation. The exclusion of the solvent molecules provides a significant reduction of the number of atoms, but it is still necessary to utilise an appropriate theoretical framework. For this purpose, we briefly describe the Density Functional Theory (DFT) in section 2.

In this work, we used DFT to obtain the two stable configurations of a relatively small peptide, namely, polylysine of 150 residues. Such structures are then utilised as a reference in a proposed model of circular dichroism (CD), which is an experimental technique that takes advantage of the chiral nature of biological molecules, allowing the identification of structural changes. Once calibrated, we make use of additional approximations in order to compare the results of our model with experimental data, as explained in the following sections.

2. Theoretical background

2.1. Density functional theory

The DFT is a computational method that allows to perform QM modelling of divers systems including condensed-matter physics, quantum chemistry, and the ground-state of a variety of molecules that could be of biological origin. A good review on the subject can be consulted in Ref. [1]. The electron density $\rho(\vec{r})$ is the central quantity in DFT, since it determines the probability of finding any of the N electrons of the system within the volume element $d\vec{r}$. It is defined as the integral over the spin coordinates of all electrons,

$$\rho(\vec{r}_1) = N \int \cdots \int |\Psi(\vec{x}_1, \vec{x}_2, ..., \vec{x}_N)|^2 ds_1 d\vec{x}_1 \cdots d\vec{x}_N, \tag{1}$$

together with the normalisation condition, $\int \rho(\vec{r})d\vec{r} = N$. The electron density is a positive function that can be experimentally measured and it shows an exponential decay for large distances, $\rho(\vec{r}) \sim \exp[2\sqrt{2I}|\vec{r}|]$, where I is the exact ionization energy. The determination of $\rho(\vec{r})$ is based on two fundamental theorems established by Hohenberg and Kohn (H-K) in 1964 [2]. For an external potential $v(\vec{r})$ acting on the N electrons of the system, the first H-K theorem states that $v(\vec{r})$ is determined (to within a constant) by the electron density $\rho(\vec{r})$. In other words, $v(\vec{r})$ is a unique functional of $\rho(\vec{r})$. The second H-K theorem states that for all v-representable densities, $E_v[\rho] \geq E_v[\rho_0]$, with $E_v[\rho_0]$ being the ground-state energy and ρ_0 is the corresponding ground-state density. The second theorem is of course a direct consequence of the variational principle [3].

Besides the H-K theorems, the Kohn-Sham (K-S) equations can be determined as a function of N orbitals ψ_i [4]. Starting with the expression for the energy functional $E[\rho]$ that depends on the interactions between the N electrons and M nuclei of the system,

$$\begin{aligned} E[\rho] &= -\frac{1}{2}\sum_i^N \langle\psi_i|\nabla^2|\psi_i\rangle + \frac{1}{2}\sum_i^N\sum_j^N \int\int |\psi_i(\vec{r}_1)|^2 \frac{1}{r_{12}} |\psi_j(\vec{r}_2)|^2 d\vec{r}_1 d\vec{r}_2 \\ &+ E_{XC}[\rho] - \sum_i^N \int \sum_A^M \frac{Z_A}{r_{1A}} |\psi_i(\vec{r}_1)|^2 d\vec{r}_1, \end{aligned} \tag{2}$$

with Z_A being the valence of nucleus A, $r_{12} = |\vec{r}_2 - \vec{r}_1|$, and the explicit form of the term $E_{XC}[\rho]$ is unknown in general. Applying the usual constraint $\langle\psi_i|\psi_j\rangle = \delta_{ij}$, leads to the set of equations,

$$\left(-\frac{1}{2}\nabla^2 + V_S(\vec{r}_1)\right)\psi_i = \epsilon_i\psi_i, \tag{3}$$

where,

$$V_S(\vec{r}_1) = \int \frac{\rho(\vec{r}_2)}{r_{12}} d\vec{r}_2 + V_{XC}(\vec{r}_1) - \sum_{A}^{M} \frac{Z_A}{r_{1A}},$$

ϵ_i are the energy eigenvalues of orbitals ψ_i, and $V_{XC} = \delta E_{XC}/\delta\rho$. At this point, it is necessary to provide an approximation for E_{XC}. In the generalised gradient approximation (GGA) there is a dependence on densities and their gradients,

$$E_{XC}^{GGA}[\rho_\alpha, \rho_\beta] = \int f(\rho_\alpha, \rho_\beta, \nabla\rho_\alpha, \nabla\rho_\beta) d\vec{r}. \tag{4}$$

This term can be combined with the K-S contribution to obtain a hybrid functional, namely,

$$E_{XC}^{hyb} = aE_{XC}^{KS} + (1-a)E_{XC}^{GGA}, \tag{5}$$

here, a is a fitting parameter. In the K-S scheme, the orbitals are expanded in a set of L predefined basis functions η_μ,

$$\psi_i = \sum_{\mu=1}^{L} c_{\mu i}\eta_\mu. \tag{6}$$

The Gaussian-type-orbitals are a convenient choice for the basis functions, i.e., $\eta^l \sim r^l \exp[-\kappa r^2]$.

We obtained the ground-state of a poly-L-lysine (also denoted as polylysine, for short) of 150 residues in vacuum by using the package Orca [5] and the density functional B3LYP [6]. This peptide has two stable configurations, namely, α-helix and β-sheet, which are employed to determine two constants parameters of our CD model. On the other hand, modelling a polylysine in aqueous solution is nowadays out of the scope of DFT not only from the computational point of view. There are still some difficulties in describing certain kinds of intermolecular interactions such as dispersion forces. Instead, we made use of a classical force-field to model our peptides dissolved in water.

2.2. *Circular dichroism*

An important feature of biological molecules is that a considerable number of them are chiral, in other words, they are not identical to their mirror images. Chiral structures can be distinguished and characterised by polarised light [7]. CD is the differential absorption of left and right circularly polarised light or the difference between their respective extinction coefficients,

$$\Delta\varepsilon = \varepsilon_L - \varepsilon_R. \tag{7}$$

The rotational strength of a transition $A \leftarrow 0$ is the integrated intensity of a single wavelength band in the CD spectrum

$$R_{0A} \sim \int (\Delta\varepsilon/\lambda) d\lambda, \qquad R_{0A} = Im(\langle\Psi_0|\vec{\mu}_e|\Psi_A\rangle \cdot \langle\Psi_0|\vec{\mu}_m|\Psi_A\rangle), \tag{8}$$

with $\vec{\mu}_e$ and $\vec{\mu}_m$ being the electronic and magnetic transition dipole moments. There are theoretical calculations of CD spectra within this QM formalism that are rather involved, but they are only in qualitative agreement with the corresponding experimental data [8] and that is a consequence of the used approximations. In this paper, we present an approach that combines the accuracy of the structures obtained by means of DTF together with experimental data and the results of all-atom (classical) molecular dynamics (MD). The resulting model is computationally tractable even for proteins of hundreds of residues.

Fifty years ago, DeVoe proposed a classical theory of CD [9]. For an applied electric field $\vec{\mathbf{E}}_i$, the induced dipole moment $\vec{\mu}_i$ at each residue of polarisability tensor $\hat{\alpha}$ and components α_i, is given by

$$\vec{\mu}_i = \alpha_i \left[\vec{\mathbf{E}}_i + \sum_{i \neq j} T_{ij} \vec{\mu}_i \right], \tag{9}$$

where the components of the dipole interaction tensor $\hat{T}$ are defined as,

$$T_{ij} = -\left[3p_i q_j - \delta_{pq} r_{ij}^2\right] / r_{ij}^5, \tag{10}$$

and $(p, q = x, y, z)$. The formal solution of Eq. (9) is $\hat{\mu} = \hat{B}\hat{\mathbf{E}}$ with $\vec{\mu}_i = |\vec{\mu}_i|\vec{\mathbf{e}}_i$. It can be assumed that an average polarisability α is able to approximate the three principal components of the monomer polarisabilities [10]. An advantage of the polarisability α is that it can be extracted from experimental data, as explained in the following lines. According to the Beer-Lambert law, the monomer extinction coefficient ε is proportional to the absorbance [7], which is a measurable quantity. Additionally, we use a first order approximation to relate the extinction coefficient with the imaginary part of α, i.e.,

$$\varepsilon \simeq \frac{k_1}{\lambda} \sum_{i=1}^{m} Im(\alpha_i). \tag{11}$$

Here, k_1 is a constant, m is the total number of identical monomers (residues, in our case), and $\alpha_i = \alpha(\omega)$ with ω being the angular frequency. The real part of $\alpha(\omega)$ is obtained through a Kramers-Kronig transform,

$$Re(\alpha(\omega)) = -\frac{2}{\pi} \int_0^\infty \frac{Im(\alpha(\omega'))\omega' d\omega'}{[\omega'^2 - \omega^2]}. \tag{12}$$

On the other hand, CD can be determined through the ellipticity $[\theta] \sim \Delta\varepsilon$,

$$[\theta] = \frac{k_2}{\lambda^2} \sum_{i,j=1}^{m} (\vec{\mathbf{e}}_i \times \vec{\mathbf{e}}_j) \cdot (\vec{r}_j - \vec{r}_i) A_{ij}, \tag{13}$$

where $\hat{A} = [\hat{\alpha}^{-1} + \hat{G}]^{-1}$, $G_{ij} = T_{ij}\vec{\mathbf{e}}_i \cdot \vec{\mathbf{e}}_j$, and the constant k_2 has to be determined. The results of this model are presented in the next section.

3. Experimental part

In this section, we first depict the synthesis of our peptides and the experimental techniques used to describe them. These results together with the DFT and MD calculations are the main ingredients of our model of CD.

3.1. Poly-L-lysine synthesis

The monodisperse polymers of L-lysine (of 7, 8, and 20 residues) were manually synthesised by using Fmoc chemistry on 0.22 mmoles of Rink amide 4-Methylbenzhydrylamine hydrochloride copoly (styrene-1% Divinylbenzene) resin (100-200 mesh, loading 0.74 mmol/g, Merck-Millipore). Synthesis was carried out only with L amino acids protected with N-fluorenylmetoxycarbonyl group (Fmoc). The resins were solvated by successive washing steps with dichloromethane (DCM, 3×1 min), 37% piperidine and 0.07% Triton®X-100 in DMF (3×5 min), N,N-Dimethylformamide DMF (6×1 min), and DCM (3×1 min). Fmoc protected amino acids (Merck-Millipore) were incorporated using standard procedures and N,N'-Diisopropylcarbodiimide / 1-Hydroxybenzotriazole hydrate (1:1) in DMF (0.3 M of Fmoc-amino acid) as coupling agents. Fmoc group was removed by treatment with 37% piperidine and 0.07% Triton®X-100 in DMF. After each deprotection, peptidyl resins were washed with DMF (7×1 min) and DCM (2×1 min).

A ninhydrin test was used to monitor amino acid couplings and deprotections along the dipeptide synthesis [11]. Finally, the peptides on Rink-MBHA resin were cleaved by treatment with 3 ml of a mixture of Trifluoroacetic acid/water (95:5 v/v) at 0°C for 15 min and at room temperature for 1.5 hrs. The acid was removed by evaporation and the crude peptide was precipitated with tert-butyl methyl ether/cyclohexane mixture (1:1). The products were characterised by HPLC-MS and ESI mass spectrometry. UV detection was performed at 220 nm, in an Agilent Technologies SQ LC/MSD G6120B instrument. The polymers were purified by HPLC using a Symmetry C18 Preparative Column (100 Å, 5 μm, 19 mm × 150, Waters®) in a Waters instrument with a 2487 Dual λ Absorbance Detector, Waters 600 controller, Waters Delta 600 pump, degasser and Waters Fraction collector III with a manual injector Rheodyne®3725i controlled by the Empower II software.

3.2. Experimental techniques

We obtained a series of UV and CD spectra of our peptides in aqueous solution. The types of spectra were simultaneously acquired using a Jasco J815 spectropolarimeter interfaced with a Peltier unit. Far-UV measurements were performed in 1 mm pathlength quartz cells with PTFE stopper (Hellma). In all cases, three scans were done with a response time of 2 s, a band width of 0.1 nm, and a scan velocity of 20 nm/min.

A range of concentrations between 25-200 μM was used for poly-L-lysine of 7 and 8 residues solution, 10-100 μM for poly-L-lysine of 20 residues and 0.25-2 μM for commercial poly-L-lysine. The experiments in aqueous solution were acquired at 295 K, in NaOH (1 mM, pH 10.8) storaged under nitrogen to avoid pH change during the experiments. For each experiment, a blank solution spectrum was subtracted. The spectra were collected between 187-250 nm. The raw ellipticity was converted to molar ellipticity, $[\theta]$. Samples were left 30 min at 278 K to allow equilibration before each experiment.

4. Results and discussion

The Far-UV spectra are a measure of the absorbance of our peptides in solution, which is equal to the extinction coefficient multiplied by the pathlength and the molar concentration. From the analysed spectra of the three synthesised peptides, we found better results with the polylysine of 7 residues. In the inset of Fig. 1, we present the extinction coefficient ε of this peptide. Assuming an identical polarisability for each residue and using Eq. (11), we obtain the imaginary part of α of a single residue and its real part is computed through Eq. (12). Both components of α are shown in Fig. 1.

In the following stage, the complex polarisability is used together with the structural information of the peptide under study to determine its corresponding ellipticity from Eq. (13). Starting with the optimised polylysines obtained by DFT, we obtain the positions of the centres of polarisability (CP) of each monomer, which are analogous to the centres of mass (instead of using the atomic masses, we employ the dipole polarisabilities of neutral atoms [12]). The electric field $\vec{\mathbf{E}}_i$ is null in the z-direction and the vectors $\vec{\mathbf{e}}_i$ come from Eq. (9). In order to improve the statistics of $[\theta]$, we rotate the peptide along a spiral on the unit sphere, as suggested by Park *et al.* [13]. The constants k_1 and k_2 are determined by comparing our spectra with

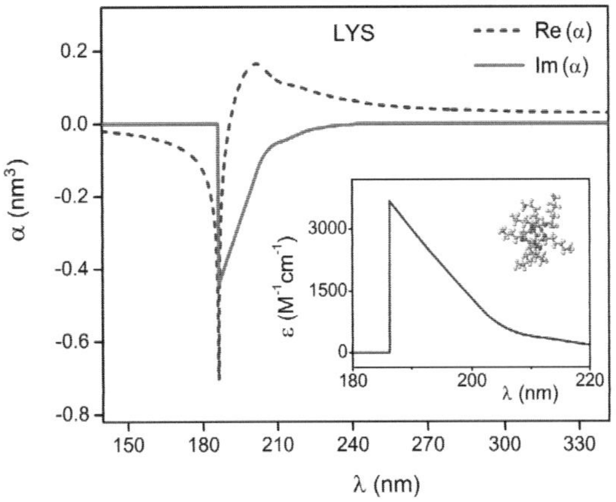

Figure 1. Real and imaginary parts of the polarisability α per residue of lysines belonging to a peptide. Both components were extracted from the extinction coefficient ε of a polylysine of 7 residues and the measured ε is shown in the inset of the figure (the details of the calculation can be consulted in the text).

the well-established profiles of known structures [7]. As a result, in Fig. 2 we draw the spectra of two polylysines of 150 residues in the configurations of α-helix and β-sheet. As it can be observed in such figure, the two spectra are at least in qualitative agreement with the known spectral shapes.

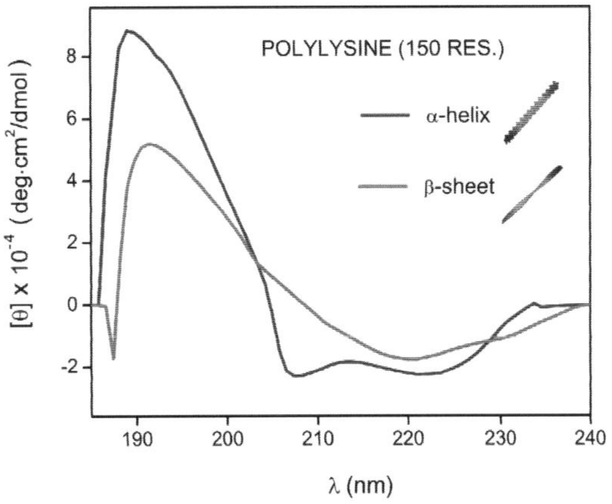

Figure 2. Molar ellipticity $[\theta]$ of two polylysines of 150 residues each one that are in different structural configurations, namely, an α-helix and a β-sheet.

We finally performed a comparison with experimental data. The commercial polylysine has an estimate of 230 residues and we carried out the corresponding MD simulation with the package Gromacs with the Gromos 53A6 force-field and using explicit solvent (three-point SPC/E representation of water molecules). In Fig. 3, we compare the experimental results with the prediction of our model for the molar ellipticity, provided that both systems have similar conditions of pH and temperature. As it can be observed in such figure, there is a fairly good agreement between the modelled spectrum and the experimental one, especially in the region of $\lambda > 210$ nm.

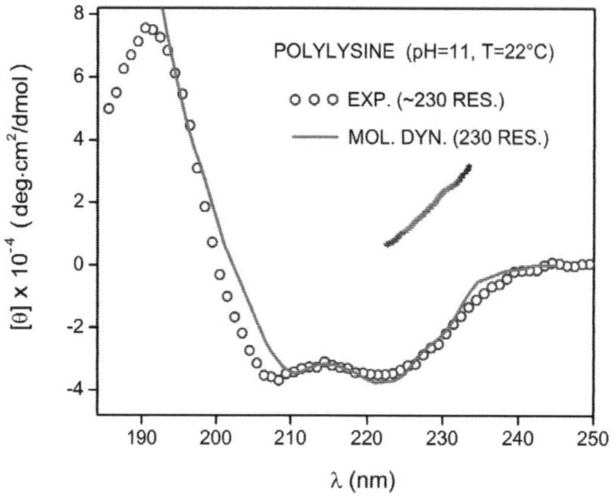

Figure 3. Molar ellipticity $[\theta]$ of an experimental system and the resulting spectrum of our model that uses, as an input, a simulated structure of 230 residues.

5. Conclusions

We have presented a model of circular dichroism that is able to reproduce, at least in a qualitative way, the known spectra of secondary structures like the α-helix and the β-sheet. In order to calibrate our model, we made use of structures obtained with DFT and experimental data such as the absorptivity and the CD spectra of small peptides that were synthesised by us. These experimental quantities are intended to capture subtle effects that include the interaction between the solvent and the peptide as well as other possible physical interactions (electrostatic, van de Waals, etc.) that still cannot be properly described with QM. More important, the result of our model is in reasonable agreement with an experimental CD spectrum and it was obtained in a reduced computational time.

Acknowledgments

We appreciate the aid of D. Jacinto-Méndez for providing the MD simulation performed at the cluster Xiuhcoatl of Cinvestav. This work was partially supported by CONACyT through Grant No. 152532. C.G.G.-R. wants to thank the programme *Tiempo de Volver* of Colciencias (Colombia) and C.G.B-C. thanks SIP-IPN Grant No. 20160239 and Medix.

References

[1] Parr R G and Yang W 1989 *Density-Functional Theory of Atoms and Molecules* (New York: Oxford University Press)
[2] Hohenberg P and Kohn W 1964 *Phys. Rev.* **136** B864
[3] Sakurai J J 1994 *Modern Quantum Mechanics* (New York: Addison-Wesley Publishing Co.)
[4] Kohn W and Sham L J 1965 *Phys. Rev.* **140** A1133
[5] Hoičevar T and Demšar J 2014 *Bioinformatics* **30** 559
[6] Stephens P J, Devlin F J, Chabalowski C F and Frisch M J 1994 *J. Phys. Chem.* **98** 11623
[7] Tinoco I *et al.* 2014 *Physical Chemistry* (Boston: Pearson Education Inc.)
[8] Besley N A and Hirst J D 1999 *J. Am. Chem. Soc.* **121** 9636
[9] DeVoe H 1965 *J. Chem. Phys.* **43** 3199
[10] Miller K J 1990 *J. Am. Chem. Soc.* **112** 8543
[11] Sarin V K *et al.* 1981 *Anal. Biochem.* **117** 147
[12] Anderson K and Sadlej A J 1992 *Phys. Rev. A* **46** 2356
[13] Park S, Bardhan J P, Roux B and Makowski L 2009 *J. Chem. Phys.* **130** 134114

On Quantum Microstates in the Near Extremal, Near Horizon Kerr Geometry

A Guneratne[1], L Rodriguez[2], S Wickramasekara[1,3] and T Yildirim[4]

[1]Department of Physics and Astronomy, The University of Iowa, Iowa City, IA 52242
[2]Department of Physics, Assumption College, Worcester, MA 01609
[3]Department of Physics, Grinnell College, Grinnell, IA 50112
[4]Department of Physics, Arizona State University, Tempe, AZ 85287

E-mail: ananda-guneratne@uiowa.edu,ll.rodriguez@assumption.edu, s-wickram@uiowa.edu,tuna.yildirim@asu.edu

Abstract. We study the thermodynamics of near horizon near extremal Kerr (NHNEK) geometry within the framework of AdS_2/CFT_1 correspondence. We start by shifting the horizon of near horizon extremal Kerr (NHEK) geometry by a general finite mass. While this shift does not alter the geometry in that the resulting classical solution is still diffeomorphic to the NHEK solution, it does lead to a quantum theory different from that of NHEK. We obtain this quantum theory by means of a Robinson-Wilczek two-dimensional Kaluza-Klein reduction which enables us to introduce a finite regulator on the AdS_2 boundary and compute the full asymptotic symmetry group of the two-dimensional quantum conformal field theory on the respective AdS_2 boundary. The s-wave contribution of the energy-momentum-tensor of this conformal field theory, together with the asymptotic symmetries, generate a Virasoro algebra with a calculable center, which agrees with the standard Kerr/CFT result, and a non-vanishing lowest Virasoro eigenmode. The central charge and lowest eigenmode produce the Bekenstein-Hawking entropy and Hawking temperature for NHNEK.

1. Introduction

Black hole thermodynamic quantities [1–3],

$$\begin{cases} T_H = \frac{\hbar\kappa}{2\pi} & \text{Hawking Temperature} \\ S_{BH} = \frac{A}{4\hbar G} & \text{Bekenstein-Hawking Entropy} \end{cases} \tag{1}$$

have provided a testbed for most current competing theories of quantum gravity. It is widely believed that any viable ultraviolet completion of general relativity should reproduce some variant of (1), perhaps modulo some real finite parameter that would need to be fixed by experiment. To date, there have been a range of different approaches for arriving at (1), with string theories and loop quantum gravity being the predominant competitors but no clear consensus on which approach should be preferred over the others.

Since Hawking's original analysis of the density of quantum states in terms of Bogolyubov coefficients, effective actions and their associated energy-momentum tensors for semiclassical matter fields have been explored in various settings for arriving at T_H [4, 5]. Of particular interest is the realization by Robinson and Wilczek (RW) that anomalous two-dimensional chiral theories in the near horizon of black holes are rendered unitary by requiring the black hole to radiate at temperature T_H [6–9]. The RW procedure requires a dimensional reduction yielding two-dimensional analogues (RW2DA) for various types of four-dimensional black holes, beyond the basic Schwarzschild case, coupled to two-dimensional matter fields.

The AdS/CFT correspondence, i.e., the conjecture that quantum gravity on an anti-de Sitter space is dual to a conformal field theory [10], has spawned a surge, led by Strominger [11, 12], Carlip [13–16], Park [17–19] and others, in applying CFT techniques to compute Bekenstein-Hawking entropy of various black holes. By far, the most notable example is the Kerr/CFT correspondence and its extensions [12, 20], where the general idea is that the asymptotic symmetry group (ASG), preserving certain metric boundary or fall off conditions, is generated by a Virasoro algebra with a calculable central extension:

$$[\mathcal{Q}_m, \mathcal{Q}_n] = (m-n)\mathcal{Q}_{m+n} + \tfrac{c}{12} m\left(m^2-1\right)\delta_{m+n,0}, \tag{2}$$

where $m, n \in \mathbb{Z}$. The Bekenstein-Hawking entropy is then obtained from Cardy's formula [21,22] in terms of the central charge c and the normalized lowest eigenmode $\mathcal{Q}_0$ (without Casimir shift):

$$S = 2\pi\sqrt{\frac{c\cdot \mathcal{Q}_0}{6}}. \tag{3}$$

Since surface gravity is usually employed in regulating the quantum charges of (2), thus leading to a finite $\mathcal{Q}_0$, there is some difficulty in using (3) for the extremal Kerr geometry where surface gravity, and therewith Hawking temperature, vanishes. To circumvent this, a thermal Cardy formula is often employed:

$$S = \tfrac{\pi^2}{3}\left(c_L T_L + c_R T_R\right) \tag{4}$$

where the subscripts L and R refer to the dual, two chiral CFT theories with central charges c_L and c_R and Frolov-Thorne vacuum temperatures T_L and T_R which, in the near extremal case, are

$$T_L = \frac{(GM)^2}{2\pi J} \quad \text{and} \quad T_R = \frac{\sqrt{(GM)^4-(GJ)^2}}{2\pi J}. \tag{5}$$

It follows that $T_L = \frac{1}{2\pi}$, $T_R = 0$ in the extremal case, where $GJ = M^2$. Further, since $c_L = c_R = c = 12J$ in the extremal case, we can readily obtain the near horizon extremal Kerr (NHEK) entropy from (4), which has exactly the same form as the Bekenstein-Hawking entropy:

$$S_{BH} = \tfrac{\pi^2}{3}(c_L T_L + c_R T_R) = \tfrac{\pi^2}{3} c T_L = 2\pi J. \tag{6}$$

Because of the equality of $c_L = c_R$, we can also identify a NHEK temperature,

$$T = T_L + T_R = \tfrac{1}{2\pi}, \tag{7}$$

which is in contrast to the vanishing Hawking temperature in this limit.

Note that, while the use of the thermal Cardy formula was motivated by the failure of regularization of the quantum charges (2), a finite zero mode may be nevertheless inferred from (6) by the identification

$$\frac{\partial S_{CFT}}{\partial \mathcal{Q}_0} = \frac{\partial S_{BH}}{\partial \mathcal{Q}_0} \equiv \frac{1}{T} \Rightarrow \mathcal{Q}_0 = \tfrac{\pi^2}{6} c T^2. \tag{8}$$

Turning things around, the temperature $T = \frac{1}{2\pi}$ of (7) can be obtained as a general result from (8) when the ASG of the CFT contains a proper $SL(2,\mathbb{R})$ subgroup. From the definition (8) we also see that T in general should be unitless (with $\hbar = 1$). In passing, we note that an interesting candidate for such a general temperature is the Hawking temperature scaled by the finite time regulator $1/\kappa$. This gives $T = \frac{1}{2\pi}$, which also extends smoothly to extremality (a similar identification can be found in [13, 23]).

The desired general temperature result $T = \frac{1}{2\pi}$ motivates us to consider defining a general Kerr entropy that reduces to (6) at extremality. To that end, if we combine the Frolov-Thorne temperatures (5) with (7) *and* use $c_L = c_R = c = 12J$, we can obtain the standard area law

$$S_{BH} = 2\pi\left(GM^2 + \sqrt{G^2M^4 - J^2}\right). \tag{9}$$

The drawback of this result is that it combines quantities derived separately at extremality and near-extremality. It is also not obvious that the combination of temperatures in (5), $T = T_L + T_R$, yields the value $\frac{1}{2\pi}$, except in the extremal limit. However, since the extremal result $c = 12J$ is consistent with the general expression $c = \frac{3A}{2\pi G}$, we may recast c, T_L, T_R in terms of more general variables to obtain

$$c = \frac{3A}{2\pi G},\ T_L = \frac{4(GM)^2}{A} \text{ and } T_R = \frac{4\sqrt{(GM)^4 - (GJ)^2}}{A}. \tag{10}$$

Substituting (10) into (4) yields $S_{BH} = \frac{A}{4G}$ and, assuming they smoothly extend back to non-extremality,

$$T = T_L + T_R = \frac{4(GM)^2}{A} + \frac{4\sqrt{(GM)^4 - (GJ)^2}}{A} = \frac{1}{2\pi}. \tag{11}$$

Such a procedure would provide a more wholesome calculation of near-extremal Kerr black hole entropy. This is precisely the aim of this note, i.e., to construct a CFT dual for the near horizon, near extremal Kerr (NHNEK) geometry and compute its entropy by the statistical Cardy formula (3), without mixing results derived separately at extremality and near-extremality. This will require the computation of the full ASG, which we will do within an AdS_2/CFT_1 correspondence by performing a RW two-dimensional reduction of the NHNEK geometry in a specific finite mass gauge, following similar previous constructions [24].

2. Geometry

Consider the generic Kerr metric

$$\begin{aligned} ds^2_{Kerr} = &- \frac{\Sigma\Delta}{(r^2+a^2)^2 - \Delta a^2 \sin^2\theta} dt'^2 + \Sigma\left[\frac{dr^2}{\Delta} + d\theta^2\right] \\ &+ \frac{\left((r^2+a^2)^2 - \Delta a^2\sin^2\theta\right)\sin^2\theta}{\Sigma}\left[d\phi' + \frac{2rGMa}{(r^2+a^2)^2 - \Delta a^2\sin^2\theta} dt'\right]^2, \end{aligned} \tag{12}$$

where

$$\Sigma = r^2 + a^2\cos^2\theta, \quad \Delta = (r - r_+)(r - r_-), \quad r_\pm = Gm \pm \sqrt{(Gm)^2 - a^2}, \quad a = \tfrac{J}{m}. \tag{13}$$

The NHEK geometry is a four-dimensional vacuum solution derived from the above Kerr metric by introducing the transformations

$$r = Gm + \lambda U,\ t' = \frac{t}{\lambda},\ \phi' = \phi + \frac{t}{2Gm\lambda}, \tag{14}$$

and taking the limit $\lambda \to 0$, known as the extremal near horizon limit:

$$ds^2_{NHEK} = \frac{1+\cos^2\theta}{2}\left[-\frac{U^2}{\ell^2}dt^2 + \frac{\ell^2}{U^2}dU^2 + \ell^2 d\theta^2\right] + \ell^2 \frac{2\sin^2\theta}{1+\cos^2\theta}\left(d\phi + \frac{U}{\ell^2}dt\right)^2, \tag{15}$$

where $\ell^2 = 2G^2M^2$. This extremal metric may be tuned to near-extremality via a finite temperature gauge:

$$ds^2_{NHNEK} = \frac{1+\cos^2\theta}{2}\left[-\frac{U^2-\epsilon^2}{\ell^2}dt^2 + \frac{\ell^2}{U^2-\epsilon^2}dU^2 + \ell^2 d\theta^2\right] + \ell^2 \frac{2\sin^2\theta}{1+\cos^2\theta}\left(d\phi + \frac{U}{\ell^2}dt\right)^2, \tag{16}$$

where $\epsilon = \frac{1}{2\lambda}(r_+ - r_-)$ is a finite excitation above extremality. The line elements (15) and (16) are classically diffeomerophic, but they exhibit differing quantum theories. To extract the details of these differences, we make one more tuning by adding a finite ADM mass parameter. The resulting metric representing this finite mass gauge,

$$\begin{aligned} ds^2_{NHNEK} =& \frac{1+\cos^2\theta}{2}\left[-\frac{r^2-2rGM+a^2}{r_+^2+a^2}dt^2 + \frac{r_+^2+a^2}{r^2-2rGM+a^2}dr^2 + \left(r_+^2+a^2\right)d\theta^2\right] + \\ & \frac{2\sin^2\theta}{1+\cos^2\theta}\left(r_+^2+a^2\right)\left[d\phi + \left(\frac{r-2GM}{r_+^2+a^2}\right)dt\right]^2, \end{aligned} \tag{17}$$

is still diffeomorphic to (15) and (16), and clearly exhibits global $AdS_2 \times S^2$ topology, our reasons for maintaining the same label NHNEK in (17). However, it is this form of the NHNEK metric with the mass gauge that will prove useful for the tuning purposes in our CFT construction and the calculation of the full ASG.

3. Quantum fields in NHNEK spacetime

Note that the NHNEK metric (15) is of the general form

$$ds^2 = K_1(\theta)\, {g^{(2)}}_{\mu\nu}dx^\mu dx^\nu + K_2(\theta)\, e^{-2\varphi}d\theta^2 + K_3(\theta)\, e^{-2\varphi}\left[d\phi + \mathcal{A}\right]^2, \tag{18}$$

where ${g^{(2)}}_{\mu\nu}$ is the reduced two-dimensional metric $(\mu, \nu = 0, 1)$, φ is a free scalar field and $\mathcal{A}$ is a $U(1)$ gauge field, collectively often called Kaluza-Klein fields. The two-dimensional field splitting of (18) provides a robust platform for constructing CFT duals in the near extremal case by way of the ASG of the Kaluza-Klein fields. We will now exploit the structure of (18) to examine the near horizon matter theory of (17) via the RW dimensional reduction procedure. Our goal here is to apply our previous techniques from [24–26] to study the resulting thermodynamics of the NHNEK within an AdS_2/CFT_1 formalism.

3.1. RW dimensional reduction

Our initial ansatz leading to the NHNEK solution involved a specific decomposition of our four-dimensional spacetime into a two-dimensional black hole and matter fields. It is necessary to check that these fields are the correct RW2DA useful in a holographic study of the quantum spacetime in the near horizon regime. To that end, let us consider a single free scalar field φ in the background of (17) with action:

$$\begin{aligned}S_{free} =& \frac{1}{2}\int d^4x\sqrt{-g}g^{\mu\nu}\partial_\mu\varphi\partial_\nu\varphi \\ =& -\frac{1}{2}\int d^4x\,\varphi\left[\partial_\mu\left(\sqrt{-g}g^{\mu\nu}\partial_\nu\right)\right]\varphi \\ =& -\frac{1}{2}\int d^4x\,\varphi\left[\partial_t\left(-\sin\theta\left(a^2+r_+^2\right)\frac{r_+^2+a^2}{r^2-2rGM+a^2}\partial_t\right)+\right. \\ & \partial_r\left(\sin\theta\left(a^2+r_+^2\right)\frac{r^2-2rGM+a^2}{r_+^2+a^2}\partial_r\right)+\partial_\theta\left(\sin\theta\partial_\theta\right)+ \\ & \partial_\phi\left(\left\{-\sin\theta\left(a^2+r_+^2\right)\left(\frac{r-2GM}{r_+^2+a^2}\right)^2\frac{r_+^2+a^2}{r^2-2rGM+a^2}+\frac{\left(\cos\left(2\theta\right)+3\right)^2}{16\sin\theta}\right\}\partial_\phi\right)+ \\ & \left.2\partial_t\left(\sin\theta\left(a^2+r_+^2\right)\frac{r-2GM}{r_+^2+a^2}\frac{r_+^2+a^2}{r^2-2rGM+a^2}\partial_\phi\right)\right]\varphi.\end{aligned} \tag{19}$$

The above functional is reduced to a two-dimensional theory by expanding the four-dimensional scalar field in terms of spherical harmonics

$$\varphi(t,r,\theta,\phi)=\sum_{lm}\varphi_{lm}(r,t)Y_l{}^m(\theta,\phi), \tag{20}$$

where φ_{lm} has the form of a complex interacting two-dimensional scalar field. Integrating out angular degrees of freedom, transforming to tortoise coordinates $dr^*=f(r)dr$ and considering the region very close to r_+, we find that the two-dimensional action is much reduced. This is due to the fact that all interaction, mixing and potential terms ($\sim l(l+1)\ldots$) are weighted by a factor of $f(r(r*))\sim e^{2\kappa r^*}$, which vanishes exponentially fast as $r\to r_+$. This leaves us with an infinite collection of massless charged scalar fields in the very near horizon region, with $U(1)$ gauge charge equal to the azimuthal quantum number $e=m$ and remnant functional:

$$S=-\frac{r_+^2+a^2}{2}\int d^2x\,\varphi_{lm}^*\left[-\frac{1}{f(r)}\left(\partial_t-im\mathcal{A}_t\right)^2+\partial_r f(r)\partial_r\right]\varphi_{lm}. \tag{21}$$

Thus, we arrive at the RW2DA for the NHNEK solution given by:

$$g^{(2)}{}_{\mu\nu}=\begin{pmatrix}-f(r) & 0\\ 0 & \frac{1}{f(r)}\end{pmatrix} \tag{22}$$

and $U(1)$ gauge field

$$\mathcal{A}=\mathcal{A}_t dt. \tag{23}$$

Given the initial ansatz (18), it is not surprising that the only relevant physical fields in the region $r\sim r_+$ are the above RW2DAs, which reinforces the holographic statement that we may learn much about the quantum nature of spacetime in the near horizon regime from the semiclassical analysis of $g^{(2)}{}_{\mu\nu}$, $\mathcal{A}$ and φ_{lm}.

3.2. *Effective gravitational action and asymptotic symmetries*

We would like to interpret (21) as a useful action for gravity in the near horizon of the classical four-dimensional spacetime. This can be done by considering only the s-wave contribution and making a field redefinition rendering the scalar field unitless [26]. The s-wave approximation is sensible in this scenario since we will interpret φ_{lm} as a component of the gravitational field and hence it should be real and unitless. Most of the interesting gravitational dynamics seem to be contained in this region or approximation [27]. We also note that in [24] it was shown that φ_{lm} dies exponentially fast in time by analyzing the asymptotic behavior of its field equation. However, we find the statement relating φ_{lm} to a real gravitational field component, a stronger justification to neglect higher order terms in l and m. These arguments motivate the field redefinition

$$\varphi_{00} = \sqrt{\frac{6}{G}}\psi, \tag{24}$$

where ψ is now unitless and the $\sqrt{6}$ is chosen to recover the Einstein coupling $\frac{1}{16\pi G}$ in the effective action (21) within the s-wave approximation. Using (24) in (21) gives

$$S^{(2)}[\psi, g] = \frac{3(r_+^2 + a^2)}{G} \int d^2x\sqrt{-g^{(2)}}\psi \left[D_\mu \left(\sqrt{-g^{(2)}}g_{(2)}{}^{\mu\nu}D_\nu\right)\right]\psi, \tag{25}$$

where D_μ is the gauge covariant derivative. In addition to re-dressing the scalar field, our choice of field redefinition has also rendered the effective coupling unitless, hinting towards a finite quantum theory. The effective action of this quantum theory, which may be extracted via zeta-function regularization of the functional determinant in (25), is given by the sum of two functionals [7, 28]:

$$\Gamma = \Gamma_{grav} + \Gamma_{U(1)}, \tag{26}$$

where

$$\Gamma_{grav} = \frac{(r_+^2 + a^2)}{16\pi G} \int d^2x\sqrt{-g^{(2)}}R^{(2)}\frac{1}{\Box_{g^{(2)}}}R^{(2)}, \quad \Gamma_{U(1)} = \frac{3e^2(r_+^2 + a^2)}{\pi G} \int \mathcal{F}\frac{1}{\Box_{g^{(2)}}}\mathcal{F}, \tag{27}$$

and $R^{(2)}$ is the Ricci scalar curvature obtained from $g^{(2)}{}_{\mu\nu}$, and $\mathcal{F} = d\mathcal{A}$ is the $U(1)$ invariant curvature two form. Next, let us introduce the auxiliary scalars Φ and B satisfying:

$$\Box_{g^{(2)}}\Phi = R^{(2)} \text{ and } \Box_{g^{(2)}}B = \epsilon^{\mu\nu}\partial_\mu\mathcal{A}_\nu, \tag{28}$$

which transform the functional (26) into a Liouville CFT of the form:

$$\begin{aligned} S_{NHCFT} =& \frac{(r_+^2 + a^2)}{16\pi G} \int d^2x\sqrt{-g^{(2)}}\left\{-\Phi\Box_{g^{(2)}}\Phi + 2\Phi R^{(2)}\right\} \\ &+ \frac{3e^2(r_+^2 + a^2)}{\pi G} \int d^2x\sqrt{-g^{(2)}}\left\{-B\Box_{g^{(2)}}B + 2B\left(\frac{\epsilon^{\mu\nu}}{\sqrt{-g^{(2)}}}\right)\partial_\mu A_\nu\right\} \end{aligned} \tag{29}$$

Now, we turn our attention to computing the ASG of (25). The behavior of the RW2DA fields at large r is defined by

$$g^{(0)}{}_{\mu\nu} = \begin{pmatrix} -\frac{r^2}{\ell^2} + \frac{2rGM}{\ell^2} - \frac{a^2}{\ell^2} + \mathcal{O}\left(\left(\frac{1}{r}\right)^3\right) & 0 \\ 0 & \frac{\ell^2}{r^2} + \mathcal{O}\left(\left(\frac{1}{r}\right)^3\right) \end{pmatrix}, \tag{30}$$

$$\mathcal{A}^{(0)}{}_t = \frac{r}{\ell^2} - \frac{2GM}{\ell^2} + \mathcal{O}\left(\left(\frac{1}{r}\right)^3\right), \tag{31}$$

which yield an asymptotically AdS_2 configuration with Ricci Scalar, $R = -\frac{2}{l^2} + O\left(\left(\frac{1}{r}\right)^1\right)$, where $\ell^2 = r_+^2 + a^2$. In addition, we impose the following metric and gauge field fall-off conditions:

$$\delta g_{\mu\nu} = \begin{pmatrix} \mathcal{O}\left(\left(\frac{1}{r}\right)^3\right) & \mathcal{O}\left(\left(\frac{1}{r}\right)^0\right) \\ \mathcal{O}\left(\left(\frac{1}{r}\right)^0\right) & \mathcal{O}(r) \end{pmatrix} \text{ and } \delta\mathcal{A} = \mathcal{O}\left(\left(\tfrac{1}{r}\right)^0\right), \tag{32}$$

which imply the following set of asymptotic metric preserving diffeomorphisms

$$\xi_n = \xi_1(r)\frac{e^{in\kappa(t\pm r^*)}}{\kappa}\partial_t + \xi_2(r)\frac{e^{in\kappa(t\pm r^*)}}{\kappa}\partial_r, \tag{33}$$

where r^* is the tortoise coordinate,

$$\xi_1 = Cre^{in\kappa r^*}, \quad \xi_2 = \frac{irC(r - GM)}{\kappa n\,(r^2 - 2rGM + a^2)}, \tag{34}$$

C is an arbitrary normalization constant and κ is the surface gravity of the NHNEK black hole. Under diffeomorphisms (33), the gauge field transforms as:

$$\delta_\xi \mathcal{A}_\mu = \left(\mathcal{O}\left(\left(\tfrac{1}{r}\right)^0\right),\ \mathcal{O}\left(\left(\tfrac{1}{r}\right)^1\right)\right) \tag{35}$$

and thus δ_ξ may be elevated to a total symmetry of the action, i.e.,

$$\delta_\xi \to \delta_{\xi+\Lambda}, \tag{36}$$

in accordance with (32). Switching to light cone coordinates $x^\pm = t \pm r^*$ (where large r behavior will be synonymous with large x^+ behavior), we see that the set $\xi_n^\pm$ is well-behaved on the $r \to \infty$ boundary and form a centerless Witt or $Diff(S^1)$ subalgebra:

$$i\left\{\xi_m^\pm, \xi_n^\pm\right\} = (m - n)\xi_{m+n}^\pm. \tag{37}$$

3.3. Energy-momentum and the full ASG

We define the energy-momentum tensor and $U(1)$ current of (29) in the usual way:

$$\begin{aligned}
\langle T_{\mu\nu}\rangle =& \frac{2}{\sqrt{-g^{(2)}}}\frac{\delta S_{NHCFT}}{\delta g^{(2)\mu\nu}} \\
=& \frac{r_+^2 + a^2}{8\pi G}\left\{\partial_\mu\Phi\partial_\nu\Phi - 2\nabla_\mu\partial_\nu\Phi + {g^{(2)}}_{\mu\nu}\left[2R^{(2)} - \frac{1}{2}\nabla_\alpha\Phi\nabla^\alpha\Phi\right]\right\} \\
&+ \frac{6e^2(r_+^2 + a^2)}{\pi G}\left\{\partial_\mu B\partial_\nu B - \frac{1}{2}g_{\mu\nu}\partial_\alpha B\partial^\alpha B\right\} \text{ and} \\
\langle J^\mu\rangle =& \frac{1}{\sqrt{-g^{(2)}}}\frac{\delta S_{NHCFT}}{\delta\mathcal{A}_\mu} = \frac{6e^2(r_+^2 + a^2)}{\pi G}\frac{1}{\sqrt{-g^{(2)}}}\epsilon^{\mu\nu}\partial_\nu B.
\end{aligned} \tag{38}$$

Next, solving the equations of motions for the auxiliary fields,

$$\Box_{g^{(2)}}\Phi = R^{(2)}, \quad \Box_{g^{(2)}}B = \epsilon^{\mu\nu}\partial_\mu\mathcal{A}_\nu \tag{39}$$

using the metric (22) and gauge field (23) and employing the modified Unruh vacuum boundary conditions [29]

$$\begin{cases} \langle T_{++}\rangle = \langle J_{+}\rangle = 0 & r \to \infty,\ \ell \to \infty \\ \langle T_{--}\rangle = \langle J_{-}\rangle = 0 & r \to r_{+} \end{cases}, \tag{40}$$

we determine all relevant integration constants of (38) and (39). For large r and to $\mathcal{O}(\frac{1}{\ell})^2$, which we will denote as the single limit $r \to \infty$ in the remainder of this section, the resulting energy-momentum tensor is dominated by one holomorphic component, $\langle T_{--}\rangle$. We are interested in the response of the energy-momentum tensor and the $U(1)$ current to total symmetry $\delta_{\xi_n^- + \Lambda}$, which may be obtained by the use of the boundary fields (30) and (31):

$$\begin{cases} \delta_{\xi_n^- + \Lambda} \langle T_{--}\rangle = \xi_n^- \langle T_{--}\rangle' + 2\langle T_{--}\rangle (\xi_n^-)' + \frac{r_+^2 + a^2}{4\pi G} (\xi_n^-)''' + \mathcal{O}\left(\left(\frac{1}{r}\right)^3\right) \\ \delta_{\xi_n^- + \Lambda} \langle J_{-}\rangle = \mathcal{O}\left(\left(\frac{1}{r}\right)^3\right). \end{cases} \tag{41}$$

From this, we see that $\langle T_{--}\rangle$ transforms asymptotically as the energy-momentum tensor of a one dimensional CFT with center:

$$\frac{c}{24\pi} = \frac{r_+^2 + a^2}{4\pi G} \Rightarrow c = \frac{3A}{2\pi G}. \tag{42}$$

We should also note that the above central charge is in congruence with the 2-dimensional conformal/trace anomaly [30]:

$$\langle T_{\mu}{}^{\mu}\rangle = -\frac{c}{24\pi} R^{(2)}. \tag{43}$$

3.4. *Virasoro algebra*

Next, we define the quantum generators from the conserved charges:

$$\mathcal{Q}_n = \lim_{r\to\infty} \int dx^- \langle T_{--}\rangle \xi_n^-, \tag{44}$$

The algebraic structure of these generators can be found by calculating response of (44) to a total symmetry, while compactifying the x^- coordinate to a circle from $0 \to 2\pi/\kappa$:

$$\delta_{\xi_m^- + \Lambda} \mathcal{Q}_n = [\mathcal{Q}_m, \mathcal{Q}_n] = (m-n)\mathcal{Q}_n + \frac{c}{12} m \left(m^2 - 1\right) \delta_{m+n,0}, \tag{45}$$

Hence, the quantum symmetry generators form a centrally extended Virasoro algebra with regulated/normalized zero-mode $\mathcal{Q}_0 = \frac{A}{16\pi G}$.

3.5. *AdS_2/CFT_1 and entropy of NHNEK*

The main conclusion of the foregoing discussion is that, by employing the finite mass gauge (17), it is possible to show that the near-extremal Kerr throat is holographically dual to a CFT with center

$$c = \frac{3A}{2\pi G} \tag{46}$$

and lowest Virasoro eigenmode

$$\mathcal{Q}_0 = \frac{A}{16\pi G}. \tag{47}$$

We are now free to use the above results in the statistical Cardy Formula (3),

$$S = 2\pi\sqrt{\tfrac{c\mathcal{Q}_0}{6}} = \tfrac{A}{4G} = 2\pi\left(GM^2 + \sqrt{G^2M^4 - J^2}\right), \tag{48}$$

which is in agreement with the area law (9). However, here we have derived it without mixing results computed separately at near-extremality and extremality. In addition, c and $\mathcal{Q}_0$ extend smoothly to extremality in the limit as $a \to GM$, giving

$$\lim_{a\to GM} c = 12J \text{ and } \lim_{a\to GM} \mathcal{Q}_0 = J/2. \tag{49}$$

The limiting value c here is identical to the left central charge obtained in the Kerr/CFT correspondence [12] and, together with the limit of $\mathcal{Q}_0$, it reproduces the extremal Kerr entropy by way of the statistical Cardy formula (3). In addition, our derived zero-mode in (46) is in accordance with the following assertion [24]:

Assertion 1. *The lowest Virasoro eigenmode of a quantum CFT is proportional to the irreducible mass of its dual black hole,*

$$\mathcal{Q}_0 = GM_{irr}^2. \tag{50}$$

Note that the irreducible mass is the final ADM mass state of a Kerr black hole after it has completed its Penrose process. This assertion may have broad generality with a large avenue of application, though we do not have a rigorous proof at this time.

3.6. Near extremal black hole temperature

To extract the NHNEK temperature, we will focus on the gravitational part of (29), i.e.,

$$S_{grav} = \frac{(r_+^2 + a^2)}{16\pi G}\int d^2x\sqrt{-g^{(2)}}\left\{-\Phi\square_{g^{(2)}}\Phi + 2\Phi R^{(2)}\right\}. \tag{51}$$

The energy-momentum is given by:

$$\begin{aligned}\langle T_{\mu\nu}\rangle =& \frac{2}{\sqrt{-g^{(2)}}}\frac{\delta S_{NHCFT}}{\delta g^{(2)\mu\nu}}\\ =& \frac{r_+^2 + a^2}{8\pi G}\left\{\partial_\mu\Phi\partial_\nu\Phi - 2\nabla_\mu\partial_\nu\Phi + {g^{(2)}}_{\mu\nu}\left[2R^{(2)} - \frac{1}{2}\nabla_\alpha\Phi\nabla^\alpha\Phi\right]\right\},\end{aligned} \tag{52}$$

which, as before, may be brought to the form (40) by using (38). However, on the horizon limit $r \to r_+$, we are left with just one holomorphic component,

$$\langle T_{++}\rangle = -\frac{r_+^2 + a^2}{32\pi G}f'\left(r^+\right)^2. \tag{53}$$

This is precisely the Hawking flux (HF, radiation flux $\sim {T_r}^t$) of the NHNEK metric, weighted by the central charge (42),

$$\langle T_{++}\rangle = cHF = -c\tfrac{\pi}{12}\left(T_H\right)^2 \tag{54}$$

with Hawking temperature [31, 32]

$$T_H = \frac{f'\left(r^+\right)}{4\pi}. \tag{55}$$

This is an interesting result, for it suggests that the AdS_2/CFT_1 correspondence constructed here contains information about both black hole entropy and black hole temperature. Though the $\langle T_{++}\rangle$ component in the horizon limit is not precisely the Hawking flux of the four-dimensional parent black hole, given prior knowledge of the central extension, it is possible to read off or extract the relevant information from the correspondence.

4. Concluding remarks

We have analyzed quantum near-extremal Kerr black hole properties in the near horizon regime by way of AdS_2/CFT_1 correspondence, as outlined in Table 1. This extends our previous work [24,25] to a new spacetime geometry. The main results of this work includes the derivation of the central charge $c = \frac{3A}{2\pi G}$ from a Lagrangian analysis of conserved currents of two near horizon theories and the correct quadratic two-form transformation law on the holographic renormalized boundary. It is conceivable that other $AdS_2 \times S^2$ gauges, exhibiting the field

CFT	Black Hole
Conformal Group	Asymptotic Symmetry Group
center	$\frac{3A}{2\pi G}$
Hamiltonian eigenvalue	GM_{irr}^2
Regulator	κ_{NHNEK}

Table 1. Black-Hole/Near-Horizon-CFT Duality

splitting of (18), exist with physical relevance and connections to other classical near-extremal solutions. Many analogues to the NHEK for charged rotating black holes with negative and positive cosmological constants have been shown to exist (see [20] for a comprehensive review). They suggest that there may exist similar such analogues to the NHNEK solution, to which the line of reasoning of this note may be applicable. In particular, Assertion 1 may be a useful tool in the asymptotic symmetry analysis of other extremal black holes. They have vanishing surface gravity, a complication that is generally handled by the use of the thermal Cardy formula (4). However, extremal black holes in general do have well defined horizons, which lead to a finite non-zero irreducible mass. This fact enables us to invoke Assertion 1 to implement the standard statistical Cardy formula (3) and obtain the Bekenstein-Hawking entropy for a wide class of extremal black holes.

Acknowledgments

We thank Vincent Rodgers, Jacob Willig-Onwuachi, John Baker and Shanshan Rodriguez for enlightening discussions. SW thanks the organizers of QuantumFest 2015 for their kind invitation and extraordinary hospitality during the conference. This work was supported in part by the HHMI Science Education Award 52006298 and the Grinnell College CSFS and MAP programs.

This work was completed just before the passing of Sujeev Wickramasekara on December 28th, 2015. Sujeev was our friend, mentor, teacher and collaborator. A true visionary and inspiration in his exemplification of a teacher-scholar and human being! He more than inspired us to pursue greatness in all aspects of our life goals. His teachings and mentorship have and will continue to influence and outline our lives. He is dearly missed and remembered always...

References

[1] Hawking S W 1975 *Commun. Math. Phys.* **43** 199–220
[2] Bardeen J M, Carter B and Hawking S W 1973 *Commun. Math. Phys.* **31** 161–170
[3] Bekenstein J D 1973 *Phys. Rev.* D **7** 2333–2346
[4] Mukhanov V F, Wipf A and Zelnikov A 1994 *Phys. Lett.* B **332** 283–291; hep-th/9403018
[5] Rodriguez L 2011 *Black-hole/near-horizon-CFT duality and 4 dimensional classical spacetimes.* Ph.D. dissertation University of Iowa
[6] Robinson S P and Wilczek F 2005 *Phys. Rev. Lett.* **95** 011303; gr-qc/0502074
[7] Iso S, Umetsu H and Wilczek F 2006 *Phys. Rev.* D **74** 044017; hep-th/0606018

[8] Zampeli A, Singleton D and Vagenas E C 2012 *JHEP* **1206** 097; gr-qc/1206.0879
[9] Das S, Robinson S P and Vagenas E C 2008 *Int. J. Mod. Phys.* D **17** 533–539; gr-qc/0705.2233
[10] Maldacena J M 1998 *Adv. Theor. Math. Phys.* **2** 231–252; hep-th/9711200
[11] Strominger A 1998 *JHEP* **02** 009; hep-th/9712251
[12] Guica M, Hartman T, Song W and Strominger A 2009 *Phys. Rev.* D **80** 124008; gr-qc/0809.4266
[13] Carlip S 2011 *JHEP* **1104** 076; gr-qc/1101.5136
[14] Carlip S 1999 *Phys. Rev. Lett.* **82** 2828–2831; hep-th/9812013
[15] Carlip S 2005 *Class. Quant. Grav.* **22** R85–R124; gr-qc/0503022
[16] Carlip S 1999 *Class. Quant. Grav.* **16** 3327–3348; gr-qc/9906126
[17] Park M I and Ho J 1999 *Phys.Rev.Lett.* **83** 5595; hep-th/9910158
[18] Park M I 2002 *Nucl.Phys.* B **634** 339–369; hep-th/0111224
[19] Kang G, Koga J i and Park M I 2004 *Phys. Rev.* D **70** 024005; hep-th/0402113
[20] Compere G 2012; hep-th/1203.3561
[21] Blöte H W J, Cardy J A and Nightingale M P 1986 *Phys. Rev. Lett.* **56**
[22] Cardy J A 1986 *Nucl. Phys.* B **270**
[23] Chang-Young E and Eune M 2012; arXiv:1212.3031
[24] Button B K, Rodriguez L, Whiting C A and Yildirim T 2011 *Int.J.Mod.Phys.* A **26** 3077–3090; arXiv:1009.1661
[25] Rodriguez L and Yildirim T 2010 *Class. Quant. Grav.* **27** 155003; arXiv:1003.0026
[26] Button B K, Rodriguez L and Wickramasekara S 2013 *JHEP* **10** 144; arXiv:1307.7125
[27] Strominger A 1994 Les Houches lectures on black holes, hep-th/9501071
[28] Leutwyler H 1985 *Phys. Lett.* B **153** 65
[29] Unruh W G 1976 *Phys. Rev.* D **14** 870
[30] Francesco P D, Mathieu P and Sénéchal D 1997 *Conformal Field Theory* (Springer)
[31] Jiang Q Q and Wu S Q 2007 *Phys. Lett.* B **647** 200–206; hep-th/0701002
[32] Caldarelli M M, Cognola G and Klemm D 2000 *Class. Quant. Grav.* **17** 399–420; hep-th/9908022
[33] Frolov V P and Thorne K S 1989 *Phys. Rev.* D **39** 2125–2154

Towards Precision Measurements at UASLP

S Hamzeloui, N Arias, V Abediyeh, D Martínez, M Gutiérrez, E Uruñuela, E del Rio, E Cerda-Méndez, E Gomez

Instituto de Física, Universidad Autónoma de San Luis Potosí (UASLP), San Luis Potosí 78290, México

E-mail: egomez@ifisica.uaslp.mx

V M Valenzuela

Facultad de Ciencias Físico Matemáticas, Universidad Autónoma de Sinaloa, Sinaloa 80000, México

Abstract. Atomic interferometry is a very sensitive technique to measure small forces. Here we present an overview of the progress towards interferometric measurements in our laboratory. We characterize the magnetic field noise and describe the strategies to minimize the sensitivity to magnetic field fluctuations. We introduce as well a system for Raman excitation with minimum phase noise and the frequency filtering needed to implement it. Finally, we demonstrate atomic interferometry with a frequency sensitivity of 3 Hz.

1. Introduction

Laser cooling is a widely used technique nowadays. The control achieved over the atomic sample improves the precision of different measurements in physics. The best example is that of atomic clocks where 18 digits of precision have been achieved in the determination of time [1]. The use of interferometric techniques is largely responsible for the continuous improvement of measurements. Cold atoms are well suited for such techniques since they allow for coherence times of several seconds [2,3]. In our group we are particularly interested in the measurement of small forces. Here cold atoms have been used to study the gravitational force [4], variations of gravity at short distances [5], interactions of atoms with surfaces [6,7] and the weak force [8,9].

The separated oscillatory fields method of Ramsey is at the heart of an atomic interferometer [10]. In this method a $\pi/2$ pulse creates a coherent superposition between two hyperfine levels. The system is then left to evolve for a time T, and a second $\pi/2$ pulse is applied to read the relative phase between the two levels. An ideal atomic interferometer must have high sensitivity and it should not be affected by environmental perturbations. Magnetic field variations are often a concern, but it is possible to choose energy levels that remain unperturbed by those fluctuations.

Here we present our progress on atomic interferometry. We explore different ways to suppress the sensitivity to magnetic field fluctuations. Also, the use of a phase modulator allows for the excitation of Raman transitions with a minimum of phase noise. The flexibility of the system opens the door to complex excitation schemes.

2. Atomic interferometry

In this section we present the basic concepts of an atomic interferometer applied to gravimetry. Two ground hyperfine levels ($|g_1\rangle$, $|g_2\rangle$) are connected via a Raman transition that uses vertical counter propagating lasers of frequency ω_1 and ω_2. The lasers have a detuning Δ with respect to an excited state to selectively induce a two photon transition. The counter propagating Raman transition is sensitive to the velocity of the atoms along the laser beam propagation. The velocity (v_z) dependence of the detuning given by

$$\delta = \omega_2 - \omega_1 - \omega_0 - 2k(v_z + \hbar k/m), \tag{1}$$

with ω_0 the hyperfine splitting, k the laser wave vector magnitude and m the mass of the atom. The velocity dependence originates from the Doppler effect of the atoms as they change their velocity during free fall.

To implement a gravimeter we start with the atoms in the state $|g_1 p\rangle$, where the p represents a particular value of momentum along z. A $\pi/2$ Raman pulse creates an equal superposition of the two ground states. The Raman pulse changes the internal state as well as the atomic momentum by $2\hbar k$. After some free evolution time the state is given by

$$|\Psi(t)\rangle = \left(|g_1 p\rangle - ie^{i\Phi(t)}\,|g_2 p + 2\hbar k\rangle\right)/\sqrt{2}. \tag{2}$$

The relative phase depends on the energy difference ($E_2 - E_1$) of the two levels that includes the hyperfine splitting and the gravitational potential

$$\Phi(t) = \int \frac{E_2 - E_1}{\hbar} dt = \omega_0 t + \left(\frac{mg}{\hbar}\right) \int (z_2 - z_1) dt. \tag{3}$$

A π pulse is applied after a free evolution time τ to reverse the momentum difference of the two hyperfine levels. To close the interferometer, a $\pi/2$ pulse is applied after an additional free evolution time τ. The position difference in Eq. 3 comes from the momentum kick during the splitting process. The contribution of the last term is $2gk\tau^2$ right before the second $\pi/2$ pulse and is what gives the sensitivity to the gravitational acceleration g.

3. Magnetic sensitivity of atomic levels and transitions

The sensitivity to magnetic field has some advantages. Atomic magnetometers are used in medical applications [11], quantum measurements [12] or searches for physics beyond the Standard Model [13]. Measurements in magnetic field gradients can be used to obtain spatial information, as it is the case for example in Nuclear Magnetic Resonance [14]. Here the atoms are resonant only at a particular position which can be selectively excited. The same idea can be extended for the case of cold gases where two sequential position measurements separated by a time delay provides a way to measure the velocity [15, 16].

For gravimetry applications, however, magnetic field fluctuations become a source of noise. The energy of an atomic level shifts in a magnetic field due to the Zeeman effect [15], which gives an additional contribution to the relative phase evolution of Eq. 3. The states with m_F=0 have no linear Zeeman shift and are thus preferred for interferometric measurements. The hyperfine transition between $m_F = 0$ levels (the clock transition) allows for long coherence times [3]. Also, in the case of ^{87}Rb, the hyperfine transition between $|F = 1, m_F = -1\rangle$ and $|F = 2, m_F = 1\rangle$ is quite insensitive to magnetic field fluctuations at 3.2 Gauss [17], and shows long coherence times as well [18]. Each of the levels in this last transition are sensitive to the magnetic field whereas the transition remains insensitive to magnetic fluctuations. Combining these two transitions we will look for reduced magnetic sensitivity at a tunable field.

Figure 1 shows the measured magnetic field noise at our laboratory. We measure the noise with a magneto resistance sensor (Honeywell HMC1002) and we digitize the signal using a DAQ card (NI USB-6289). The magnetic variations are dominated by line noise and its harmonics. Reducing magnetic noise using feedback is not the best strategy since it adds broadband noise. Instead we apply narrow band corrections only at the line frequency and harmonics using coils in Helmholtz configuration. Adjusting the phase and amplitude of the correcting signal we manage to selectively reduce each peak in the noise spectrum (60 Hz, 180 Hz, etc.) by more than an order of magnitude. The power supplies nearby are responsible for most of the noise we observe. We verified that the noise decays as $1/r^2$ as you move away from the supply as expected from radiative fields. We moved all the supplies more than 6 m away from the experiment where they give contributions below 10 μGauss.

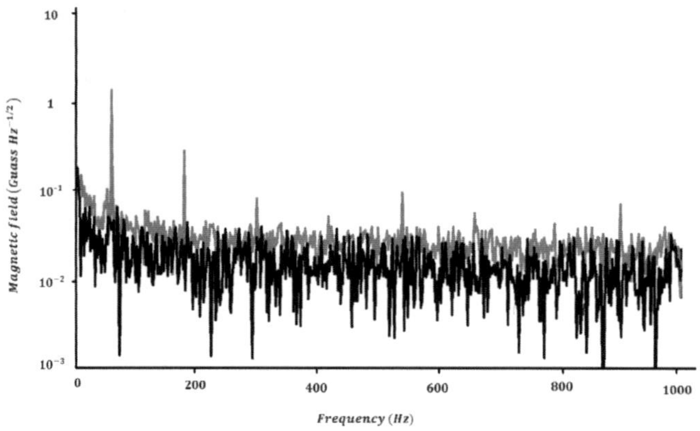

Figure 1. Magnetic field noise in the laboratory (red) and noise floor (black).

4. Phase modulator for Raman transitions

A Raman transition requires two phase locked lasers with a frequency difference close to the hyperfine splitting (Eq. 1). Having low phase noise is an important requirement for interferometry. A phase lock is usually achieved by sending both laser beams to a detector. The beat note is then compared with a reference signal in an Optical Phase Lock Loop (OPLL) to obtain the feedback that is sent to one of the lasers [19]. Achieving low phase noise with this technique is quite challenging. Alternatively one can obtain both beams from a single laser. Modulating the light either in amplitude or phase produces sidebands that can be used for the extra beam. Since both beams come from the same laser they are automatically phase locked, and the phase noise depends on the quality of the RF source feeding the modulator. This technique is easier to implement and has lower phase noise than the traditional method.

The use of extra beams from modulation has been exploited in laser cooling and trapping. By modulating at the hyperfine splitting frequency one can derive a repumper beam out of the trap laser [20]. Acousto-optical modulators only work for small hyperfine splittings and the beams obtained are spatially separated. Electro-optical modulators work for higher frequencies and the beams propagate in the same direction. This technique was only used for repumping beams since that require less power. We extended the idea for the trapping beams using a fiber modulator and a double pass tapered amplifier [21]. With that system we demonstrated the simultaneous trapping of two different isotopes of rubidium (^{85}Rb and ^{87}Rb) using a single laser and a fiber modulator. The system provided a big simplification for laser cooling since all the

Journal of Physics: Conference Series **698** (2016) 012011 doi:10.1088/1742-6596/698/1/012011

control for the beams was moved from the optical to the RF world, which is easier to work with, cheaper and more reliable.

Light modulators have also been used for exciting Raman transitions [4, 22–24]. The carrier and one sideband work as the Raman pair when the modulation frequency equals that of the transition. The power out of the modulator is small and an amplification stage is needed to boost the power. The amplifier introduces extra phase noise, and when using a tapered amplifier, there is a pedestal of spontaneous emission that limits the coherence of the sample.

Figure 2 shows the ideal system for Raman excitation. The acousto-optical modulator produces the pulses and the electro-optical modulator creates the phase locked beams. The scheme requires a modulator capable of handling high power as it is the case for the Advr WPM-P78P78-ALO that takes up to 250 mW of optical power. The power on a sideband can be as high as 50 mW requiring no amplification stage after it. The two beams follow exactly the same path, therefore vibration insensitivity is also obtained except for the last mirror. The system offers the possibility of complex excitation patterns by replacing the RF source with an arbitrary wave generator.

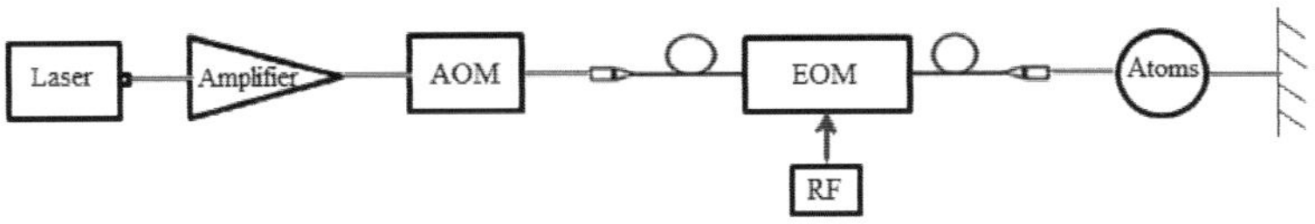

Figure 2. System for Raman excitation using a phase modulator. AOM: acousto-optical modulator, EOM: electro-optical modulator. The retro reflecting mirror is at the end on the right.

5. Frequency filtering

An amplifier is required after the modulator in the Raman system of Fig. 2 whenever the output power is not high enough. Tapered amplifiers produce a broad pedestal of spontaneous emission besides the amplified beam. The pedestal is many orders of magnitude smaller, but still produces significant scattering due to nearly resonant contributions. In our previous work with dipole traps [25], we removed the resonant contribution using heated Rb cells. Still the lifetime of the dipole trap was limited by the scattering from the pedestal.

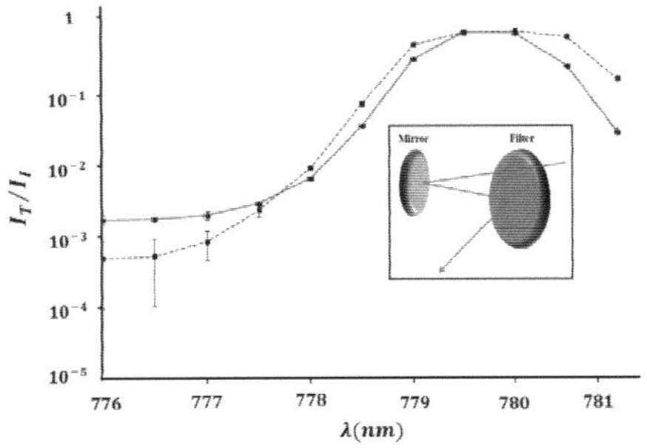

Figure 3. Fraction of transmitted light as a function of wavelength for the double pass interference filter with (solid red line) and without (dashed blue line) tapered amplifier. Inset: double pass schematic.

We have implemented two frequency filters to spectrally clean the beam out of the tapered amplifier, one based on an interference filter and the other on a diffraction grating. For the first one, we use a sharp low pass interference filter (Semrock LP02-785RU, transition width 7.9

Journal of Physics: Conference Series **698** (2016) 012011 doi:10.1088/1742-6596/698/1/012011

nm) that blocks wavelengths below 785 nm. The cut off wavelength can be tuned by changing the incidence angle. For a dipole trap in rubidium we use a beam at 785 nm. The interference filter blocks the D2 line components at 780 nm but not those on the D1 line at 795 nm. To block high frequencies we direct the beam again towards the interference filter, but we use it in reflection this second time. The double pass operation introduces a sharp tunable band pass filter (Fig. 3). We get 75% central transmission and 25 dB attenuation when we are 2.5 nm away from the band center.

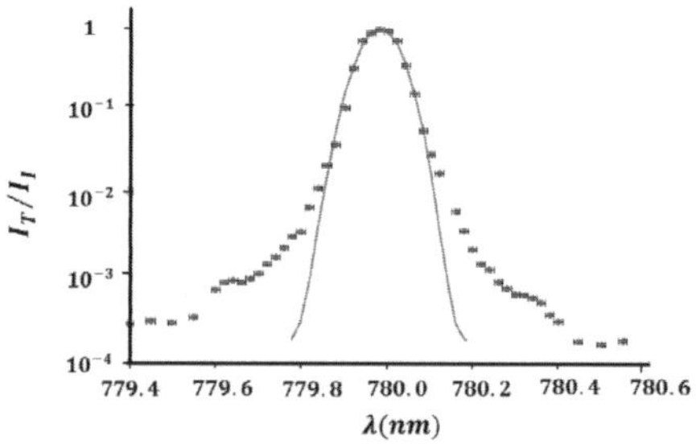

Figure 4. Fraction of transmitted light as a function of wavelength for the grating filter. Red solid curve is a Gaussian fit.

The interference filter is very easy to use and works well for beams detuned by several nm as it is the case for a dipole trap, but it might not be sharp enough for the Raman beams that require smaller detunings. There we use instead a frequency filter based on a diffraction grating, similar to a monochromator. The grating we use (Edmund Optics 43-848) gives good diffraction efficiency (90%) at 780 nm. We spatially filter the light with a pinhole, expand and collimate the beam before the grating. The diffracted beam goes through a lens (f=20 cm) that focus the beam to the frequency filtering pinhole (25 μm). Figure 4 shows the characterization of the filter. We get 77 % central transmission and 30 dB attenuation when we are 0.3 nm away from the band center.

6. New vacuum system

We have improved our vacuum system for the measurements with microwaves and Raman beams (Fig. 5). The Pyrex vacuum cell reduces problems with Eddy currents and has better optical access. The vacuum system is quite compact and it is pumped with a NexTorr D100-5 pump that combines a 100 l/s getter pump and a 5 l/s ion pump. The pump produces no vibration and introduces magnetic field gradients smaller than 20 mG/cm at the position of the atoms. We reach pressures close to 10^{-10} Torr that translate into lifetimes of around 10 s (limited by the Rb dispenser).

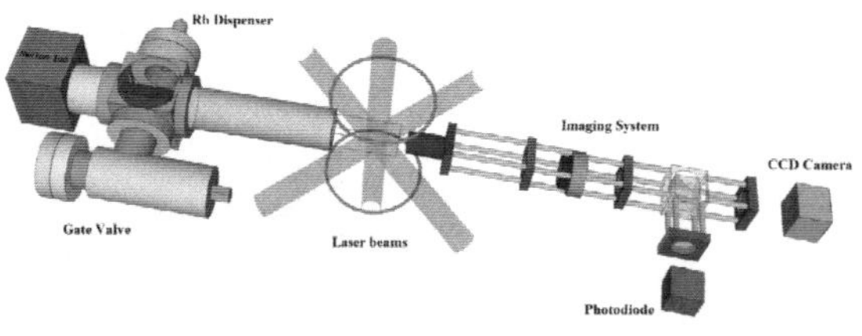

Figure 5. New vacuum and imaging system.

We monitor the atoms using fluorescence with a double relay imaging system (1:1 amplification). The system uses best form lenses to reduce aberrations and we place an iris after the first relay to minimize background light. It has a numerical aperture of 0.3, a theoretical resolution of 1.5 μm and collects 2.3 % of the light emitted by the atoms. A 50-50 splitter in the imaging system sends light to a CCD camera and a photo diode connected to a DAQ card.

7. Atomic interferometric fringes

With the new vacuum system we have been able to demonstrate molasses to 3 μK. We apply optical pumping to put more than 88% of the atoms in the $|F = 1, m_F = 0\rangle$ state of ^{87}Rb. We excite Rabi oscillations between hyperfine or Zeeman levels. Figure 6 shows the interferometry fringes we get using the clock transition. The fringes have a width of 50 Hz determined by the 10 ms separation between pulses. With this interferometer we achieve a frequency precision of 3 Hz. We have also observed interferometric fringes in the other magnetic insensitive two photon transition ($|F = 1, m_F = -1\rangle$ to $|F = 2, m_F = 1\rangle$). This requires simultaneous excitation with microwaves using a horn and RF from a homemade resonant antenna.

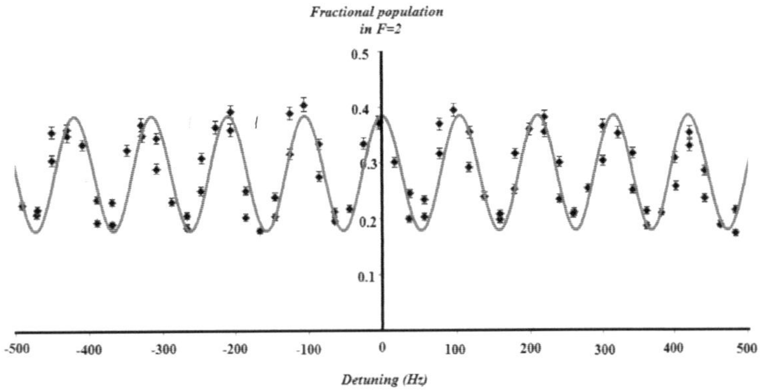

Figure 6. Interferometry fringes using microwaves in the clock transition. Each $\pi/2$ pulse is 20μs long and they are separated by 10 ms.

8. Conclusion

We demonstrated atomic interferometric in our laboratory with a frequency precision of 3 Hz. The interferometry required the construction of a new vacuum system and the implementation of different components for coherent manipulation of the atoms. We introduced a system for Raman excitation with a minimum of phase noise based on a fiber modulator. Its implementation requires frequency filtering that we achieved using a double pass interference filter with a pass band 2.5 nm wide, or alternatively with a blazed diffraction grating with a width of 0.3 nm.

Acknowledgments

This work was supported by CONACyT, UASLP, FAI and Fundación Marcos Moshinsky. We thank Jonathan Espinosa and José Rocha for their assistance with electronics.

References

[1] Bloom B J, Nicholson T L, Williams J R, Campbell S L, Bishof M, Zhang X, Zhang W, Bromley S L and Ye J 2014 An optical lattice clock with accuracy and stability at the 10^{-18} level *Nature* **506**, 71

[2] Davidson N, Lee H J, Adams C S, Kasevich M and Chu S 1995 Long atomic coherence times in an optical dipole trap *Phys. Rev. Lett.* **74**, 1311

[3] Dudin Y O and Kuzmich A 2013 Light storage on the time scale of a minute *Phys. Rev.* A **87** 031801

[4] Dickerson S M, Hogan J M, Sugarbaker A, Johnson D M S and Kasevich M A 2013 Multiaxis inertial sensing with long-time point source atom interferometry *Phys. Rev. Lett.* **111** 083001

[5] Ferrari G, Poli N, Sorrentino F and Tino G M 2006 Long lived Bloch oscillations with bosonic Sr atoms and application to gravity measurement at the micrometer scale *Phys. Rev. Lett.* **97** 060402

[6] Carusotto I, Pitaevskii L, Stringari S, Modugno G and Inguscio M 2005 Sensitive measurement of forces at the micron scale using Bloch oscillations of ultracold atoms, *Phys. Rev. Lett.* **95** 093202

[7] Harber D M, Obrecht J M, McGuirk J M and Cornell E A 2005 Measurement of the Casimir-Polder force through center-of-mass oscillations of a Bose-Einstein condensate *Phys. Rev.* A **72** 033610

[8] Gorelov A, Melconian D, Alford W P, *et. al.* 2005 Scalar interaction limits from the $\beta - \nu$ correlation of trapped radioactive atoms *Phys. Rev. Lett.* **94**, 142501

[9] Gomez E, Aubin S, Sprouse G D, Orozco L A and DeMille D P 2007 Measurement method for the nuclear anapole moment of laser-trapped alkali-metal atoms *Phys. Rev.* A **75** 033418

[10] Ramsey N F 1950 A molecular beam resonance method with separated oscillating fields *Phys. Rev.* **78** 695

[11] Kominis I K, Kornack T W, Allred J C and Romalis M V 2003 A subfemtotesla multichannel atomic magnetometer *Nature* **422** 596

[12] Trail C M, Jessen P S and Deutsch I H 2010 Strongly enhanced spin squeezing via quantum control *Phys. Rev. Lett.* **105** 193602

[13] Canè F, Bear D, Phillips D F, Rosen M S, Smallwood C L, Stoner R E, Walsworth R L and Kostelecký V A 2004 Bound on Lorentz and CPT violating boost effects for the neutron *Phys. Rev. Lett.* **93** 230801

[14] Levitt M H 2008 *Spin Dynamics: Basics of Nuclear Magnetic Resonance* (2nd edition, Wiley)

[15] Castaños L O and Gomez E 2014 Model for a phase-space selector using microwave transitions *Phys. Rev.* A **89** 013406

[16] Castaños L O and Gomez E 2014 Velocity selector with a microwave magnetic dipole transition *arXiv:1403.5018v1*

[17] Matthews M R, Hall D S, Jin D S, Ensher J R, Wieman C E, Cornell E A, Dalfovo F, Minniti C and Stringari S 1998 Dynamical response of a Bose-Einstein Condensate to a discontinous change in internal state *Phys. Rev. Lett.* **81** 243

[18] Treutlein P, Hommelhoff P, Steinmetz T, Hänsch T W and Reichel J 2004 Coherence in Microchip Traps *Phys. Rev. Lett.* **92** 203005

[19] Cheinet P, Pereira dos Santos F, Petelski T, Le Gouët J, Kim J, Therkildsen K T, Clarion A and Landragin A 2006 Compact laser system for atom interferometry, *Appl. Phys.* B **84** 643

[20] Weiner J, Bagnato V S, Zilio S and Julienne P S 1999 Experiments and theory in cold and ultracold collisions *Rev. Mod. Phys.* **71** 1

[21] Valenzuela V M, Hamzeloui S, Gutiérrez M and Gomez E 2013 Multiple isotope magneto-optical trap from a single diode laser, *J. Opt. Soc. Am. B.* **30** 1205

[22] Charrière R, Cadoret M, Zahzam N, Bidel Y and Bresson A 2012 Local gravity measurement with the combination of atom interferometry and Bloch oscillations, *Phys. Rev.* A **85** 013639

[23] Döring D, McDonald G, Debs J E, Figl C, Altin P A, Bachor H A, Robins N P and Close J D 2010 Quantum-projection-noise-limited interferometry with coherent atoms in a Ramsey-type setup *Phys. Rev.* A **81** 043633

[24] Zi-Chao Z, Rong W, Chun-Yan S, Tang L and Yu-Zhu W 2011 Magnetic field measurement based on a stimulated two-photon Raman transition, *Chin. Phys.* B **20** 034206

[25] Valenzuela V M, Hernández L and Gomez E 2012 High power rapidly tunable system for laser cooling *Rev. Sci. Instrum.* **83** 015111

An Invitation to the Mathematics of Topological Quantum Computation

E C Rowell
Mathematics Department, Texas A&M University, College Station, TX 77843-3368
E-mail: rowell@math.tamu.edu

Abstract. Two-dimensional topological states of matter offer a route to quantum computation that would be topologically protected against the nemesis of the quantum circuit model: decoherence. Research groups in industry, government and academic institutions are pursuing this approach. We give a mathematician's perspective on some of the advantages and challenges of this model, highlighting some recent advances. We then give a short description of how we might extend the theory to three-dimensional materials.

1. Introduction

In [1] we find the following convenient definition: "*Quantum computation* is any computational model based upon the theoretical ability to manufacture, manipulate and measure quantum states." A topological quantum computer is a hypothetical device that relies upon a kind of topological symmetry in topologically ordered states of matter to carry out fault-tolerant quantum computation. Usually, these topological phases are taken to be (effectively) 2-dimensional systems of anyons: point-like quasi-particles that emerge in certain condensed matter systems. Topological phases of matter were first realized through quantum Hall effects, for example in fractional quantum Hall liquids in the experiments of Tsui and Störmer in 1982 [2] which led to the Nobel prize they shared with Laughlin [3] in 1998. In the past few years [1, 4, 5] the possibility of building a quantum computer using physical systems exhibiting topological phases has motivated significant investment of resources towards realizing such a project. Both Microsoft and Alcatel-Lucent are currently pursuing topological qubits. Besides the potential commercial computational benefits, these states of matter are fascinating from both the physical and mathematical perspectives, connecting the seemingly unrelated subjects of quantum topology and condensed matter.

The mathematical foundations of topological quantum computation are usually cast in categorical language. In this survey we will avoid categorical terminology in the hopes that this will broaden the readership, without significantly sacrificing precision. There are other expository works of varying length and precision on the subject. The shortest and most elementary is [6], written for a general audience. The text [7] gives a fairly comprehensive account aimed at physicists, while the excellent survey [8] mainly focuses on topological and condensed matter themes. The short survey [1] introduces the topological model to mathematicians, while the very precise and complete [9] makes full use of the categorical language.

The main goal of this survey is to give non-experts a taste of the mathematics of topological quantum computation, illustrating how the theory arises naturally and hopefully inspiring further reading. In the first part of this survey we will present a first approximation of the mathematical model for anyons on surfaces. Some technical details will be neglected, but these can be reconciled by further reading (for example [9]). The second part will be devoted to describing some of the recent advances and open problems based upon this model.

2. Anyons on surfaces

2.1. Motivation

According to [8]: "A system is in a *topological phase* if, at low temperatures and energies and long wavelengths, all observable properties e.g., correlation functions are invariant under smooth deformations of the space-time manifold in which the system lives," or, alternatively, "if its low energy effective field theory is a topological quantum field theory TQFT, i.e., a field theory whose correlation functions are invariant under diffeomorphisms." The mathematics of anyons in 2D can be distilled down to the study of (quasi-)particles on surfaces. How such systems may arise in nature is explained in [8], we content ourselves to present Figure 1 as a mathematician's cartoon of the fractional quantum Hall effect.

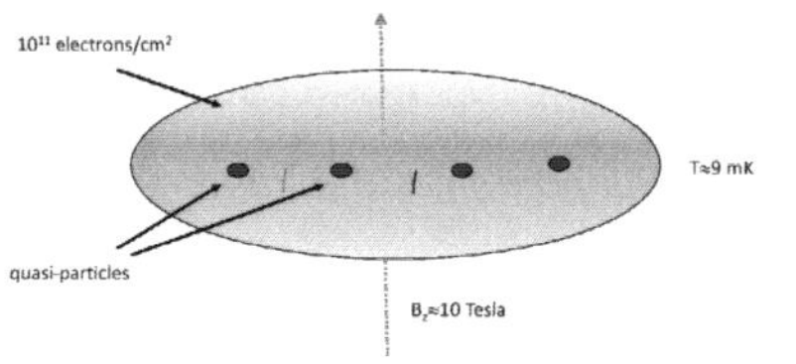

Figure 1. Electrons are effectively confined to a 2D disk due to the low energy, perpendicular magnetic field causes quasi-particles to emerge.

The spin-statistics theorem of Fierz and Pauli shows that in 3 spacial dimensions, every particle is either a fermion or a boson: particle exchange of a system of two indistinguishable particles changes the state by at most a sign: $|\psi_1\psi_2\rangle = \pm|\psi_2\psi_1\rangle$. Mathematically, this is related to the fact that the group of motions of two points in 3-dimensional space is S_2, the group of order 2. In 2 spacial dimensions the group of motions of n points in a disk is the braid group $\mathcal{B}_n$, an infinite group with generators: $\sigma_1, \ldots, \sigma_{n-1}$ obeying:

(R1) $\sigma_i\sigma_{i+1}\sigma_i = \sigma_{i+1}\sigma_i\sigma_{i+1}$ for $1 \leq i \leq n-2$ and

(R2) $\sigma_i\sigma_j = \sigma_j\sigma_i$ if $|i-j| > 1$

The suggests that *exchange statistics of point-like particles in 2 spacial dimensions are richer than in 3 dimensions.* Wilczek [10] coined the term "anyons" to describe particles obeying exchange statistics $|\psi_1\psi_2\rangle = e^{2\pi\theta i}|\psi_2\psi_1\rangle$ for any θ, where $\theta = 0$ and $\theta = 1/2$ correspond to bosons and fermions, respectively. Since the time evolution of a closed system is unitary, particle exchange for n indistinguishable anyons in the disk induces a unitary representation of the braid group $\mathcal{B}_n$, on the (Hilbert) state space of such configurations.

Assuming we can manufacture, manipulate and measure 2-dimensional topological phases of matter, we have a scheme for topological quantum computation (see Figure 2).

Our first problem is now clear: how can we model a quantum mechanical system of anyons on a surface? Two mathematical facts emerge from the quantum computational scheme in Figure 2: (1) to each surface with labeled points we must assign a Hilbert space in a consistent way, i.e. respecting the topological invariance and quantum mechanical constraints, and (2) particle exchange must correspond to a unitary operation on the associated Hilbert space.

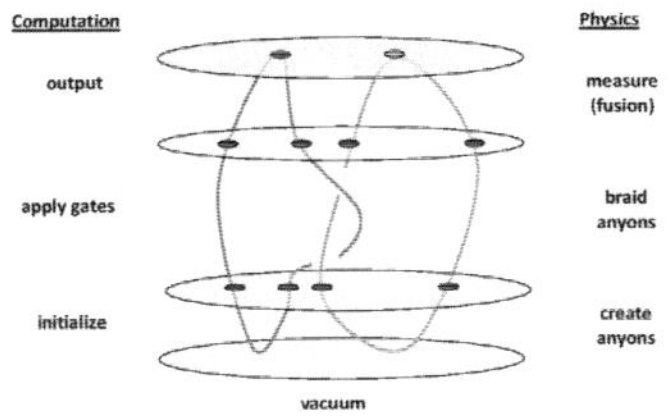

Figure 2. A topological quantum computation scheme: particle-antiparticle pairs are drawn from the vacuum, particle exchange is performed followed by a measurement of the resulting particle type(s).

2.2. State spaces for anyons on surfaces

Let us explore the problem of assigning a Hilbert space $\mathcal{H}(M, \ell)$ to a system of anyons ℓ on a surface M, in a way consistent with quantum mechanical principles and the topological nature of anyons. What we will be describing is essentially the 2-dimensional part of a $(2+1)$-TQFT, see [11] for an excellent online resource. Suppose our system supports exactly r distinguishable indecomposable anyon types, which we label by a set $\mathcal{L} = \{0, \ldots, r-1\}$ of "colors" or "particle types". We include the vacuum as a sort of invisible anyon type which we label by 0, or sometimes **1**. As we interpret the anyons as (quasi)-particles, each anyon a must have an antiparticle, which we denote by $\hat{a}$. Notice that the vacuum is its own antiparticle so $\hat{0} = 0$. Anyons on a surface are point-like, so we may imagine them as small boundary circles on the surface labeled by $a \in \mathcal{L}$. A convenient topological interpretation of anti-particles is that a positively oriented circle labeled by a is the same as a negatively oriented circle labeled by $\hat{a}$. Topological invariance means that any topologically allowed operations such as anyon exchange, or twisting part of a surface must correspond to a unitary operator on the corresponding Hilbert space. The underlying Hilbert space must remain the same, provided the operation leaves the topology of the surface M with labeled punctures the same.

The notions of *entanglement*, (particle-anti-particle) *duality* and *locality* lead us to the following:

- *Disjoint Union Axiom* The Hilbert space of two disjoint systems (surfaces with labeled boundary) (M_1, ℓ_1) and (M_2, ℓ_2), considered as a composite system is $\mathcal{H}(M_1, \ell_1) \otimes \mathcal{H}(M_2, \ell_2)$.
- *Duality Axiom* $\mathcal{H}(M, \ell)^\dagger \cong \mathcal{H}(\overline{M}, \hat{\ell})$, where $\overline{M}$ is the surface M with opposite orientation and $\hat{\ell}$ means apply $\hat{}$ to each $x \in \ell$.
- *Gluing Axiom* The global Hilbert space is determined by local Hilbert spaces. More precisely, if a surface (M, ℓ) with boundary labels ℓ is obtained from $(M_g, x, \hat{x}, \ell)$ by gluing the boundary circles labeled by x and $\hat{x}$ together, then

$$\mathcal{H}(M, \ell) = \bigoplus_{x \in \mathcal{L}} \mathcal{H}(M_g, x, \hat{x}, \ell).$$

Taken together, these axioms point us to a cut-and-paste procedure to determine the Hilbert space associated to a surface with labeled boundary from "initial conditions", i.e. the Hilbert spaces of a few less complicated surfaces. The first three initial conditions are:

- *Empty Set Axiom* $\mathcal{H}(\emptyset) \cong \mathbb{C}$
- *Disk Axiom* $\mathcal{H}(D^2, x) \cong \delta_{0x}\mathbb{C}$
- *Annulus Axiom* $\mathcal{H}(A, x, y) \cong \delta_{x\hat{y}}\mathbb{C}$

As axioms these are meant to be unquestioned–however they can be justified as consistency. For example, the Disk and Annulus Axioms can be understood as particle-anti-particle creation rules. The last initial condition(s) are meant to determine the Hilbert spaces for all surfaces with labeled boundary, subject to compatibility with the previous axioms:

- *Pants Axiom* $\mathcal{H}(P(z_1, z_2, z_3), x, y, z) = \mathbb{C}^{N^z_{xy}}$ for some $N^z_{x,y} \geq 0$.

Here $P(z_1, z_2, z_3) := S^2 \smallsetminus \{z_1, z_2, z_3\}$ is the sphere with three boundary components ("a pair of pants") labeled by x, y and z. The problem now is to choose the N^z_{xy} for all triples of labels $(x, y, z) \subset \mathcal{L}^3$ compatible with all of the axioms. The alert reader will realize that the pants axiom does not uniquely assign a Hilbert space to labeled pair of pants. Indeed, there is a 6-fold ambiguity: one for each permutation of the labels x, y and z! This technicality is dealt with by replacing surfaces M by so-called *m-surfaces* or *extended surface* which have a bit more structure. The idea is that one should fix a particular "standard" surface with boundary and use it to parametrize all surfaces with boundary that are homeomorphic the the standard surface. The parametrization then removes the ambiguity. This is carefully addressed in [12, Chapter 4] or [13, Chapter 4]. For a sphere with punctures it amounts to choosing some labels as "inputs" and others as "outputs" and carefully ordering the labels along two parallel lines. For two input labels a and b and one output label c these give us *fusion channels* H^c_{ab}: a Hilbert space of dimension N^c_{ab}, see Figure 3.

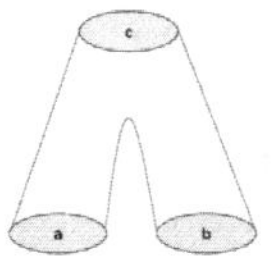

Figure 3. The labeled surface associated with the fusion channel H^c_{ab}, which is to be read upwards, and interpreted as the number of independent ways that particles a and b can fuse to give particle c.

The interpretation of the Hilbert space associated with the surface in Figure 3 as a fusion channel suits our purposes very well: we would like to imagine these as processes. Dual to fusion channels are the *splitting channels* with 1 input (say c) and 2 outputs (say a and b) denoted H^{ab}_c. Using the axioms above, *every* state space can be obtained via tensor products and direct sums of splitting and fusion channels. Which fusion/splitting channels correspond to the annulus axiom? By choosing the fusion output or splitting input to be the vacuum 0 label we get $H^0_{ab} \cong \delta_{b,\hat{a}}\mathbb{C} \cong H^{ab}_0$.

Exercise 1. *Use the gluing axiom and the annulus axiom to show that the state space with input and output a has:* $H^a_{a0} \cong \mathbb{C} \cong H^{a0}_a$.

The precise sense in which these space are dual will be described below.

In general $N^z_{xy} \neq N^y_{xz}$: this is due to the technicality mentioned above. However, using the annulus, gluing and disjoint union axioms one can show that $N^{\hat{y}}_{x\hat{z}} = N^z_{xy}$: one simply glues cylinders (annuli) with labels z and y to change them from inputs to outputs and vice versa. Similarly the duality axiom implies that $N^z_{xy} = N^{\hat{z}}_{\hat{x}\hat{y}}$. In fact, if we assume that every particle is indistinguishable from its anti-particle (i.e. $\hat{x} = x$), the *dimension* of $\mathcal{H}(M, \ell)$ can be computed unambiguously: for then the N^z_{xy} are fully symmetric in the three indices.

To get a better intuition for how these axioms allow one to determine (at least the dimension of) the Hilbert space of any surface with labeled boundary we suggest trying the following:

Exercise 2. *Show that* $\dim \mathcal{H}(T^2) = |\mathcal{L}|$*, where* T^2 *is the 2-dimensional torus.*

Topological invariance provides many useful constraints on the numbers N^z_{xy}, since these are interpreted as dimensions of Hilbert spaces associated with a thrice punctured sphere. For example, interchanging labels x and y (i.e. braiding, but keeping both x and y as inputs) is a topological (commutativity) operation, so $N_{xy}^z = N_{yx}^z$. To give algebraic interpretations we define, for each label a, a *fusion* matrix N_a by $(N_a)_{c,b} = N_{ab}^c$. In this formalism the above calculations imply that $N_{\hat{x}} = N_x^\dagger$. The *fusion rules* express the fusion channels as a superposition of particle types that can occur as an output with inputs x and y: $x \times y = \sum_z N_{xy}^z z$. A cascade of fusion channels can then be interpreted as a matrix product. A topological associativity constraint and a consequence for matrices is illustrated in Figure 4.

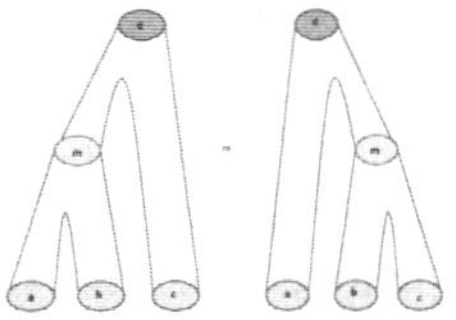

Figure 4. Equivalent (extended) labeled surfaces. The dimension of each can be computed via the gluing (over m) and disjoint union axioms, and then interpreted as a matrix product. For example the the LHS is: $\sum_m N_{ab}^m N_{\hat{m}c}^d$. As a consequence one finds $N_a N_c = N_c N_a$.

2.3. Initialization, transformations and measurement

So far we have only discussed the spaces of states. We can now proceed to explore the unitary operations (quantum gates), processes and further constraints in our model. To do this, we need to establish notation for states themselves. How do we describe a specific state in, for example, the fusion channel H_{ab}^c as in Figure 3? It is convenient to denote such a vector by the skeleton of the corresponding space, i.e. the trivalent graph ⅄ with the three extremal vertices labeled by a, b and c and the degree three vertex labeled to distinguish it from other states in H_{ab}^c. For example, we might choose a basis for H_{ab}^c, so that there are $\dim(H_{ab}^c)$ labels. Similarly, we use appropriately labeled graphs Y to denote states in the dual space H_c^{ab}. It is tempting, and indeed can be justified mathematically using the gluing axiom, to stack these graphs to represent a cascade of splitting/fusion operations. For example, if we compose compatibly labeled Y (input c, outputs a, b) and ⅄ (inputs a, b, output c) the result is a vector in the 1-dimensional state space $H_{c0}^c \cong \mathbb{C}$. One typically choosing the bases so that this pairing coincides with the inner product on the Hilbert space H_{ab}^c. A complete treatment of this diagrammatic yoga of *graphical calculus* involving such pictures can be found in [14, Appendix E] and [9, Section 4.2]. See also the discussion of the topological twist below and Figure 5 for the picture associated with the braiding operators. Some calculations of this form are illustrated in Figures 7 and 8.

The vacuum state corresponds to a disk with boundary labeled by 0, which we can use as an invisible input or output without changing the state. The creation of a particle-antiparticle from the vacuum corresponds to a disk with boundary labeled by 0 and two interior boundary circles labeled by a and $\hat{a}$. This process translates to a linear (Hermitian) operator on state spaces: $b_a : H_{00}^0 \to H_0^{a\hat{a}}$, with a corresponding (dual) annihilation $d_a : H_0^{a\hat{a}} \to H_{00}^0$. Indeed, we assume we can create (via some physical process) any number of particle-antiparticle pairs, from which we obtain, from the vacuum state, a state in $H_0^{a_1\hat{a}_1\cdots a_n\hat{a}_n}$. This corresponds to *initialization* in the quantum computational model. On the other hand, if we are given a system with state vector in $H_{a_1\cdots a_n}^{b_1\cdots b_m}$ we assume we can measure the total charge (i.e. the label) of a pair of adjacent particles, perhaps by bringing them together and measuring the energy. This is the *measurement* stage of the quantum computation, a Hermitian operator represented graphically as composing with a fusion operator Y.

The time evolution of the space of states must be a unitary operator. In particular, a sequence of particle exchanges corresponding to a braid β induces a unitary transformation $|\psi\rangle \mapsto U_\beta|\psi\rangle$. In the topological model of [1] these are (all of) the *quantum circuits* [15].

Figure 5. Interchanging the positions of two identical particles induces a quantum gate–the image of σ_i under a unitary representation on the state space.

If we consider a simple case where $a = \hat{a}$, then the state space $H_0^{a,\dots,a}$ corresponding to n particles of type a supports a unitary representation of the braid group $\mathcal{B}_n$ via particle exchange, as illustrated in Figure 5. More generally, we always obtain a unitary representation of the small group of *pure braids* consisting of those braids with each strand beginning and ending in the same position. The computational strength of the model is hidden in this unitary representation of $\mathcal{B}_n$. A particle a is called *non-abelian* if the image of the $\mathcal{B}_n$ representation on the state space of n type a particles is non-abelian (for some n). To have a reasonable computational model this is a bare minimum. An anyon a is called *(braiding) universal* if any unitary operator can be approximately achieved as the image of some braid β acting on a state space of n type a particles via particle exchange (plus some technical "no-leakage" condition that we ignore). The search for non-abelian and universal anyons is a major thrust of experimental condensed matter physics.

To summarize the processes we assume are available: 1) we can create any number of particle-antiparticle pairs, 2) we may exchange these particles to rotate our initial state and 3) we may measure the particle type of any pair of neighbouring particles. One key is that after braiding the particles' world lines, a neighbouring particle-antiparticle pair may have obtained a different total charge (besides 0, i.e. the vacuum). To get meaningful information from this process we must repeat the same process several times, taking a tally of the outputs (particle types). The topological degrees of freedom ensure that slight variations in the process (e.g. small deviations in the trajectory of a particle in space-time) do not influence the output. The empirically computed probability distribution of output particle types constitutes the result of the quantum computation.

3. Fundamental Questions

In the remainder of this survey we would like to address a few fundamental questions:

(i) How can we distinguish indecomposable particle types?

(ii) Is there a "periodic table" of topological phases of matter?

(iii) How can we detect non-abelian and universal anyons in (idealized) experiments?

3.1. Distinguishing particles

For any two particles types a and b, we must have an (idealized) quantum process that distinguishes the particle types. Essentially, we need to be able to determine some unknown particle type using creation, braiding and measurement.

Figure 6 illustrates the process, which leads to a certain non-degeneracy constraint on the braiding. The columns of the S-matrix can be seen to be simultaneous eigenvectors for the (commuting) fusion matrices $\{N_a : a \in \mathcal{L}\}$, so that S diagonalizes the N_a. This leads to the famous *Verlinde formula*: $N_{ij}^k = \frac{1}{D^2}\sum_r \frac{S_{ir}S_{jr}S_{hatkr}}{S_{0r}}$ where D^2 is an overall normalization constant.

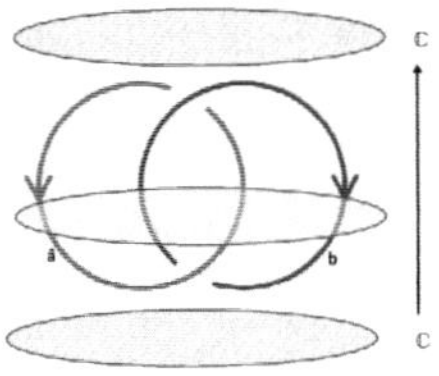

Figure 6. Creating particle-antiparticle pairs of types a and b, braiding and then measuring the amplitude of the vacuum output produces a constant map $S_{ab} \in \mathbb{C}$ for each $a, b \in \mathcal{L}$. For fixed a, we require that the vector of outputs over all $b \in \mathcal{L}$ be linearly independent with (in fact, orthogonal to) that of any other a'. That is, the matrix S must be orthogonal (up to an overall phase).

Anyons may sometimes also be distinguished by their *topological spin*, the phase acquired on the (1-dimensional) state space of the cylinder labeled by a upon twisting by 2π. For this one should imagine that the particle a is a small line segment so that the world lines are really ribbons rather than 1-dimensional curves. Then if we twist a by 2π radians, its trajectory traces out a narrow ribbon with a twist. Depicting trajectories as curves we have

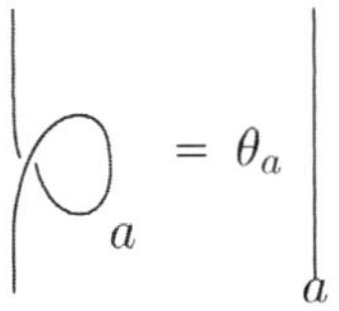

where $\theta_a = e^{2\pi i h_a}$ and $h_a \in \mathbb{Q}$ is the topological twist. We must take care to remember that the picture on the left acquires a phase θ_a when pulled straight.

3.2. Periodic table

How many distinct models of anyonic systems with exactly $n = |\mathcal{L}|$ distinct particle types are there? More generally, is there a classification of such models? Using extensive algebraic constraints we [16] recently proved that, for fixed n, there are finitely many possible models. The precise asymptotics of the number of distinct theories as $n \to \infty$ are unknown, but it can be shown that it grows faster than any polynomial. A classification up to $|\mathcal{L}| \leq 5$ is known (see [17] and references) with constructions coming from quantum groups and finite groups, see Table 1.

$\|\mathcal{L}\|$	Models
1	Vec
2	Fib, $\mathbb{Z}_2$
3	$\mathbb{Z}_3, PSU(2)_7$, Ising
4	products, $\mathbb{Z}_4$, $PSU(2)_9$
5	$\mathbb{Z}_5, PSU(2)_{11}, SU(3)_4/\mathbb{Z}_3, SU(2)_4$

Table 1. $SU(N)_k$ are the level k representations of the affine Kac-Moody algebra of type A_{N-1}, and $PSU(2)_k$ consists of the "integer spin" representations. The $\mathbb{Z}_n$ models are abelian–each fusion channel is 1-dimensional, with fusion rules like the multiplication in $\mathbb{Z}_n$.

The following are two explicit examples:

Example 1. *The Fibonacci theory has two labels* $\mathcal{L} = \{\mathbf{1}, f\}$ *with fusion rules* $f \times f = \mathbf{1} + f$. *The* S*-matrix and topological twists are:* $S = \begin{pmatrix} 1 & \frac{1+\sqrt{5}}{2} \\ \frac{1+\sqrt{5}}{2} & -1 \end{pmatrix}$ *and* $\theta_f = e^{4\pi i/5}$. *The name comes* *from the fact that* $f^n = F_{n-1}\mathbf{1} + F_n f$ *where* F_i *is the well-known Fibonacci sequence:* $0, 1, 1, \ldots$.

Example 2. *The Ising theory has three labels* $\mathcal{L} = \{\mathbf{1}, \sigma, \psi\}$ *and fusion rules* $\sigma \times \sigma = \mathbf{1} + \psi$, $\sigma \times \psi = \sigma$, $\psi \times \psi = \mathbf{1}$. *The S-matrix and twists are* $S = \begin{pmatrix} 1 & \sqrt{2} & 1 \\ \sqrt{2} & 0 & -\sqrt{2} \\ 1 & -\sqrt{2} & 1 \end{pmatrix}$ *and* $\theta_\sigma = e^{\pi i/8}$, $\theta_\psi = -1$. *The* ψ *particle is the famous Majorana fermion.*

3.3. Detecting Non-abelian and universal anyons

The *quantum dimension* $\dim(a)$ of a particle type a is the maximal eigenvalue of the fusion matrix N_a. By the Perron-Frobenius theorem in matrix theory, this eigenvalue is real and positive. In fact, it can be shown that $\dim(a) \geq 1$, since no power of N_a is 0. The Fibonacci particle f has $\dim(f) = \frac{1+\sqrt{5}}{2}$ whereas the Ising particle σ has $\dim(\sigma) = \sqrt{2}$. If $\dim(a) > 1$ we say that a is *non-degenerate*: in this case $a \times \hat{a} = \mathbf{1} + b$ where $b \neq \mathbf{1}$. Recently we [18] showed that non-degeneracy of a implies a is non-abelian. The essence of the argument is illustrated in Figures 7 and 8. If we suppose that the braiding operators commute then we may simultaneously diagonalize them. Restricting to each irreducible sector we may further assume that they act as scalar multiples of the identity.

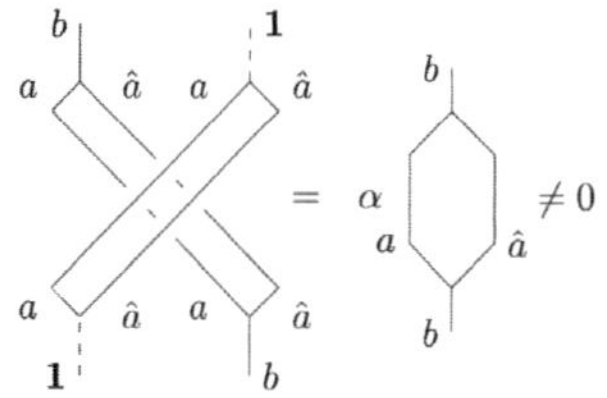

Figure 7. The top loop may be deleted at the expense of a non-zero scalar α, yielding the non-zero state on the right.

Thus it is impossible that the braiding operators commute, since this contradicts the calculation in Figure 7 of a non-zero state. This result show that, in principle, experimentalists can detect non-abelian anyons by measuring the quantum dimension.

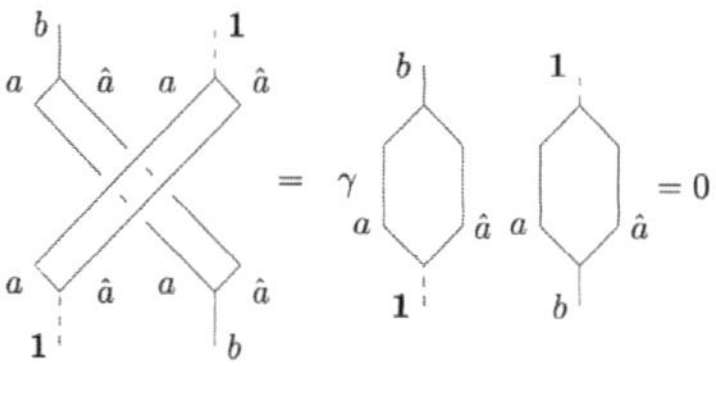

Figure 8. If the exchange of a pair of a (respectively $\hat{a}$) particles and the full position interchange of a a-$\hat{a}$ pair are each multiples of the identity operation we produce a zero state by the annulus axiom.

It is known [19] that the Fibonacci anyon is universal, whereas the Ising anyon is not, despite the fact that both are non-degenerate (and hence non-abelian). The braid group image corresponding to an array of Ising anyons is non-abelian, but finite. How can we distinguish these models? Over the last few years we have found significant evidence for the following:

Conjecture 1. *The anyon a is (braiding) universal if, and only if,* $\dim(a)^2$ *is not an integer.*

Journal of Physics: Conference Series **698** (2016) 012012 doi:10.1088/1742-6596/698/1/012012

One strong piece of evidence for this conjecture is that, for models associated with quantum groups, the braid group image is infinite if and only if $\dim(a)^2$ is not an integer. This latter weaker version of the conjecture goes by the name *property F* (see [20]).

4. Three-dimensional generalizations

Can we generalize our model for 2-dimensional systems to 3-dimensions in a meaningful way? By the above-mentioned spin statistics theorem, point-like particles in 3 dimensions do not admit interesting braiding statistics. However, the motions of *loop-like* particles (e.g. vortices) in 3-dimensional space is mathematically interesting, and physical realizations are being studied as well [21]. Consider a collection of n identical oriented loops (circles) inside a ball. There are two obvious local symmetries *Loop interchange* $\bigcirc \leftrightarrow \bigcirc$ and *Leapfrogging*, see Figure 9.

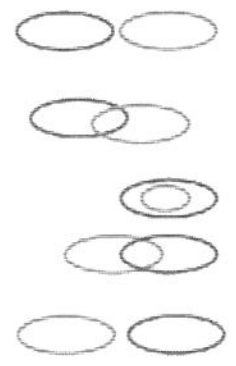

Figure 9. Reading upwards, the left (green) loop passes under and through the right (brown) loop with the final positions of the two loop interchanged.

As we assume the loops are oriented we do not permit a flip of a single loop, since this reverses the orientation. If we denote by s_i the interchange of loops i and $i+1$ and by σ_i the leapfrogging operation on loops i and $i+1$ the corresponding trajectories are in $3+1$-dimensional space-time. We can visualize them as in Figure 10.

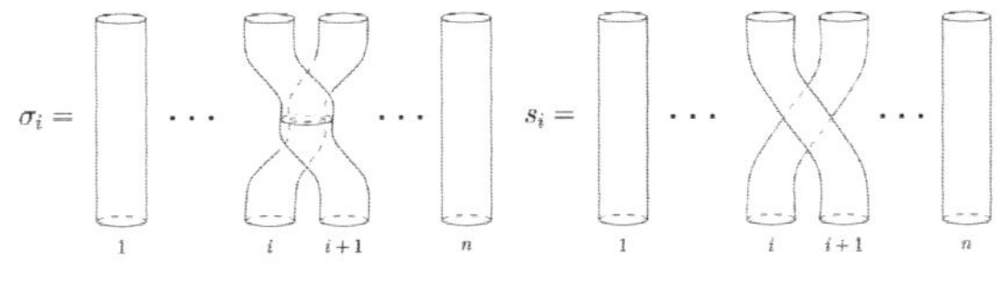

Figure 10. The trajectories of the leapfrogging σ_i and loop interchange s_i operations on n loops take place in $(3+1)$-dimensional space-time. Imagining the time axis as moving from bottom to top, we visualize σ_i and s_i through their projections into 2-dimensions.

The group generated by σ_i and s_i for $1 \leq i \leq n-1$ is called the *Loop Braid Group*, $\mathcal{LB}_n$, defined abstractly as the group satisfying:

Braid relations:

(R1) $\sigma_i\sigma_{i+1}\sigma_i = \sigma_{i+1}\sigma_i\sigma_{i+1}$
(R2) $\sigma_i\sigma_j = \sigma_j\sigma_i$ if $|i-j| > 1$

Symmetric Group relations:

$(S1) s_is_{i+1}s_i = s_{i+1}s_is_{i+1}$
(S2) $s_is_j = s_js_i$ if $|i-j| > 1$
(S3) $s_i^2 = 1$

Mixed relations:

(M1) $\sigma_i\sigma_{i+1}s_i = s_{i+1}\sigma_i\sigma_{i+1}$
(M2) $s_is_{i+1}\sigma_i = \sigma_{i+1}s_is_{i+1}$
(M3) $\sigma_is_j = s_j\sigma_i$ if $|i-j| > 1$

This is a relatively new area of development, for which many questions and research directions remain unexplored. A first mathematical step is to study the unitary representations of the loop braid group, which is already underway [22, 23]. It might also be reasonable to consider other configurations, such loops bound concentrically to an auxillary loop or knotted loops.

5. Conclusions

We have briefly illustrated how modeling the physical properties and computational applications of anyons on surfaces leads to a rich mathematical theory. This theory, in turn, can be used to probe fundamental questions and guide experiments in 2-dimensional topological phases of matter. Moreover, topological considerations suggest that 3-dimensional materials might also be studied in an analogous way, using loop-like excitations.

Acknowledgments

This article is based upon two lectures given at QuantumFest 2015 held at Monterrey Tec, Estado de Mexico campus. I would like to thank that institution and the organizers for a stimulating conference and wonderful hospitality. The author was partially supported by US NSF grant DMS-1410144.

This article is dedicated to the memory of our friend Sujeev Wickramasekara.

6. References

[1] Freedman M, Kitaev A, Larsen M J and Wang Z 2003 Bull. *Amer. Math. Soc.* **40** 31
[2] Tsui D C, Stormer H L and Gossard A C 1982 *Phys. Rev. Lett.* **48** 1559
[3] Laughlin R B 1983 *Phys. Rev. Lett.* **50** 1395
[4] Kitaev A 2003 *Annals Phys.* **303** 2
[5] Freedman M 1998 *Proc. Natl. Acad. Sci. USA* **95** 98
[6] Collins G 2006 *Computing with Quantum Knots* Sci. Amer. **294** issue 4 pp. 57–63
[7] Pachos J 2012 *Introduction to topological quantum computation* (Cambridge: Cambridge University Press)
[8] Nayak C, Simon S H, Stern A, Freedman M and Das Sarma S 2008 *Rev. Mod. Phys.* **80** 1083
[9] Wang Z 2010 *Topological quantum computation.* (Providence: Amer. Math. Soc.)
[10] Wilczek F 1982 *Phys. Rev. Lett.* **49** 957
[11] Walker K 1991 On Witten's 3-manifold invariants *Preprint* http://canyon23.net/math
[12] Turaev V 1994 *Quantum invariants of knots and 3-manifolds* (Berlin: Walter de Gruyter & Co.)
[13] Bakalov B and Kirillov Jr. A 2001 *Lectures on tensor categories and modular functors* (Providence: Amer. Math. Soc.)
[14] Kitaev A 2006 *Ann. Phys.* **321** 2
[15] Recently some models employing partial measurement [24] have been explored, but for the sake of simplicity we will only consider braiding operators as our quantum circuits.
[16] Bruillard P, Ng S-H, Rowell E and Wang Z 2013 Rank-finiteness for modular categories *Preprint math.QA/1310.7050*
[17] Bruillard P, Ng S-H, Rowell E and Wang Z 2015 On classification of modular categories by rank *Preprint math.QA/1507.05139*
[18] Rowell E and Wang Z 2015 Degeneracy implies non-abelian statistics *Preprint cond-mat.str-el/1508.04793*
[19] Freedman M, Larsen M J and Wang Z 2002 *Comm. Math. Phys.* **228** 177
[20] Naidu D and Rowell E 2010 *Algebr. Represent. Theory* **14** 837
[21] Wang C and M. Levin M 2014 *Phys. Rev. Lett.* **113** 080403
[22] Kadar Z, Martin P, Rowell E and Wang Z 2014 Local representations of the loop braid group *Preprint math.QA/1411.3768*
[23] Bruillard P, Chang L, Hong S-M, Plavnik J Y, Rowell E and Sun M Y 2015 *J. Math. Phys.* **56** 111707
[24] Cui X and Wang Z 2015 *J. Math. Phys.* **56** 032202

Journal of Physics: Conference Series **698** (2016) 012013 doi:10.1088/1742-6596/698/1/012013

Interaction-Free Measurements of Optical Semitransparent Objects

Z Blanco-Garcia and O Rosas-Ortiz

Physics Department, Cinvestav, A.P. 14-740, 07000, México D.F, Mexico

E-mail: `iblanco@fis.cinvestav.mx,orosas@fis.cinvestav.mx`

Abstract. We substitute the fully absorbing obstacle in the Elitzur-Vaidman experiment by a semitransparent object and show that the probabilities of detection can be manipulated in dependence of the transparency of such an object. Then, we connect our results with the delayed choice experiment proposed by Wheeler. It is found that the transparency of the obstacle determines either a particle-like or a wave-like behaviour of a test photon.

1. Introduction

In quantum mechanics the pure states describe the dynamics of single particles and the measuring devices are macroscopic apparatuses that obey the rules of the classical theory [1]. One of the most controversial elements of the theory is the *axiom of the wave-packet reduction* suggested by the Born's probabilistic interpretation (see e.g. [2] and references quoted therein for details). Indeed, before a measurement the particle is in a superposition of all the possible states associated with a given observable A. After measuring A, the state $|\psi\rangle$ of the particle is not arbitrary anymore, it has been *reduced* to the state $|\alpha\rangle$ that corresponds to the measured quantity α. Before such a process we have no means to know the specific value of α, so the act of measurement involves an unpredictable perturbation of the particle state. To be more precise, let us follow [2] and consider a light beam χ_a polarized in the direction $\vec{e}_a$. If the beam impinges on a polarization filter $A^{\perp}$ that selects polarization $\vec{e}_a^{\perp}$ perpendicular to $\vec{e}_a$, no photon will cross $A^{\perp}$. The situation changes if the beam is first transmitted trough a filter B selecting the polarization $\vec{e}_b$ ($\vec{e}_a \cdot \vec{e}_b = \cos\theta$) because, depending on θ, a part of the photons emerging from B will cross the filter $A^{\perp}$. Clearly, the measuring device B not only selects but also affects the state of the photons.

An intriguing question is if the superposed quantum states are attainable to microscopic systems only. The consequences of assuming that mesoscopic and macroscopic systems can be described by wave-functions were first indicated by Schrödinger and gave rise to the concept of quantum entanglement [3]. However, is it possible to identify entanglement in a system as large and complex as the one composed by a living cat and a decaying atom? Although this last is still an open problem, objects as large and classical as the fullerene molecules, the 'soccer-ball-shaped carbon cages C_{60}', have the ability of forming interference patterns [4–6]. On the other hand, the unavoidable influence of the measuring device on the state of a given particle has been harnessed in surprising form by Elitzur and Vaidman [7]. They designed an experiment in which a single photon is sent towards a Mach-Zehnder interferometer that includes a fully absorbing obstacle B in one of the arms (see Figure 1). The quantum state of the photon is divided by a beam splitter (BS_1) into two new quantum states represented by wave-packets propagating along the separated space trajectories defined by the arms of the interferometer. The obstacle represents an apparatus that measures the presence of the photon so that this last is either absorbed in the obstructed path or its wave-packet collapses to the free path. In the second case, the photon will be registered by the detector D_2 with probability $1/4$. This last result is remarkable because in absence of the obstacle the superpositions are fully destructive in the position of D_2, so that the probability of registering the photon in D_2 is zero. Thus, the presence of the obstacle is revealed if D_2 is activated, no matter that the obstacle was never illuminated by the single photon in the circuit. A phenomenon called 'quantum seeing in the dark' [8].

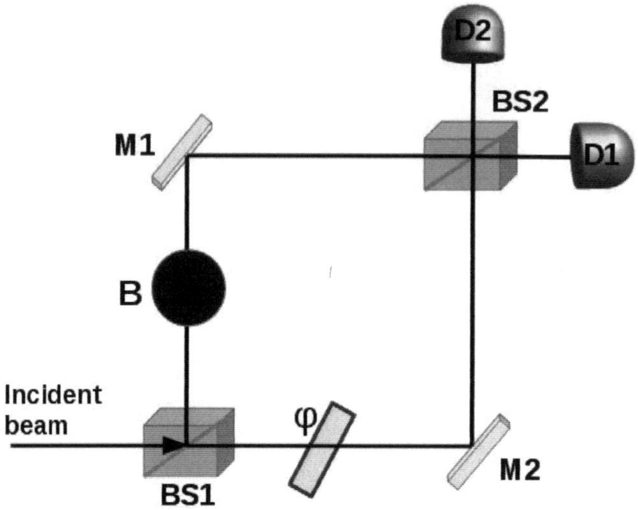

Figure 1. The Elitzur-Vaidman experiment for interaction-free measurement. An absorbing object B in one of the arms of the Mach-Zehnder interferometer is 'seen in the dark' as a consequence of the quantum correlations between the wave-packets in the interferometer circuit.

In this contribution we explore the consequences of substituting the fully absorbing obstacle B by a semitransparent one in the experimental arrangement of Elitzur and Vaidman. Although some research has been already developed in this trend, see e.g. [9–11], we are interested in the connection between the Elitzur-Vaidman approach and the 'delayed choice' experiment proposed by Wheeler [12]. Such an experiment includes a conventional Mach-Zender array in which the observer has the choice of removing the second beam splitter to perform a which-path measurement (i.e., a particle-like experiment). The most interesting part of the Wheeler's proposal is that the removing of BS_2 can be done once the photon is already inside the interferometer. In this form, the photon faces either an interference or a which-path experimental setup just when it cannot, in principle, 'adjust' its behaviour (wave-like or particle-like) consistently. Although such arrangement was originated as a *gedanken* experiment, its quantum predictions have been verified in actual laboratories by using diverse systems, see recent results in e.g. [13, 14].

Next, after a brief survey of the Mach-Zehnder interferometer generalities, we recover the Elitzur-Vaidman results by using the Hubbard representation of the operators associated with the devices in the optical bench. Then we substitute the full absorbing obstacle of Elitzur and Vaidman by a semitransparent one and show that the probabilities of registering the photon in one of the two detectors can be manipulated accordingly. Finally, we show that these results are in connection with the delayed choice experiment of Wheeler.

2. Quantum states of light in a Mach-Zehnder interferometer

The spatial states of the photons in the interferometer are elements of the two-dimensional vector space that is generated by the orthonormal basis $|r_H\rangle$, $|r_V\rangle$, where the sub-label H (V) stands for the horizontal (vertical) orientation defined in Figure 1. The occupation states of the photons are defined by the Fock vectors $|0\rangle$ and $|1\rangle$, meaning zero photons and a single photon respectively. Therefore, the quantum states in the circuit of the interferometer are elements of the four-dimensional vector space

$$\mathcal{H} = \text{Span}\{|r_H\rangle \otimes |0\rangle, |r_H\rangle \otimes |1\rangle, |r_V\rangle \otimes |0\rangle, |r_V\rangle \otimes |1\rangle\}, \tag{1}$$

where $A \otimes B$ is the direct product between A and B. The mirror M, beam splitter BS, and phase shifter φ indicated in Figure 1 are two-channel operators expressed in Hubbard representation:

$$M = i(|r_H\rangle\langle r_V| + |r_V\rangle\langle r_H|), \quad BS = \frac{1}{\sqrt{2}}(\mathbb{I}_s + M), \quad \varphi = e^{i\phi}|r_H\rangle\langle r_H| + |r_V\rangle\langle r_V|. \tag{2}$$

Here $\mathbb{I}_s$ is the identity operator in the vector space $\mathcal{H}_s = \text{Span}\{|r_H\rangle, |r_V\rangle\}$, and $\phi \in [0, 2\pi)$. The dyads $X^{k,j} = |r_k\rangle\langle r_j|$, $r, j, = H, V$, are square matrices of order 2 that have entry 1 in position (k, j) and zero in all other entries (see [15] for details). These operators are promoted to act in the space $\mathcal{H}$ as follows

$$M \leftrightarrow M \otimes \mathbb{I}_f, \qquad BS \leftrightarrow BS \otimes \mathbb{I}_f, \qquad \varphi \leftrightarrow \varphi \otimes \mathbb{I}_f, \tag{3}$$

with $\mathbb{I}_f$ the identity operator in the vector space $\mathcal{H}_f = \text{Span}\{|0\rangle, |1\rangle\} \subset \text{Span}\{|n\rangle\}_{n\geq 0}$.

In the conventional Mach-Zehnder interferometer the entire circuit corresponds to the action of the operator $A = (BS_2)(M\varphi)(BS_1)$ on a given initial state $|\psi_{in}\rangle$ which, without loss of generality, will be taken as $|\psi_{in}\rangle = |r_H\rangle \otimes |1\rangle \equiv |r_H, 1\rangle$. That is, only one photon (in the horizontal channel) is sent towards the interferometer. After some calculations one arrives at the state

$$|\psi\rangle = A|\psi_{in}\rangle = \tfrac{i(e^{i\phi}-1)}{2}|r_V, 1\rangle - \tfrac{(e^{i\phi}+1)}{2}|r_H, 1\rangle. \tag{4}$$

Therefore, the photon will be registered in either D_1 (horizontal arm) or D_2 (vertical arm) with probability

$$P_{D_1} = \tfrac{1}{2}(1 + \cos\phi), \qquad P_{D_2} = \tfrac{1}{2}(1 - \cos\phi). \tag{5}$$

If the optical path difference of the two arms in the interferometer is zero we can take the phase $\phi = 0$ to recover the well known result $P_{D_1} = 1$ and $P_{D_2} = 0$.

2.1. The Elitzur-Vaidman model

To include the presence of a perfect absorbing object B in the first vertical arm we have to identify the appropriate operator. The boson annihilation operator $a|n\rangle = \sqrt{n}|n-1\rangle$, $n \geq 0$, is useful in this matter. After promoting it as $a \leftrightarrow X^{V,V} \otimes a + X^{H,H} \otimes \mathbb{I}_f$, one gets

$$\begin{aligned}
|\psi_{in}\rangle &\xrightarrow{BS_1} \tfrac{1}{\sqrt{2}}(|r_H, 1\rangle + i|r_V, 1\rangle) \xrightarrow{\varphi} \tfrac{1}{\sqrt{2}}(e^{i\phi}|r_H, 1\rangle + i|r_V, 1\rangle) \\
&\xrightarrow{B} \tfrac{1}{\sqrt{2}}(e^{i\phi}|r_H, 1\rangle + i|r_V, 0\rangle) \xrightarrow{M} \tfrac{i}{\sqrt{2}}(e^{i\phi}|r_V, 1\rangle + i|r_H, 0\rangle) \\
&\xrightarrow{BS_2} \tfrac{1}{2}(ie^{i\phi}|r_V, 1\rangle - e^{i\phi}|r_H, 1\rangle - |r_H, 0\rangle - i|r_V, 0\rangle) = |\psi_{EV}\rangle.
\end{aligned} \tag{6}$$

So that we recover the Elitzur-Vaidman prediction $P_{D_1} = P_{D_2} = \frac{1}{4}$, with probability 1/2 of finding that the photon is absorbed by the obstacle B.

3. Seeing in the dark a semi-transparent object

Let us assume that the obstacle B is charaterized by its capability of transmitting or absorbing photons. If Λ_t and Λ_a are the related (complex) transmission and absorption coefficients we can represent them in polar form $\Lambda_t = \beta e^{i\theta}$ and $\Lambda_a = \alpha e^{i\gamma}$, with β and α nonnegative numbers such that $\alpha^2 + \beta^2 = 1$, and the phases θ and γ defining the related principal branches. Therefore, the obstacle will be fully absorbent for $\beta = 0$ and transparent for $\beta = 1$. Its presence in the interferometer can be associated to the action of the operator $X^{V,V} \otimes (\Lambda_a a + \Lambda_t \mathbb{I}_f) + X^{H,H} \otimes \mathbb{I}_f$ on the vector states representing the photon in the circuit. The operation is similar to the one in (6) and gives

$$|\psi_{abs}\rangle = \tfrac{i(e^{i\phi} - \Lambda_t)}{2} |r_V, 1\rangle - \tfrac{(e^{i\phi} + \Lambda_t)}{2} |r_H, 1\rangle - \tfrac{\Lambda_a}{2}(|r_H, 0\rangle + |r_V, 0\rangle). \qquad (7)$$

Besides the probability $\alpha^2/2$ of absorption we have

$$P_{D_1} = \tfrac{1}{4}(1 + \beta^2 + 2\beta\cos\Delta), \quad P_{D_2} = \tfrac{1}{4}(1 + \beta^2 - 2\beta\cos\Delta), \quad \Delta = \theta - \phi. \qquad (8)$$

Clearly, depending on β and Δ, we can manipulate the probabilities in order to increase the registering of our test photon in D_2 over the possibility of destroying it by the action of the obstacle B. As we can appreciate in Figure 2(a), there is no way to get success if $\Delta = 0$. That is, if the optical path in the interferometer arms is such that $\theta = \phi$, the major probability will be associated with either destroying the photon (black dotted curve in the figure) or registering it in D_1 (blue continuous curve in the figure), this last giving no information about the presence of B as discussed above. However, if $\theta = \phi + \pi/2$ we have success for $\beta > 0.57$ because $P_{D_1} = P_{D_2} > \alpha^2/2$, see Figure 2(b).

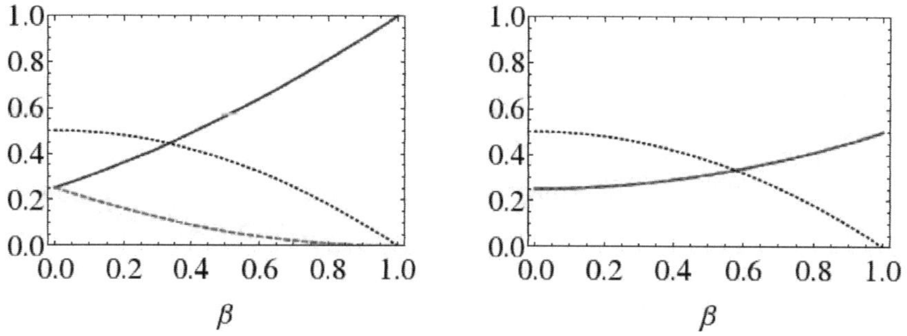

(a) $\Delta = 0$ (b) $\Delta = \pi/2$

Figure 2. Probability of registering a test photon in D_1 (blue, continuous) and D_2 (red, dashed) as a consequence of adding an obstacle of transmission coefficient $\Lambda_t = \beta e^{i\theta}$ in the first vertical arm of a Mach-Zehnder interferometer. In (a) and (b) the optical path is such that $\theta = \phi$ and $\theta = \phi + \pi/2$ respectively. Notice that $P_{D_1} = P_{D_2}$ in (b). The dotted black curve has been included as a reference and corresponds to the probability of absorption $\alpha^2/2$.

In general, for a fully absorbing obstacle ($\beta = 0$) we recover the Elitzur-Vaidman results, no matter the value of Δ. On the other hand, for complete transparency $\beta = 1$ and $\Delta = 2n\pi$, $n = 0, 1, \ldots$, only the detector D_1 is activated. Therefore we cannot, in principle, distinguish if the obstacle is present or not. However, in this case there is still an arbitrary value of the phase θ which could be different from zero. Such a phase is inherited to the final state and can be calculated (in principle) by measuring the delay time of the test photon with respect to a similar wave-packet propagating freely and traveling the same optical distance. Our point is that even if the obstacle is transparent we would know about its presence by measuring time-delays in the photons arriving at the zone of D_1. If by chance, besides $\beta = 1$ we have $\theta = 0$, then the obstacle is practically invisible whenever the photon is registered by D_1.

3.1. Approaching the Wheeler gedanken-experiment

Figure 3 includes a graphic of the probabilities (8) as a function of β and Δ. For $\beta = 0$ (i.e., a fully absorbing obstacle) the probabilities of registering the photon by D_1 and D_2 are the same and equal to 1/4. This last because the photon wave-packet collapses to the free path in the interferometer and the beam splitter BS_2 decouples it in a pair of coherent packets. Thus, the photon exhibits a particle-like behaviour because the result corresponds to a which-path experiment. A different situation arises if the obstacle is transparent ($\beta = 1$) because, depending on the optical path of the interferometer (i.e., depending on ϕ), the interference of the outgoing packets is such that the major probability of registering the photon alternates between D_2 and D_1. In this case the photon exhibits a wave-like behaviour as the result corresponds to an interference experiment. For other values of β and Δ there is not a clear distinction between the particle-like and wave-like behaviour of the photon. Our results are in qualitative agreement with the recent (experimental and theoretical) results reported in e.g. [13,14], where the authors use a controlled-Hadamard operation to simulate a movable beam splitter BS_2.

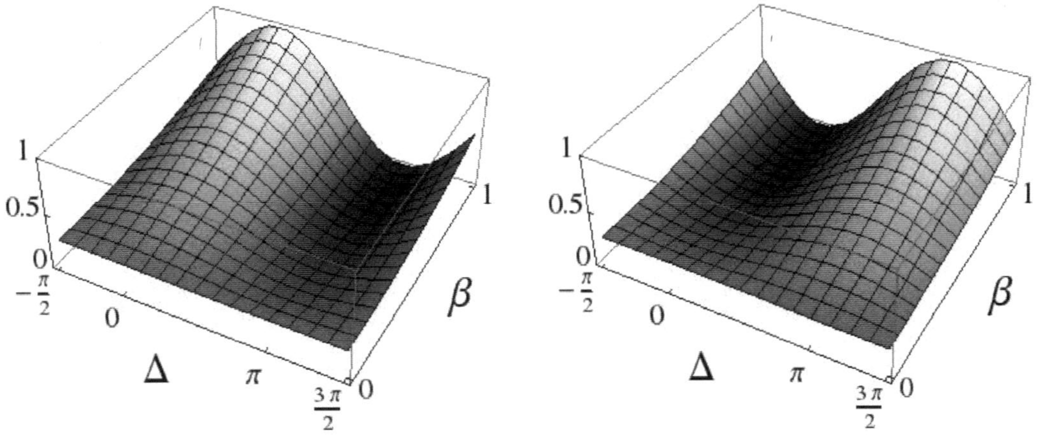

(a) $P_{D_1}(\beta, \Delta)$ (b) $P_{D_2}(\beta, \Delta)$

Figure 3. The probabilities (8) as a function of Δ and β. For $\beta = 0$ the photon exhibits particle-like behaviour while $\beta = 1$ is associated to wave-like behaviour. Then, depending on β, the results correspond either to a which-path or to an interferometer experiment.

4. Conclusions

We have shown that the Elitzur-Vaidman model of interaction-free measurements can be connected with the delayed choice experiment suggested by Wheeler. The main point is the substitution of the fully absorbing obstacle of Elitzur and Vaidman by a semitransparent one. Actual implementations of experimental arrays oriented to verify in the laboratory our quantum predictions are in progress (for details see [16]).

Acknowledgments
The comments and suggestions by the anonymous referees are acknowledged. ZBG gratefully acknowledges the funding received through a CONACyT Scholarship.

References

[1] von Neumann J 1932 *Mathematische Grundlagen der Quanten-mechanik* (Springer-Verlag Berlin) (English translation: Princeton University Press, Princeton, 1955)
[2] Mielnik B and Rosas-Ortiz O 2007 *Quantum Mechanical Laws, a chapter in Fundamentals of Physics*, Ed. J L Morán López (Encyclopedia of Life Support (EOLSS), Eolss Publishers, Oxford)
[3] Schrödinger E 1935 Die Naturwissenschaften **23** 807 (English translation in Ref [12])
[4] Arndt M, Nairz O, Vos-Andreae J, et. al. 1999 Wave-particle duality of C_{60} molecules *Nature* **401** 680
[5] Nairz O, Arndt M and Zeilinger A 2002 Experimental verification of the Heisenberg uncertainty principle for fullerene molecules *Phys. Rev.* A **65** 032109
[6] Nairz O, Arndt M and Zeilinger A 2003 Quantum interference experiments with large molecules *Am. J. Phys.* **71** 319
[7] Elitzur A C and Vaidman L 1993 Quantum mechanical interaction-free measurements *Found. Phys.* **23** 987
[8] Kwiat P, Weinfurter H and Zeilinger A 1996 Quantum Seeing in the Dark *Sci. Am.* **275** 72
[9] Azuma H 2006 Interaction-free measurement with an imperfect absorber *Phys. Rev.* A **74** 054301
[10] Jang J-S 1999 Optical interaction-free measurement of semitransparent objects *Phys. Rev.* A **59** 2322
[11] Thomas S, Kohstall C, Kruit P and Hommelhoff P 2014 Semitransparency in interaction-free measurements *Phys. Rev.* A **90** 053840
[12] Wheeler J A and Zurek W H (Eds.) 1983 Quantum Theory and Measurement (Princeton University Press, Princeton) pp 182-213
[13] Peruzzo A, Shadbolt P, Brunner N, et. al. 2012 A Quantum Delayed-Choice Experiment *Science* **338** 634
[14] Kaiser F, Coudreau T, Milman P, et. al. 2012 Entanglement-Enabled Delayed Choice Experiment *Science* **338** 637
[15] Enríquez M and Rosas-Ortiz O 2013 The Kronecker product in terms of Hubbard operators and the Clebsch-Gordan decomposition of $SU(2) \times SU(2)$ *Ann. Phys.* **339** 218
[16] Blanco-Garcia Z 2004 *Mediciones no destructivas en mecánica cuántica* (MSc Thesis (in Spanish), Physics Department, Cinvestav, Mexico)

Geometry of the Quantum States of Light in a Mach-Zehnder Interferometer

A Romero-Osnaya
Physics Department, Cinvestav, A.P. 14-740, 07000, México D.F, Mexico
E-mail: aromero@fis.cinvestav.mx

Abstract. A geometric approach to the quantum states of light that is injected into a Mach-Zehnder interferometer is discussed. It is based on the Jordan-Schwinger representation of angular momentum and permits the identification of the photon states with cones in the three-dimensional space that are rotated by the action of the optical devices. Some cases of the *evolution loops*, dynamical processes for which any state evolves cyclically, are presented.

1. Introduction

Diverse techniques for manipulating the state of single quantum systems have been developed in recent years, mainly addressed to precision measurements. This is because the controlled evolution of the quantum state is as relevant as the measurement by itself. In the theoretical scenery, for instance, the so called *inverse method* (see e.g. [1, 2]) has been used to find a way of controlling the free evolution of non-relativistic charged particles by time-independent magnetic fields [3, 4] (see also [5, 6]). The method has been also applied to determine the *evolution loops* [7, 8], dynamical processes for which any state evolves cyclically, associated with two-level quantum systems [9, 10]. On the other hand, using the above ideas, the experimental possibilities of steering an energy state to a given destination have been discussed [11]. In this paper we present preliminary results of our research addressed to control the quantum states of light in a Mach-Zehnder interferometer [12]. The main motivation of such a trend is that our theoretical predictions can be verified in the optical bench of any well equipped laboratory of quantum optics. With this in mind, in Section 2 we revisit the properties of the space of states associated with the conventional two-channel optical devices that integrate the interferometer. Then, in Section 2.1 we use the Jordan-Schwinger approach [13, 14] to get a set of angular momentum operators in term of the photon states in Heisenberg representation. The description in terms of the Stokes parameters is then used in Section 3 to identify the photon states with cones in the three-dimensional space that are rotated by the action of the optical devices. Some implementations to get evolution loops are discussed in Section 3.1 and final remarks are given in Section 4.

2. Vector space of states

Let us consider the Mach-Zehnder interferometer depicted in Figure 1. We are interested in the case when n_1+n_2 photons are sent towards the interferometer, with n_1 (n_2) photons propagating along the horizontal (vertical) direction defined in the figure. As usual, the interferometer is composited by two beam splitters and two mirrors that define two possible optical paths for each photon that may differ in a quantity $\Delta\ell$. This last can be corrected by using phase shifters in both the theoretical calculations and the experimental setup. The initial state is then defined by the tensor product $|n_1, n_2\rangle = |n_1\rangle \otimes |n_2\rangle$, where $|n_{1,2}\rangle$ is a Fock state belonging to the Hilbert space $\mathcal{H}_{1,2}$. To represent the interchange of photons between the vertical and the horizontal configurations as the action of an optical device we shall use the boson operators

$$[a_i, a_j] = 0, \quad [a_i^\dagger, a_j^\dagger] = 0, \quad [a_i, a_j^\dagger] = \delta_{ij}, \quad i, j = 1, 2, \tag{1}$$

which are defined to act on $\mathcal{H}_i$ only. Therefore, we promote them to act on the entire space $\mathcal{H} = \text{span}\{|n_1\rangle \otimes |n_2\rangle\}_{n_1,n_2 \geq 0}$ as follows [15]:

$$a_1 \leftrightarrow a_1 \otimes \mathbb{I}_2, \quad a_1^\dagger \leftrightarrow a_1^\dagger \otimes \mathbb{I}_2, \quad a_2 \leftrightarrow \mathbb{I}_1 \otimes a_2, \quad a_2^\dagger \leftrightarrow \mathbb{I}_1 \otimes a_2^\dagger, \tag{2}$$

where $\mathbb{I}_i$ stands for the identity operator in $\mathcal{H}_i$. For instance, the combined action of the annihilator operators is given by

$$\begin{aligned} (a_1 \otimes a_2)|n_1, n_2\rangle &= (a_1|n_1\rangle) \otimes (a_2|n_2\rangle) \\ &= \sqrt{n_1 n_2}\,|n_1 - 1, n_2 - 1\rangle\,. \end{aligned} \tag{3}$$

For the sake of simplicity, in the sequel we shall omit the symbol '$\otimes$', so that we will use the simpler notation $A_1 \otimes A_2 = A_1 A_2$ for the direct product of the operators A_1 and A_2.

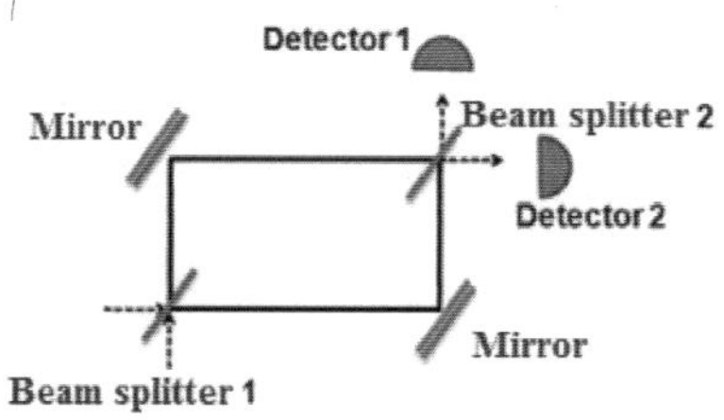

Figure 1. The Mach-Zehnder interferometer. Dotted arrows represent incoming and outcoming beams.

The majority of the optical devices are two-channel in the sense that two incoming signals can be processed simultaneously to produce either a single or a double output signal. In the Heisenberg picture the signal is represented by the boson operators rather than by the Fock vectors, so that the matrix array

$$\begin{pmatrix} a_1' \\ a_2' \end{pmatrix} = \begin{pmatrix} u_{11} & u_{12} \\ u_{21} & u_{22} \end{pmatrix} \begin{pmatrix} a_1 \\ a_2 \end{pmatrix} \tag{4}$$

represents the most general transformation that is suffered by the incoming signal. That is, we shall work with the operator-like vectors $|\hat{a}\rangle = (a_1, a_2)^T$ that are transformed by the operators $U = [u_{ij}]$. To ensure the symmetry properties of the optical devices we know that the matrices $[u_{ij}]$ are the unitary representation of the operators U in the space of the vectors $|\hat{a}\rangle$. For instance, to represent the beam splitters appearing in Figure 1 and the phase shifters that are necessary to correct the optical path difference $\Delta\ell$ we use respectively the matrices

$$D(\alpha) = \begin{pmatrix} \cos\frac{\alpha}{2} & -i\sin\frac{\alpha}{2} \\ -i\sin\frac{\alpha}{2} & \cos\frac{\alpha}{2} \end{pmatrix}, \quad F(\phi_1, \phi_2) = \begin{pmatrix} e^{i\phi_1} & 0 \\ 0 & e^{i\phi_2} \end{pmatrix}. \tag{5}$$

Another useful matrix, leading to essentially the same outputs as $D(\alpha)$ but with an additional phase in the reflected beam, is as follows

$$E(\alpha) = \begin{pmatrix} \cos\frac{\alpha}{2} & -\sin\frac{\alpha}{2} \\ \sin\frac{\alpha}{2} & \cos\frac{\alpha}{2} \end{pmatrix}. \tag{6}$$

The entire circuit in the interferometer is represented by the action of the operator $D(\alpha)F(\phi_1,\phi_2)D(\gamma)$ on the initial state $|in\rangle = |\hat{a}\rangle$. In particular, for $\gamma = -\alpha = \pi/2$ and $\phi_2 = -\phi_1$ the output state $|out\rangle = |\hat{a}'\rangle$ is given by $|out\rangle = M_z(\phi_1)\,|in\rangle$, where

$$M_z(\phi_1) = \begin{pmatrix} \cos\phi_1 & -\sin\phi_1 \\ \sin\phi_1 & \cos\phi_1 \end{pmatrix}. \tag{7}$$

2.1. Jordan-Schwinger representation of angular momentum

In the Jordan-Schwinger representation [13, 14], we can use the boson operators to construct the components of the angular momentum operator

$$J_x = \tfrac{1}{2}(a_1^\dagger a_2 + a_2^\dagger a_1), \quad J_y = \tfrac{-i}{2}(a_1^\dagger a_2 - a_2^\dagger a_1), \quad J_z = \tfrac{1}{2}(a_1^\dagger a_1 - a_2^\dagger a_2), \tag{8}$$

as well as the number operator

$$N = a_1^\dagger a_1 + a_2^\dagger a_2 = N_1 + N_2. \tag{9}$$

In the above expressions we are adhereing to the notation of Ref. [16]. It is easy to verify that N commutes with all the J-operators as well as the following commutation rules

$$[J_x, J_y] = iJ_z, \quad [J_y, J_z] = iJ_x, \quad [J_z, J_x] = iJ_y. \tag{10}$$

In turn, the operator $J^2 = J_x^2 + J_y^2 + J_z^2$ can be written as

$$J^2 = \tfrac{N}{2}\left(\tfrac{N}{2}+1\right), \quad J_z = \tfrac{1}{2}\left(N_1 - N_2\right). \tag{11}$$

Thus, the total number of photons $N = N_1 + N_2$ gives information about the total angular momentum while the difference in the number of incoming photons $N_1 - N_2$ is associated with the z-projection of $\mathbf{J}$. After an straightforward calculation we get

$$\begin{gathered} \langle J_x\rangle_{n_1,n_2} = \langle J_y\rangle_{n_1,n_2} = 0, \quad \langle J_z\rangle_{n_1,n_2} = \tfrac{1}{2}(n_1 - n_2), \\ \left\langle J^2\right\rangle_{n_1,n_2} = \tfrac{n}{2}(\tfrac{n}{2}+1). \end{gathered} \tag{12}$$

This last permits a geometrical description of the state $|n_1, n_2\rangle$ in $\mathbb{R}^3$, see Figure 2(a). Indeed, depending on n_1 and n_2, the expectation values (12) represent a cone of height $h = |\langle J_z\rangle|$ and generatrix $s = \sqrt{\langle J^2\rangle}$ that is aligned with the vertical. The positiveness of $\langle J_z\rangle$ determines the orientation of the cone.

3. Geometric action of the optical devices

In the description of the quantum Stokes parameters [17] (see also [18]), the operators J_i can be identified with the cartesian coordinates that localize the axis of the cone:

$$J_i = \tfrac{1}{2}\,\langle \hat{a}\,|\,\sigma_i\,|\,\hat{a}\rangle\,. \tag{13}$$

After the transformation generated by the unitary operator U one gets [16]:

$$J_i' = \left\langle \hat{a}\,\middle|\,\sigma_i'\,\middle|\,\hat{a}\right\rangle = \left\langle \hat{a}\,\middle|\,U^\dagger \sigma_i U\,\middle|\,\hat{a}\right\rangle, \tag{14}$$

In particular, the action of the beam splitter $D(\alpha)$ on the components of $\mathbf{J}$ gives

$$J_z' = \cos\alpha J_z + \sin\alpha J_y, \quad J_y' = \cos\alpha J_y - \sin\alpha J_z, \quad J_x' = J_x. \tag{15}$$

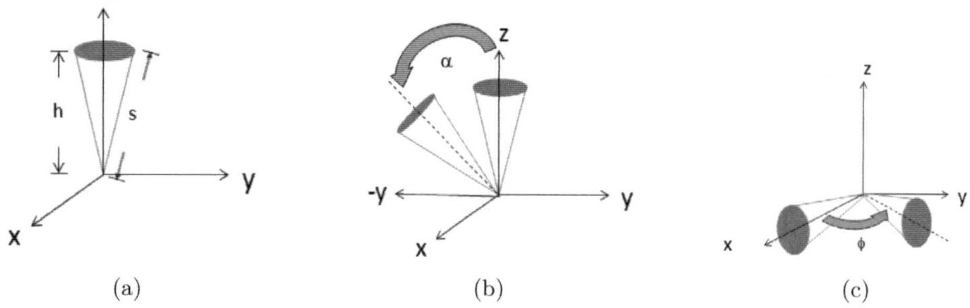

Figure 2. (a) The quantum state of n_1 photons in the horizontal arm and n_2 photons in the vertical arm of a Mach-Zehnder interferometer can be represented as a cone (oriented along the z-axis) of height $h = \frac{1}{2}|n_1 - n_2|$ and generatrix $s = \frac{n}{2}(\frac{n}{2}+1)$ (b) The action of the beam splitter $D(\alpha)$ produces rotations of the cone by an angle α around the x-axis (c) The action of the phase shifter $F(\phi_1, \phi_2)$ produces rotations by an angle $\phi = \phi_2 - \phi_1$ around the z-axis.

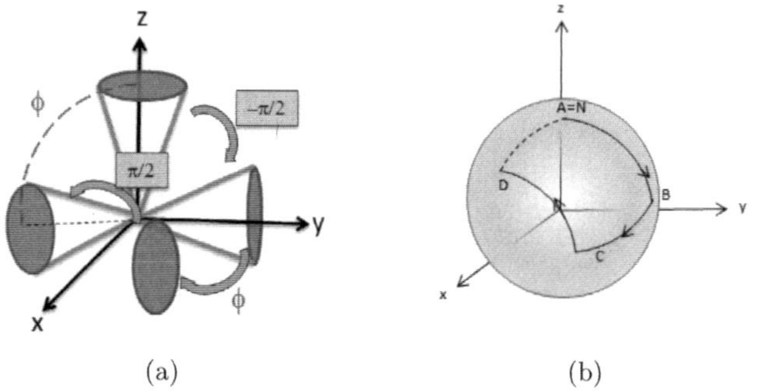

Figure 3. (a) The net effect of the optical devices that integrate the Mach-Zehnder interferometer is reduced to a finite rotation around the y-axis (b) As the genneratrix s is a constant (the number of photons $n = n_1 + n_2$ is preserved), the axis of the cone describes a path on a sphere of radius h.

Or, in matrix form, $\mathbf{J}' = R_x(\alpha)\mathbf{J}$. Thus, the action of the beam splitter produces rotations of the vector $\mathbf{J}$ by an angle α around the x-axis, see Figure 2(b). In turn, the phase shifter $\mathrm{F}(\phi_1, \phi_2)$ produces rotations by an angle $\phi = \phi_2 - \phi_1$ around the z-axis: $\mathbf{J}' = R_z(\phi)\mathbf{J}$, see Figure 2(b). In this form, the entire circuit of the interferometer is associated to the operator

$$U_{global} = D(\pi/2)F(\phi_1, \phi_2 = -\phi_1)D(-\pi/2) = M_z(\phi_1), \tag{16}$$

and gives $\mathbf{J}' = R_y(2\phi_1)\mathbf{J}$. In other words, the net effect of the entire circuit produces rotations of $\mathbf{J}$ by an angle $2\phi_1$ around the y-axis (see Figure 3). This last leads to the interference term $\langle J_z' \rangle = \frac{1}{2}(n_1 - n_2)\cos 2\phi_1$, which clearly accounts for the difference of photons in the vertical and horizontal outputs of the interferometer. It can be shown that the net rotation (16) can be also produced by acting $E(\alpha)$ on the initial state with $\alpha = 2\phi_1$ [12].

The axis of the cone describes a path on a sphere of radius h because the generatrix s is a constant, as it is shown in Figure 3(b). It is well known that the points on such a surface (i.e. the points on the 2-sphere S^2) are in a one-to-one correspondence with the pure states of any qubit (see e.g. [19] and references quoted therein). The dynamics of the two-level pure states is in this form reduced to rotations and reflections of the 2-sphere [20]. Therefore, the path depicted in Figure 3(b) should correspond to the successive transitions of pure states that are required to connect the signal outputs (points B, C and D on the sphere) associated with the optical devices involved.

3.1. Manipulating the quantum states in the interferometer

The above approach is useful in sculpturing the final quantum state of the light that is injected into the interferometer. For instance, making the rotations indicated in Figure 3 in the reverse order we arrive at another state. Namely, if we rotate around y, z and y by $\frac{\pi}{2}$, ϕ_1 and $-\frac{\pi}{2}$ respectively then the net effect is a rotation by ϕ_1 around the x-axis, see Figure 4(a). To produce such a circuit in the optical bench we only require the substitution of the beam splitters D by the ones of type E. The combination of these results is useful to produce close paths on the h-sphere. For instance, the net operator

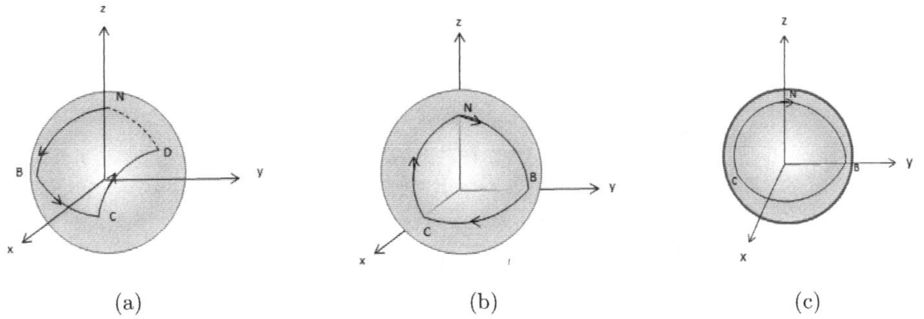

Figure 4. (a) The path described by the axis of the cone on the sphere corresponds to the rotations shown in Figure 3 in reverse order (b) Close path in which the initial state is distinguishable from the final state (c) Close path in which the initial and final states coincide.

$$U_c(\phi_1) = E(-\tfrac{\pi}{2})F(\phi_1, \phi_1 + \tfrac{\pi}{2})D(-\tfrac{\pi}{2}) = e^{i\phi_1}\mathbb{I} \tag{17}$$

produces the adding of a global phase on any given initial state, as it is shown in Figure 4(b). However, notice that even for $\phi_1 = 0$, this last result does not mean that the initial and final states coincide. The difference lies on the fact that the tangent vectors in the paths from the, and arriving at, the point N are not the same. A different situation arises if one uses

$$U_{loop}(\phi_1) = D(-\tfrac{\pi}{2})F(\phi_1, \phi_1 - \pi)D(-\tfrac{\pi}{2}) = \mathbb{I} \tag{18}$$

because this produces close paths in which the initial and final tangent vectors are the same, no matter the initial state, see Figure 4(c). This kind of path on the 2-sphere is known as *evolution loop* [9, 10] and lead, in natural form, to the notion of *geometric phases* [21].

4. Concluding remarks

We have presented preliminary results of our research addressed to control the quantum states of light in a Mach-Zehnder interferometer. Using a geometric approach, we described the action of the optical devices on the photon states as paths described on a sphere of radius $h = \frac{1}{2}|n_1 - n_2|$, with n_1 (n_2) the number of photons that is injected into the horizontal (vertical) arm of the interferometer as initial condition. Similar geometric descriptions have been already reported in connection with the dynamical manipulation of spin-1/2 systems by using magnetic fields [20,22] and with linear combinations of $j = 1/2$ angular momentum eigenkets (mesoscopic *kittens*) that manifest quantum interference effects analogous to those occurring for linear combinations of coherent light waves [23]. In our case, the paths on the 2-sphere correspond to the successive transitions of pure states that are required to connect the signal outputs associated with the optical devices involved. In particular, close paths in which the initial and final states coincide have been found to describe the so called *evolution loops* [8, 9]. The calculation of the corresponding geometric phase is in progress and will be reported elsewhere.

Acknowledgements
The author is grateful to Prof. Oscar Rosas-Ortiz for useful comments and suggestions. The support of CONACyT is acknowledged.

References

[1] Mielnik B 2013 Quantum Operations: technical or fundamental challenge? *J. Phys. A: Math. Theor.* **46** 385301; arXiv:1305.3664
[2] Mielnik B 2014 Quantum Control: discovered, repeated and reformulated ideas mark the progress *J. Phys. Conf. Ser.* **512** 012035
[3] Fernández D J and Mielnik B 1994 Controlling quantum motion *J. Math. Phys* **35** 2083
[4] Mielnik B and Ramirez A 2011 Magnetic operations: a little fuzzy physics? *Phys. Scr.* **84** 045008; arXiv:1006.1944
[5] Cruz y Cruz S and Mielnik B 2005 The parity phenomenon of the Floquet spectra *Phys. Lett.* A **352** 36
[6] Cruz y Cruz S and Mielnik B 2006 Quantum control with periodic sequences of non resonant pulses *Rev. Mex. Fis.* **S 53** 37
[7] Mielnik B 1977 Global mobility of Schrödinger's particle *Rep. Math. Phys.* **12** 331
[8] Mielnik B 1986 Evolution loops *J. Math. Phys.* **27** 2290
[9] Fernández D J and Rosas Ortiz O 1997 Inverse techniques and evolution of spin 1/2 systems *Phys. Lett.* A **236** 275
[10] Fernández D J and Rosas-Ortiz O 2001 Evolution loops and spin-1/2 systems in *Coherent States, Quantization and Gravity*; eds M. Schlichenmaier et al (Warsaw University Press, Poland) pp 81-88; arXiv:quant-ph/9910008
[11] Emmanoulidou A, et. al 2000 Steering an Eigenstate to a Destination *Phys. Rev. Lett.* **85** 1626
[12] Romero A 2015 *Sobre la geometría de estados cuánticos de la luz en un inteferómetro Mach-Zehnder* (M.Sc. Thesis (in Spanish), Physics Department, Cinvestav, Mexico)
[13] Jordan P 1935 Der Zusammenhang der symmetrischen und linearen Gruppen und das Mehrkörperproblem *Z. Phys.* **94** 531
[14] Schwinger J 1965 On angular momentum In *Quantum Theory of Angular Momentum* eds L C Biedenharn and H Van Dam (Academic Press, New York)
[15] Enríquez M and Rosas-Ortiz O 2013 The Kronecker product in terms of the Hubbard operators and the Clebsh-Gordan decomposition of $\mathcal{SU}(2) \times \mathcal{SU}(2)$ *Ann. Phys.* **339** 218; arXiv:1302.4797
[16] Yurke B, Mccall S L and Klauder J R 1986 SU(2) and SU(1,1) interferometers *Phys. Rev.* A **33** 4033
[17] Korolkova N, Leuchs G, Loudon R, Ralph T C and Silberhorn Ch 2002 Polarization squeezing and continuous-variable polarization entanglement *Phys. Rev.* A **65** 052306
[18] Leonhardt U 2003 Quantum physics of simple optical instruments *Rep. Prog. Phys.* **66** 1207
[19] Enríquez M, Reyes R and Rosas-Ortiz O 2014 A note on the geometry of the quantum states of a two-level atom in external radiation fields *J. Phys. Conf. Ser.* **512** 012021
[20] Rosas-Ortiz O 2003 Quantum control of two-level systems in *8th International Conference on Squeezed States and Uncertainty Relations*; eds H. Moya-Cessa et al (Rinton Press, USA) 360-365
[21] Aharonov Y and Anandan J 1987 Phase change during a cyclic quantum evolution *Phys. Rev. lett.* **58** 1593
[22] Fernández D J and Rosas-Ortiz O May 1997 in *Fifth International Conference on Squeezed States and Uncertainty Relations*, Balatonfured, Hungary; eds D Han, J Janszky, Y S Kim and V I Man'ko. NASA/CP-1998-206855, pp 583-588; arXiv:quant-ph/9706044
[23] Enríquez-Flores M and Rosas-Ortiz O 2010 Atomic Schrödinger cat-like states *AIP Conf. Proc.* **1287** 74

Journal of Physics: Conference Series **698** (2016) 012015 doi:10.1088/1742-6596/698/1/012015

Quantum Anatomy of the Classical Interference of n-Photon States in a Mach-Zehnder Interferometer

N Ramírez-Cruz[1], M A Bastarrachea-Magnani[2] and V Velázquez[1]

[1]Departamento de Física, Facultad de Ciencias, UNAM, A.P. 70-543, México, D.F., Mexico

[2]Instituto de Ciencias Nucleares, UNAM, A.P. 70-543, México, D.F., Mexico

E-mail: nramirez@ciencias.unam.mx, miguel.bastarrachea@nucleares.unam.mx, vicvela@ciencias.unam.mx

Abstract. In this work we present the theory for the quantum interference of states with an arbitrary number of photons in a Mach-Zehnder interferometer. We express the mathematical description of the interference in an algebraic way. We show the interference pattern of an average of a n photons input state corresponds to the classical interference pattern, which tells us the last comes from a quantum interference statistical average. Then, we propose to use this scheme to study the statistical transition from quantum to classical interference.

1. Introduction

The idea of classical interference is deep-rooted in our intuition. For example, the double-slit experiment made by Young with light [1] can be elegantly explained by thinking the light as a wave. However, when we consider a quantum description of the same experiment it is necessary to include photon interference. A study of the transition between quantum interference of photons and a wave description of this interference could help to better understand the transition between quantum and classical optics. The fact that a train of single photons can produce an interference pattern represents the main difference between both descriptions of the interference. Therefore, there is not much difference among them except for the auto-interference [2–4], a paradigm for quantum interference.

In this work we discuss the problem of recovering the classical interference of light from the quantum interference of $|n\rangle$ photon states in a Mach-Zehnder interferometer (MZI). First, we show there is no difference between the quantum and classical geometric patterns of interference due to the randomness of the photon output states $|n-k\rangle$ and $|k\rangle$, the combination produced by the beam splitters in the MZI. We will see that what we know as classical interference is no more than the statistical average of the different quantum interferences.

Finally, the output photon statistics can be selected 'on demand' among poissonian, sub-poissonian and super-poissonian quantum statistics as a function of the dephasing parameter.

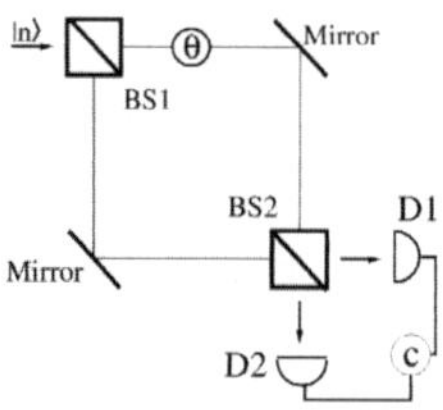

Figure 1. Schematics of a Mach-Zehnder interferometer. The n photons enter the left port through the first beam-splitter BS1.

2. n-photon interference in a Mach-Zehnder interferometer

The quantum interference's hearth lies in the state of a photon going through a beam-splitter. This state is represented by

$$|1\rangle \rightarrow \frac{1}{\sqrt{2}}(|1\rangle^T + i|1\rangle^R), \tag{1}$$

where T and R mean transmitted and reflected, respectively. We have selected the reflection phase as $\pi/2$. For n photons the quantum state is limited to the quantum components resultant of expanding a n-dimensional binomial,

$$|n\rangle \rightarrow \frac{1}{\sqrt{2^n}}(|1\rangle^T + i|1\rangle^R)^n. \tag{2}$$

If the transmitted and reflected states are redirected to a second beam-splitter, the expression for the normalized probability amplitudes of $n-k$ photons arriving to detector D_1 and k photons arriving to detector D_2 (Fig. 1) is

$$|n\rangle \rightarrow \frac{1}{(\sqrt{2})^n}\sum_{k=0}^{n} i^k \left(\frac{n!}{k!\,(n-k)!}\right)^{\frac{1}{2}} \left(e^{i\theta}-1\right)^{n-k}\left(e^{i\theta}+1\right)^{k} |n-k\rangle_{D1}|k\rangle_{D2}. \tag{3}$$

For $n = 1$ we recover the well-known case where $P_{D1} = \frac{1}{2}[1+\cos(\theta)]$ and $P_{D2} = \frac{1}{2}[1-\cos(\theta)]$ [5, 6]. See Figure 2. The photon itself interferes. The output photons arrive to the detectors with a sinusoidal probabilistic distribution.

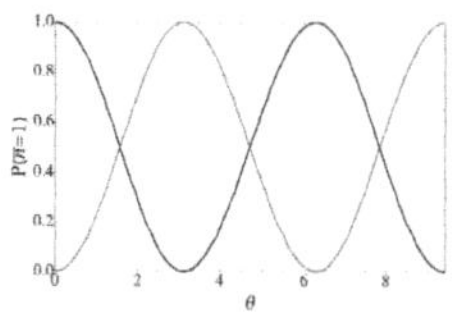

Figure 2. Interference of single photons: (Black) Probability of one photon on D1 an zero photons on D2 and (Red) probability of one photon on D1 an zero photons on D2.

This case represents a paradigm linking quantum and classic interference. A time series of single photon states $|1\rangle$, has been performed experimentally many times in a MZI [7,8], validating the shape of the interference. If we increase the photon number to the state $|2\rangle$, we note an important change in the probabilities. See Figure 3. Now, in the expansion of (3), the probability amplitude implies a cosine sum, whose arguments are θ and 2θ. For this reason, the probability amplitude has an asymmetry between the maximum and minimum among the components (2,0)

Journal of Physics: Conference Series **698** (2016) 012015 doi:10.1088/1742-6596/698/1/012015

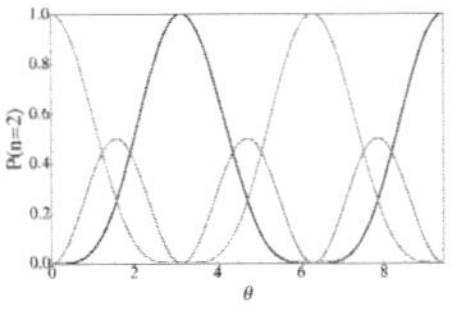

Figure 3. Interference of two photons: (Purple) Probability of two photons on D1 and zero photons on D2, (Black) probability of zero photons on D1 and two photons on D2 and (Red) probability of one photon on D1 and one photon on D2.

and (0,2). The minimum is more widespread than the maximum, which becomes narrower. Note that we draw the interference patterns for both ports.

Even though it is very complicated to obtain experimentally a quantum state with $n \geq 2$ [9], however, it is possible to validate the theory in the lower limit by using a two-photon source. In reference [8] a two-photon source coming from spontaneous parametric down-conversion (SPDC) it is used. Both photons are pumped collinearly into a MZI. In this way, the asymmetry of both states (2,0) and (0,2) can be tested. We can predict the interference pattern for quantum states with many photons, even if nowadays there are no sources which can produce those states. Figure 4 shows the probability distribution for a photon state $|10\rangle$ distributed in the different possible outcomes. As we can observe from the figure, the asymmetries between maximum and minimum increases as the photon number increases. The mixed probabilities for P_{n-k} and P_k outline an envelope which amplitude decreases with n i. e. the individual interference patterns from different combinations begin to form an envelope.

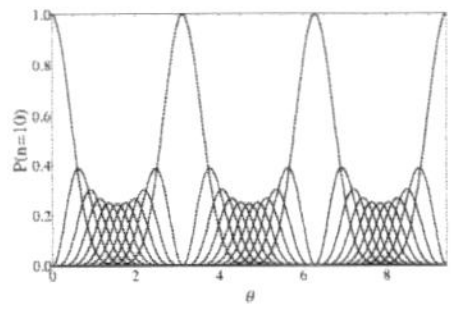

Figure 4. Interference of ten photons. It shows the probabilities of ten photons on D1 and zero photons on D2 and the combinations.

On the other hand, there are well-defined phase shifts where the state is conserved as fully quantum, $\theta = 0, \pi, 2\pi...$ in the detector D_1 direction and $\theta = \pi/2, 3\pi/2, 5\pi/2...$ in the detector D_2 direction. Outside these angles, the states lose the phase coherence rapidly building statistical mixes from the states with combinations that sum up ten photons. In order to appreciate better the fore mentioned effect, we calculate the interference of quantum states with one hundred photons at a time.

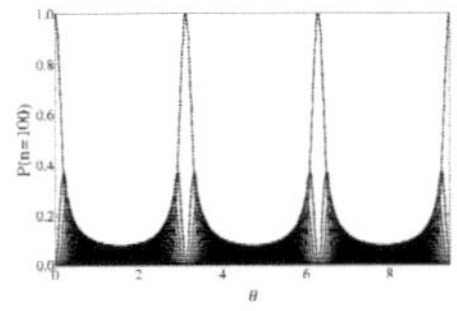

Figure 5. Interference of one hundred photons. It shows the probabilities of one hundred photons on D1 and zero photon on D2 and the combinations.

Figure 5 shows the probability distribution of finding $n = 100$ photons in both detectors, as a function of the dephasing angle θ. We can say the set of individual interference patterns should built up classical interference. We will show it in the next section.

3. Mixture of states and the the passage to classical interference

The 50 : 50 beam splitter is a random photons generator. Since the sum of photons in both outputs should be always the same number, a dephasing element in one arm can control the statistics between quantum and poisson behavior. In order to prove this, we calculate the expectation value for the photon number $N(\theta)$ in the D_1 outcome of the interferometer,

$$N_1(\theta) = \sum_{n_1'=1}^{n} n_1' P_{n_1'}(\theta), \qquad (4)$$

where $N_1(\theta)$ represents the photon number average in the port D_1 of the interferometer and $P_{n_1'}$ represents the probability of n_1' photons coming out from this port.

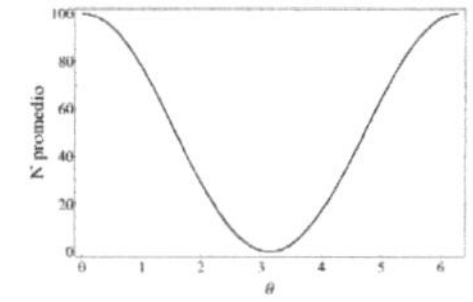

Figure 6. Average number of photons N_1 in the D_1 outcome of the IMZ. The pattern is fitted by the classic form $\frac{100}{2}[1+\cos(\theta)]$.

In Figure 6 we show the average photon numbers in the D_1 port as a function of θ. It is straightforward to recognize in this figure the classical interference pattern $P_{D_1} = \frac{N_1}{2}[1+\cos(\theta)]$, and $P_{D_2} = \frac{N_1}{2}[1-\cos(\theta)]$ [10]. Now, we discuss the meaning of the transition between a quantum state with n photons to a classical state with an average N of n photons in one of the interferometer's output. Evidently, the sum of the photons on both outcomes must be always n. In this part, we realized that the interference of a n-photon quantum state in the MZI depends on the interferometer's characteristics. First, the n photon train is forced to divide itself and reunite again randomly through de two option optical elements with the same probability. This creates two complementary interference patterns, each one composed of the statistical sum of the quantum interference patterns of the different probabilities. We can say this represents some kind of quantum decoherence originated in the geometry of the imposed trajectories of the quantum photon beam.

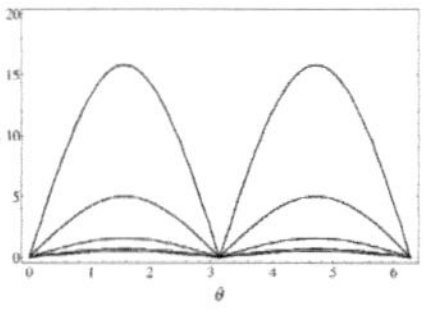

Figure 7. Widths as a function of n and θ for $n = 1$, 2, 10, 100 and 1000 photons, from small to bigger curves.

Now, we analyze the photon statistics. There is an average photon number and a standard deviation around this average, because for every dephasing angle, several probability curves of different amounts of photons converge. We show in Figure 7 the behavior of the standard deviation (distribution's width) $\sigma(n,\theta)$ as a function of the dephasing angle and the photon number [10, 11].

As it can be observed, the width is minimum for the angles where we have certainty of the photon counting and maximum for angles where the counting becomes random. If we normalize this quantity with the average photon square root, we find the graphics for every photon number

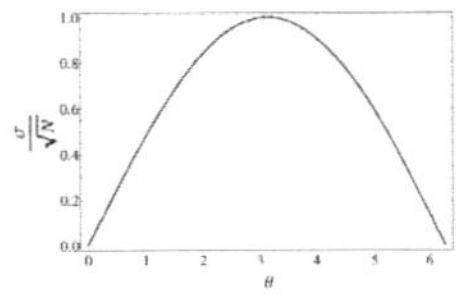

Figure 8. Photon statistics from $\sigma/\sqrt{N}$ as a function of θ.

have the same behaviour. Figure 8 shows the behavior of $F = \sigma/\sqrt{N}$. From the photon statistics theory we know if $F = 1$ the photon statistics is poissonian, in the other hand, if $F < 1$ the photon statistics is sub-poissonian.

The limit $F = 0$ corresponds to a pure quantum states $|n> |0>$ or $|0> |n>$, because for some values of θ the interferometer only has deviated the incoming photons to only one of the two outputs. The maximal unpredictability corresponds to the other limit $F = 1$, i. e. combination $|n-k>_{D_1} |k>_{D_2}$.

4. Discussion and Conclusions

We have obtained the quantum interference of the n-photon quantum states different combinations in a Mach-Zehnder interferometer. Given the randomly splitting of the quantum state in two states $|n-k\rangle|k\rangle$, the interference pattern is a statistical mixture with different levels of anticorrelation in terms of the dephasing parameter between both interferometer's arms. In order to contrast this result with a classical source's one, we know that if we introduce a coherent state $|\alpha\rangle$ in the same interferometer, we obtain as a result

$$P_{D1} = \frac{\bar{n}}{2}[1 + \cos(\theta)], \quad P_{D2} = \frac{\bar{n}}{2}[1 - \cos(\theta)], \tag{5}$$

where $\bar{n}$ represents $|\alpha|^2$, the average photon number of the coherent state [10]. In this case $F_\alpha(\theta) = 1$. The coherent state's statistics for any dephasing angle is constant and equal to the Poisson statistics. Then, an important result is that, even if the interference patters of quantum light and classical light look very similar, their punctual statistics are different. Evidently, both statistics agree on the point of maximum mixture of the quantum states. We can consider this analysis as an exercise of controlled decoherence, because the quantum states come and go as the interferometer's phase changes.

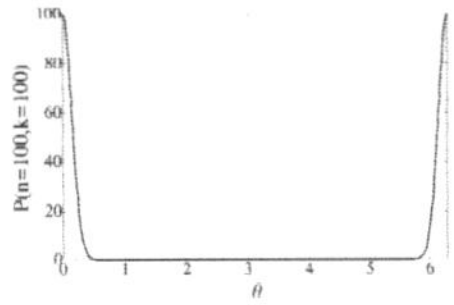

Figure 9. Quantum interference of the combination $|100\rangle|0\rangle$ as a function of θ.

The fact the quantum interference pattern is similar to the classical interference pattern tells us that we cannot control each of the multiple components of the interference at the same time, because that is what happens with the poissonian light. As we cannot control the statistics, we cannot control the interference. Here, we have another important result: to introduce quantum sates in a Mach-Zehnder interferometer produces a classical interference pattern (in both outcomes), unless we are capable of isolate some of the interference components of the combination $|n-k\rangle|k\rangle$. In Figure 9 it is shown the probability curve of finding at the MZI output the photon combination $|100\rangle|0\rangle$. As this one is not mixed with another combination, when we multiply it by 100 (the photon number at this output) it gives us directly the interference pattern for that individual combination. In this sense, this pattern represents one of the 101 quantum interference patterns. Experimentally it could be obtained through energy detectors with enough resolution (still under investigation [12]). Another option, as an state of the art, it could be an arrangement of beam splitters.

Finally, if we compare the interference patterns from Figs. 6 and 9, we can say that if there is a difference between the classical interference and the quantum interference, such difference grows with the photon number. The quantum interference has more spacial resolution, as it was expected.

In conclusion we study the interference of N photons in a IMZ. In order to determinate the output states of the interferometer we analyze the factor $F = \sigma/\sqrt{N}$, and we have quantum behaviour for some angles and a behaviour close to classic for others angles. That is, we can study whether it is possible to output states with some statistics on demand. On the other hand, we are interested in the transition from quantum to classical world. Our conclusion at this point is that classical interference is formed by the average of the quantum interference, this as seen in the average number of photons (Figure 6). This brings us to better understand this transition. The mixing of states is inevitable if we use classical detectors, but is possible to isolate some quantum quantum states using some arrays of beam splitters. This kind of isolated states can be so useful like NOON states [13–15], in quantum metrology.

Acknowledgments

We want to thank Raúl Viurquez for his comments and support to this contribution.

References

[1] Young Thomas 1804 *Philosophical Transactions of the Royal Society of London*, **94** 2.
[2] Geoffrey Ingram Taylor 1909 *Proc. Cam. Phil. Soc.* **15** 114.
[3] Dirac P 1930 *The Principles of Quantum Mechanics* (New York: Oxford University Press)
[4] Feynman R P, Leighton R B and Sands M 1965 *The Feynman Lectures on Physics Vol III* (MA: Addison-Wesley).
[5] Dunningham J and Kim T 2006 *Journal of Modern Optics* **53** 557-571.
[6] Rarity J, Larchuk T and Saleh B 1990 *Phys. Rev. Lett.* **65** 1348.
[7] Alain Aspect, Philippe Grangier and Gerard Roger 1981 *Phys. Rev. Lett.* **47** 7
[8] Galvez E, Holbrow C, Pysher M, Martin J, Courtemanche N, Heilig L and Spencer J 2005 *Am. J. Phys.* **73** 127-140.
[9] Walther P et al., 2004 *Nature* **429**, 158161.
[10] Gerry C and Knight P 2004 *Introductory Quantum Optics*, (New York: Cambridge University Press).
[11] Fox Mark 2006 *Quantum Optics, An Introduction* (New York: Oxford University Press)
[12] Natarajan C, Tanner M, and Hadfield R 2012 *Superconductor Science and Technology* **25**.
[13] Mitchell M, Lundeen J and Steinberg A 2004 *Nature* **161** 429.
[14] Dowling J 2008 *Contemp. Phys.* **49** 125.
[15] Afek I, Ambar O and Silberberg Y 2010 *Science* **328** 879.

Quantum Fest 2015 IOP Publishing
Journal of Physics: Conference Series **698** (2016) 012016 doi:10.1088/1742-6596/698/1/012016

Generation of Sources of Light with Well Defined Orbital Angular Momentum

S Cruz y Cruz[1], N Escamilla[1] and V Velázquez[2]

[1]Unidad Profesional Interdisciplinaria en Ingeniería y Tecnologías Avanzadas del Instituto Politécnico Nacional, Av. Instituto Politécnico Nacional 2580, Barrio la Laguna Ticomán, Gustavo A. Madero, 07340, Ciudad de México, Mexico.
[2]Facultad de Ciencias, Universidad Nacional Autónoma de México, Apartado Postal 70-543, 04510 Ciudad de México, Mexico.

E-mail: sgcruzc@ipn.mx,nescamillab0400@alumno.ipn.mx,vicvela@ciencias.unam.mx

Abstract. In this work, a technique to produce spatial electromagnetic modes with definite orbital angular momentum is presented. The method is based in the construction of binary diffractive gratings generated by computer. In the classical regime the gratings produce the well known Laguerre-Gaussian modes distributions when illuminated by a plane wave. In the quantum regime the grating is placed in the signal path of a spontaneous parametric down conversion layout and the diffraction pattern, observed in the coincidence count rate, shows that the single photons are projected onto spatial states consistent with a Laguerre-Gaussian modes distribution.

1. Introduction

Quantum engineering is a branch of science that emerged recently. Its objective is to understand how to improve the manipulation and control of physical systems for the development of new technologies approaching the quantum regime. Of the infinity of processes that occur at the quantum level, the interaction of radiation with matter is of fundamental importance. This is why the interest in understanding the properties of radiation and how to apply them in handling quantum systems has been growing in recent years. One of such properties is the ability to transmit energy and momentum to material systems at a microscopic scale. The energy and linear momentum of radiation are well studied concepts. However, a complete analysis of the optical orbital angular momentum and its potential applications is still under development. The concept itself has gained much interest since its discovery in 1992 [1–3] due to the fact that beams with well defined orbital angular momentum can be easily produced in practice by means of different methods [4–8]. At the same time, the extraordinary perspectives of application of their properties in the manipulation and trapping of matter, quantum information and photonics were immediately evident [9–12]. Experimental studies involving electromagnetic orbital angular momentum in nonlinear optics began shortly in the generation of frequency doubling [13, 14], where the phase matching conditions for helical beams proven to be the same as for plane waves. A similar phenomenon occurs in more general cases of frequency mixing [15, 16]. Since then, too much interest has been paid to the study of the conservation of the orbital angular momentum, spatial correlation and entanglement, of single photons produced by spontaneous parametric down conversion [17–24]. Studies of these phenomena intends to give a better understanding of the transfer of angular momentum between matter and light at the quantum level. The purpose of this work is to present a procedure to generate sources of light in definite orbital angular momentum states in the classical as well as in the quantum regime. In section 2, the Laguerre-Gaussian modes are considered, from the theoretical point of view, as solutions of the paraxial Helmholtz equation with definite orbital angular momentum. In section 3, the procedure used to generate Laguerre-Gaussian modes distributions is presented. Section 4 contains the experimental setup used to generate single photons in definite spatial states consistent with a Laguerre-Gaussian modes distribution. A spontaneous parametric down conversion (SPDC) layout is used as the source of single photons. The article closes with some conclusions.

2. Orbital angular momentum of light

The fundamental Gaussian beams of light are characterized by their plane wavefronts and Gaussian transverse intensity distributions (see Figure 1). For those modes the linear momentum density $\epsilon_0 \mathbf{E} \times \mathbf{B}$ points along the direction of propagation, meaning that there is no component of the angular momentum in that direction.

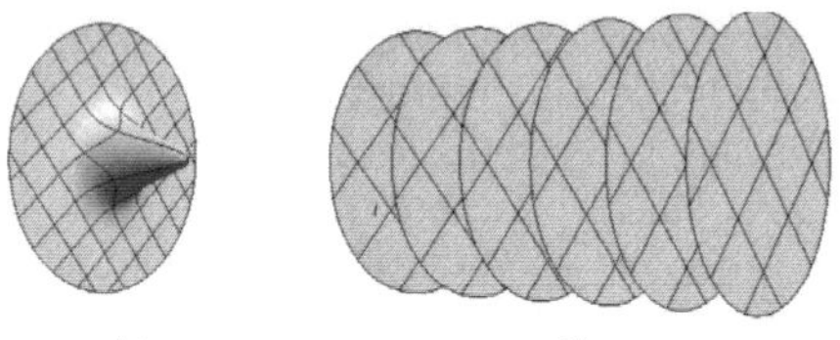

(a) (b)

Figure 1. The Gaussian transverse intensity distribution (a), and the plane wavefront (b) for a fundamental Gaussian mode.

For higher-order modes, as *e.g.* Laguerre-Gaussian beams, the points of constant phase are delayed with respect to each other and the wavefronts are no longer plane surfaces [25]. The transverse intensity distributions in those cases have also complex structures (see Figure 2).

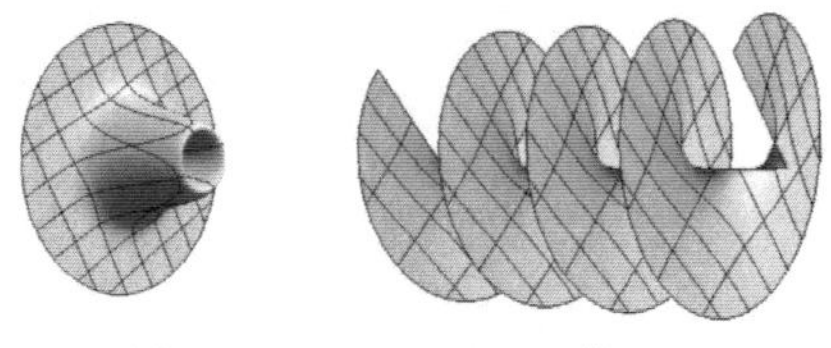

(a) (b)

Figure 2. The "annular" profile of the transverse intensity distribution (a) and the helical wavefront (b) for a Laguerre-Gaussian mode with orbital angular momentum $\ell = 1$.

These spatial modes are predicted by the Maxwell wave description of light as solutions of the wave equation in the paraxial approximation [26]. In particular, the cylindrically symmetric solutions known as the Laguerre-Gaussian modes have the form [27]

$$E(\mathbf{r}) = \frac{\mathbf{E}_0}{\sqrt{z^2 + z_R^2}} \left(\frac{2r^2}{\omega^2(z)} \right)^{\frac{|\ell|}{2}} L_n^{(|\ell|)} \left(\frac{2r^2}{\omega^2(z)} \right) e^{-\frac{ikr^2}{2R(z)}} e^{-\frac{r^2}{\omega^2(z)}} e^{i(2n+|\ell|+1)\arctan z/z_R} e^{-i(kz+\ell\theta)}, \quad (1)$$

where $\mathbf{E}_0$ is a constant vector, ℓ is an integer number, z_R is the Rayleigh range, $\omega(z)$ is the transverse width of the beam (assuming that the waist is placed at $z = 0$), $R(z)$ is the radius of curvature of the wavefront and $L_n^{(|\ell|)}$ is the associated Laguerre polynomial [27]. These modes form a complete set of orthogonal solutions to the paraxial Helmholtz equation. The phase $\ell\theta$ is responsible for the helical shape of the wavefronts (see Figure 2) and is directly related to the orbital angular momentum of the beam.

Considering the local angular momentum density of the electromagnetic field [28]

$$\mathbf{l} = \epsilon_0 \mathbf{r} \times (\mathbf{E} \times \mathbf{B}), \tag{2}$$

and using the expression (1), it is possible to show that the ratio of orbital angular momentum to energy in the direction of propagation is ℓ/ω [29], suggesting that the Laguerre-Gaussian modes posses definite orbital angular momentum proportional to ℓ. In this way, the integer number ℓ determines the number of sheets in the helical surface of the wavefront and the amount of orbital angular momentum of each axial mode of the electromagnetic field.

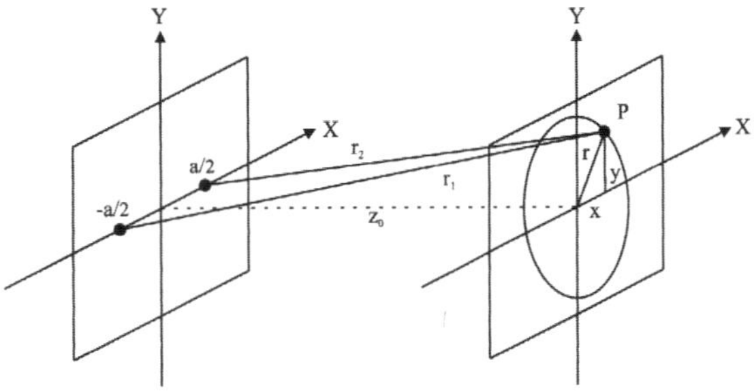

Figure 3. Geometry of two beams interference.

3. Production of light with orbital angular momentum

Different techniques have been developed in order to generate structured beams in practice. For example, it is possible to transform Hermite-Gaussian beams into Laguerre-Gaussian ones using a set of two cylindrical lenses called mode converter [1, 7, 8]. The converting process is based on the decomposition of a Laguerre-Gaussian and a diagonally oriented Hermite-Gaussian modes into Hermite-Gaussian ones. By changing the separation of the set of lenses it is possible to revert the helicity of the resulting beam. However, the most common technique in the production of helical wavefronts with any value of orbital angular momentum is the use of diffractive gratings generated by computer [4–6]. In this case the interference pattern of a plane wave and the mode to be generated is recorded in a photographic film. The resulting grid has an $\ell-$fold dislocation (a fork) at the center of the pattern. When a plane wave illuminates this pattern, the diffracted beam acquire the desired distributions of phase and intensity. Spatial light modulators (SLMs) have also been used to generate beams of a wide range of phase and intensity distributions [10, 23]. They consist in pixelated liquid crystal devices instead of the photographic film, in which different holographic patterns can be displayed.

In this work the Laguerre-Gaussian beams were produced by means of binary diffraction gratings. In order to make the design of these holograms, the interference process of a plane wave and a Laguerre-Gaussian beam of the same frequency and amplitude is considered. The total amplitude of the electric field at a point P in the far field is (see Figure 3)

$$E(\mathbf{r}, t) = E_0 e^{i(kr_1 - \omega t)} + E_0 e^{i(kr_2 + \ell\theta - \omega t)}, \tag{3}$$

where the first term is the plane wave and the second one the Laguerre-Gaussian mode. Note the additional term in the phase of this field corresponding to the phase angle that characterizes the helical beams. The total irradiance of the electromagnetic field at point P is given by

$$I(\mathbf{r}) = |E(\mathbf{r})|^2 = 2|E_0|^2(1 + \cos[k(r_1 - r_2) - \ell\theta]). \tag{4}$$

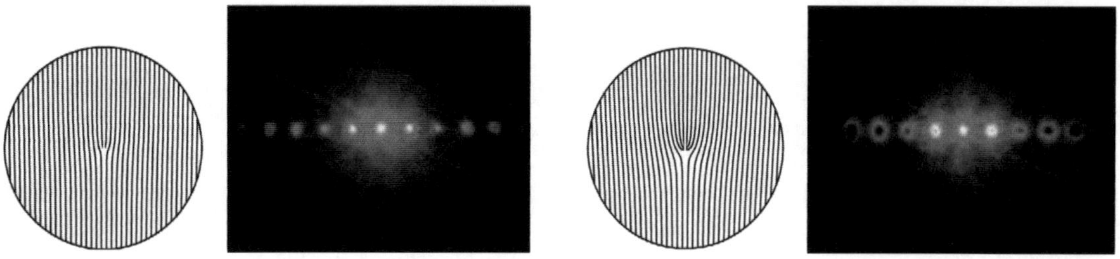

(a) (b)

Figure 4. Fork diffractive gratings with (a) $s = 2$ and (b) $s = 6$ dislocations (left) and their corresponding Laguerre-Gaussian modes intensity distributions (right).

Assuming that r_1 and r_2 are much greater than the separation a between the sources one obtains (Figure 3)

$$I(\mathbf{r}) = 2|E_0|^2 \left(1 + \cos\left[\frac{ka}{z_0} r\cos\theta - \ell\theta\right]\right). \tag{5}$$

Next, it is assumed that the boundaries between the clear and dark fringes of the interference pattern are found whenever the argument of the periodic term of (5) crosses integer multiples of π, *i.e.*

$$\ell\theta - \frac{ka}{z_0} r\cos\theta = m\pi, \quad m = 0, \pm1, \pm2, \ldots. \tag{6}$$

Observe that the distance between two clear (or dark) fringes becomes $\Lambda = 2z_0/ka$ as $r \to \infty$. Thus, the algebraic relation giving the desired grating turn out to be

$$\frac{s\theta}{\pi} = m + \frac{2r}{\Lambda}\cos\theta, \qquad s = 1, 2, \ldots \tag{7}$$

where (r, θ) are the polar coordinates, s is the number of dislocations in the grid, m is an integer number and Λ is the period of the grid at large distances from the dislocation [6, 9]. By assigning different values to these parameters one may generate diverse grids that produce particular spacial light distributions. In Figure 4 we present two diffractive gratings constructed with $s = 2$ (a) and $s = 6$ (b). As they were illuminated by a plane wave, the well known distributions of Laguerre-Gaussian modes were observed. In order to evaluate the vorticity of these modes we use the self-interference method [1], which consists in superposing two modes of the same charge ℓ and opposite handedness. Then, an interference pattern with $N = 2\ell$ regions of constructive interference can be observed. In Figure 5 we show the interference pattern of the Laguerre-Gaussian mode distribution generated by a grating with charge $s = 2$ superposed with itself but rotated πrad. We can observe a central mode of order 0, a second order mode composed by a distribution of four maxima, a third order one with eight maxima and so on. This means that the corresponding Laguerre-Gaussian modes are associated to $\ell = 0, \pm2, \pm4, \ldots$ Similar experiments with different spatial distributions allow to conclude that Laguerre-Gaussian modes associated to $\ell = sn$, $n = 0, 1, 2, \ldots$ can be generated by means of grids with $s-$fold dislocations.

Figure 5. Interference pattern of the Laguerre-Gaussian mode distribution, generated by a grating with charge $s = 2$, superposed with itself but rotated πrad. The regions of constructive interference lead to the conclusion that this grids produce Laguerre-Gaussian modes with $\ell = 0, \pm 2, \pm 4, \ldots$

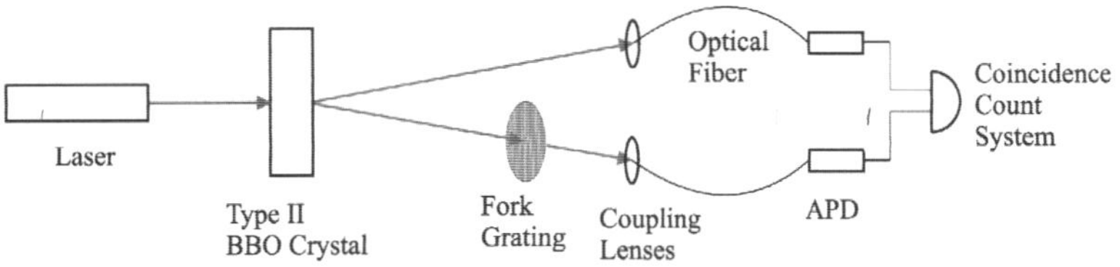

Figure 6. General outline of the setup used to observe the spatial correlation of down converted photons with orbital angular momentum.

4. Spatial correlation of light with orbital angular momentum

In this section the experimental setup used to test the generation of light with definite orbital angular momentum in the quantum regime is presented. A SPDC layout is used as a source of single photons. In our setup the pump beam is generated by a violet diode laser, of wavelength $\lambda = 405nm$ and a bandwidth of $0.78nm$, with horizontal polarization. A type I BBO crystal cut at 30° with respect to the optical axis and a semi-cone angle of 5° is placed to $13cm$ from the laser output as shown in Figure 6. A beam with a spot of transversal diameter $1.5mm$ and power of $100mW$ is incident on the BBO crystal. Two light collectors are placed at a distance of $80cm$ from the crystal, on the signal and idler arms of the SPDC setup. The collectors consist on coupling lenses connected to optical fiber to send the down converted light to the detection system. For the measurement of count rates, two avalanche photodiodes (APD), with a detection efficiency for $810nm$ single photons of about 60%, are used. When the APDs detect a photon, a $3.5V$ TTL pulse of $15ns$ is triggered. Then the coincidence system detects either individual or coincidence counts.

A diffractive grating to produce Laguerre-Gaussian modes with $\ell = 0, \pm 2, \pm 4, \ldots$ (see Figure 4(a)) is set on the signal arm of the SPDC source at a distance of $27cm$ from the BBO crystal. An He-Ne laser of wavelength $633nm$ is used to track the course of signal photons in order to align the grid. The first collector is then placed at the position of maximum counts of idler photons while the second collector make an horizontal scanning on the signal arm searching for coincidences. The aim of each idler photon is to localize in both, time and position, the corresponding signal photon. In Figure 7 it is shown the diffraction pattern observed in coincidence count rate.

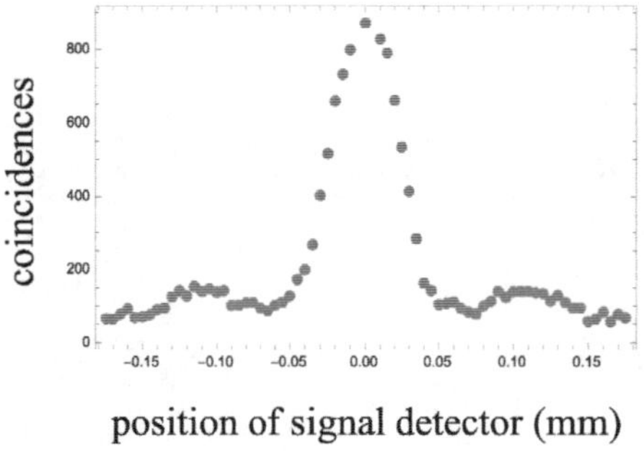

Figure 7. Coincidence count rate between the idler and signal photons as a function of the signal detector horizontal position.

Observe that this distribution shows that the signal photons are projected onto spacial states consistent with the three lowest orders of a Laguerre-Gaussian modes distribution, suggesting that they posses definite orbital angular momentum.

5. Concluding remarks

A technique for the generation of Laguerre-Gaussian modes with different values of angular momentum was established. This technique consist in the design of diverse types of diffractive gratings by means of the interference pattern of a plane wave and different Laguerre-Gaussian modes. The angular momentum of the resulting modes was determined by observing the interference pattern of two modes with opposite helicity. The designed holograms shown to produce helical modes with orbital angular momentum proportional to the number of dislocations in the grids. A spontaneous parametric down conversion source was used to produce single photons and the coincidence count rate diffraction pattern, when a fork grid is placed in the signal path was observed. The experiment allowed to conclude that the parametric down converted photons were projected onto spacial states consistent a Laguerre-Gaussian modes distribution.

Acknowledgments

This work was supported by the Projects SIP-20160527 (Instituto Politécnico Nacional, México) and PAPIIT-UNAM 116914. The authors thank the the anonymous referee for his comments that improved the presentation of this paper, and also to Gerardo Fernández and Mathieu Hautefeuille for their invaluable help in the experimental aspects of this work.

References

[1] Allen L, Beijersbergen M W, Spreeuw R J C and Woerdman J P 1992 Orbital angular momentum of light and the transformation of Laguerre-Gaussian laser modes *Phys. Rev.* A **45** 8185-8189
[2] van Enk S J and Nienhius G 1992 Eigenfunciton description of laser beams and orbital angular momentum of light *Opt. Commun.* **94** 147-158
[3] Beijersbergen M W, Allen L, van der Veen H and Woerdman J P 1993 Asigmatic laser mode converters and transfer of orbital angular momentum *Opt. Commun.* **96** 123-132
[4] Bazhenov V Yu, Vasnetsov M V and Soskin M S 1990 Laser beams with screw dislocations in their wavefronts *JETP Lett.* **52** 429-431
[5] Heckenberg N R, McDuff R, Smith C P, Rubinsztein-Dunlop R and Wegener M J 1992 Laser beams with phase singularities, *Opt. Quant. Electronics* **24** S951-S962
[6] Arlt J, Dholakia K Allen L and Padgett M J 1998 The production of multiringed Laguerre-Gaussian modes by computer-generated holograms *J. Mod. Opt.* **45** 1231-1237.
[7] Padgett M, Courtial J and Allen L 2004 Light's orbital angular momentum *Phys. Today* **57** 35-40

[8] Yao A M and Padgett M 2011 Orbital angular momentum: origins behavior and applications *Adv. Opt. Photonics* **3** 161-204
[9] Andrews L D 2008 *Structured light and its applications: an introduction to phase-structured beams and nanoscale optical forces* (USA: Academic Press)
[10] Grier D G 2003 A revolution in optical manipulation *Nature* **424** 810-816
[11] Leach J, Dennis M, Courtial J and Padgett M2004 Laser beams: knotted threads of darkness *Nature* **432** 165
[12] Gibson G, Courtial J and Padgett M J 2004 Free-space information transfer using light beams carrying orbital angular momentum *Optics Express* **12** 5448-5456
[13] Dholakia K, Simson N, Padgett M and Allen L 1996 Second-harmonic generation and the orbital angular momentum of light *Phys. Rev.* A **54** R3742-R3745
[14] Courtial J, Dholakia K, Allen L and Padgett M 1997 Second-harmonic generation and the conservation of orbital angular momentum with high-order LaguerreGaussian modes *Phys. Rev.* A **56** 41934196
[15] Grice W P and Walmsley I A 1997 Spectral information and dintinguishability in type-II down-conversion with a broadband pump *Phys. Rev.* A **56** 1627-634
[16] Beranskis A, Matijoius A, Piskarskas A, Smilgeviius V and Stabinis A 1997 Conversion of topological charge of optical vortices in a parametric frequency converter *Opt. Commun.* **140** 273276
[17] Arlt J, Dholakia K, Allen L and Padgett M J 1999 Parametric down-conversion for light beams possesing orbital angular momentum *Phys. Rev.* A **59** 3950-3952.
[18] Chunqing G A O, Guanghi W E I and Weber H 2000 Orbital angular momentum of the laser beam and the second order intensity moments *Science in China* **43** 1306-1311
[19] Mair A, Vaziri A, Weihs G and Zeilinger A 2001 Entanglement of the orbital angular momentum states of photons *Nature*, **412** 3123-316
[20] Franke-Arnold S, Barnett S M, Padgett M J and Allen L 2002 Two-photon entanglement of orbital angular momentum states *Phys. Rev.* A **65** 033823 5 pages
[21] Barbosa G A and Arnaut H H 2002 Twin photons with angular momentum entanglement: phase matching *Phys. Rev.* A **65** 053801 6 pages
[22] Huguenin J A O, Martinelli M, Caetanos D P, Coutinho dos Santos B, Almeidas M P, Souto Ribeiros P H, Nussenzveig P and Khoury A Z 2006 Orbital angular momentum exchange in parametric down conversion *J. of Mod. Opt.* **53** 647-658
[23] Yao E, Franke-Arnold S, Courtial J and Padgett M J 2006 Observation of quantum entanglement using spatial light modulators *Optics Express* **14** 1389-13094
[24] Osorio C I, Molina-Terriza C and Torres J P 2008 Correlations in orbital angular momentum of spatially entangled paired photons generated parametric down-conversion *Phys. Rev.* A **77** 015810 4 pages
[25] Galvez E J 2006 Gaussian beams in the optics course *Am. J. Phys.* **74** 355-361
[26] Bandres M A and Gutierrez-Vega J C 2008 Circular beams *Opt. Let.* **33** 177-179
[27] Siegman A E 1986 *Lasers* (USA: University Science Books)
[28] Jackson J D 1962 *Classical electrodynamics* (USA: Wiley)
[29] Allen L, Padgett J M and Babiker M 1999 The orbital angular momentum of light *in E. Wolf, Progress in Optics XXXIX* Elsevier Science B. V. 291-372

Propagation of Ultra-Short Higher-Order Solitons in a Photonic Crystal Fiber

D Bermudez

Departamento de Física, Cinvestav, A.P. 14-740, 07000 México D.F., Mexico

E-mail: dbermudez@fis.cinvestav.mx

Abstract. We study the propagation of ultra-short optical pulses (~10 fs) traveling through a photonic crystal fiber. For this, we develop a realistic model of the fiber dispersion by numerical solution and a rational fit. Then, we model the laser pulses as higher-order solitons, and we show that they present rich dynamics along their propagation inside a fiber from where the steepest pulses are of great interest in the search of optical analogues of Hawking radiation.

1. Introduction

Optical solitons are localized waves that are formed by the opposite effects of nonlinearity and dispersion in optical fibers. The understanding of such pulses has come through theoretical studies and numerical simulations mainly from the nonlinear Schrödinger equation (NLSE) [1,2], which has been successful at explaining some features of nonlinear pulse dynamics, including the emission of dispersive waves [3], soliton fission [4, 5] and supercontinuum generation [6]. Nevertheless, if we consider ultra-short pulses (~10-100 fs in the visible spectrum), several higher-order nonlinear and dispersive effects come into play and the approximations used to derive the NLSE are not valid anymore [7]. Additionally, if we consider higher-order solitons, the nonlinear effects take further importance, specially the soliton fission, which immediately brakes down the pulses by emission of dispersive waves (DW), also known as non-solitonic radiation (NSR) or Cherenkov radiation [3].

On the other hand, photonic crystal fibers (PCFs) or microstructured fibers, have a solid core surrounded by a regular array of air holes in the propagation direction that changes the dispersion relation of the fiber core. These holes can be fabricated with certain degree of freedom to engineer this relation beyond material dispersion [7–9]. The PCFs made of silica have been very successful at shedding *supercontinuum* light that has found a wide range of applications [6].

Another feature of the PCFs is that the ability to support a wide range of frequencies opens the possibility of creating very steep pulses, close to the single-cycle regime. These pulses are useful in the search of optical analogues of Hawking radiation because the effective temperature and the emission rate depend crucially on the rate of change of the refractive index due to the Kerr effect, which ultimately depends on the steepness of the pulses. See, e.g., Refs. [10, 11].

In this work, we will study the propagation of ultra-short higher-order solutions inside PCFs. In Section 2 we will present the equations that model the pulse propagation. Then, in Section 3 we will study the PCFs and obtain a rational approximation for its dispersion. In Section 4 we will write the definitions concerning higher-order solitons. Then, in Section 5 we will be able to present a study of the propagation by analyzing four slightly different pulses with three degrees of approximation in the dispersion. Finally, in Section 6 we will present our conclusions.

2. Model equations

We consider a single-mode fiber using the scalar approximation for pulses, i.e., the field structure in the radial direction is fixed and the signal $E(z,t)$ only changes in the propagation direction z. We describe the system in the co-moving frame $\tau = t - \beta_1 z$, where β_1 is the inverse of the group velocity v_g. The basic model for the pulse propagation in fibers is given by the NLSE:

$$i\partial_z \mathcal{E} + \frac{\beta_2}{2}(i\partial_\tau)^2 \mathcal{E} + \gamma |\mathcal{E}|^2 \mathcal{E} = 0, \tag{1}$$

where we have introduced the analytic signal $\mathcal{E}$ of the electric field $E = \mathrm{Re}(\mathcal{E})$, β_2 is the group velocity dispersion (GVD), γ is proportional to the third-order nonlinear susceptibility [1]. Eq. (1) is also known as cubic Schrödinger equation or Gross-Pitaevskii equation.

Nevertheless, we are interested in describing the pulse dynamics in a real fiber, in order to do that, we have to include higher-order dispersion terms, i.e., we have to replace the second term by the full-dispersion operator, usually written as a Taylor expansion as:

$$\mathcal{D}(i\partial_\tau)\mathcal{E} = \sum_{m=2}^{M} \frac{\beta_m}{m!}(i\partial_\tau)^m \mathcal{E}. \tag{2}$$

In this case, the pulse carrier frequency is shifted and not well-defined. Furthermore, it is convenient to consider the so-called *self-steepening effect* and the nonlocal terms or, more generally, we can include the dispersion inside the nonlinear term of the NLSE:

$$i\partial_z \mathcal{E} + \mathcal{D}(i\partial_\tau)\mathcal{E} + \gamma \frac{\beta_0}{\omega_0^2} \frac{(i\partial_\tau)^2}{\beta(i\partial_\tau)} |\mathcal{E}|^2 \mathcal{E} = 0. \tag{3}$$

We can study this equation in the Fourier-transformed space of frequencies ω, i.e.,

$$i\partial_z \mathcal{E}_\omega + \mathcal{D}(\omega)\mathcal{E}_\omega + \gamma \frac{\beta_0}{\omega_0^2} \frac{\omega^2}{\beta(\omega)} (|\mathcal{E}|^2 \mathcal{E})_\omega = 0, \tag{4}$$

where $\mathcal{E}_\omega$ and $(|\mathcal{E}|^2\mathcal{E})_\omega$ are signals in the Fourier space. It is not clear if we can ignore the effects of the stimulated Raman scattering (SRS) even if we are using ultra-short pulses because we are interested in the temporal shape of the pulse and not in its spectrum. Therefore, we included it in Fourier space by replacing:

$$(|\mathcal{E}|^2\mathcal{E})_\omega \to \mathcal{E}_\omega \int_{-\infty}^{\tau} R(\tau - \tau')|\mathcal{E}_\omega|^2 \mathrm{d}\tau' = 0, \tag{5}$$

where $R(\tau)$ is called the *response function*, which comes from delayed material vibrations caused by the optical field. In the case of silica fibers, an analytic approximation for this function has been recently developed for ultra-short pulses [1, 12]:

$$R(\tau) = (1 - f_2 - f_3)\delta(\tau) + f_2 \frac{\tau_1^2 + \tau_2^2}{\tau_1 \tau_2^2} \exp(-\tau/\tau_2)\sin(\tau/\tau_1) + f_3 \frac{2\tau_3 - \tau}{\tau_3^2} \exp(-\tau/\tau_3), \tag{6}$$

where $f_2 = 0.194$ is the fractional contribution of the delayed Raman response to the nonlinear polarization and $f_3 = 0.051$ is the fractional contribution of a boson peak, $\tau_1 = 12.2\,\mathrm{fs}$, $\tau_2 = 32\,\mathrm{fs}$ and $\tau_3 = 96\,\mathrm{fs}$ are time parameters to fit $R(\tau)$ with experimental data.

In the case of ultra-short pulses propagating in PCFs, the usual approximation for the dispersion given by Eq. (1) is not enough but some generalization is needed. Usually, the model given by Eqs. (4-6) works well for ultra-short pulses, but it depends critically on the shape of the dispersion $\beta(\omega)$. In the next section we will address this issue.

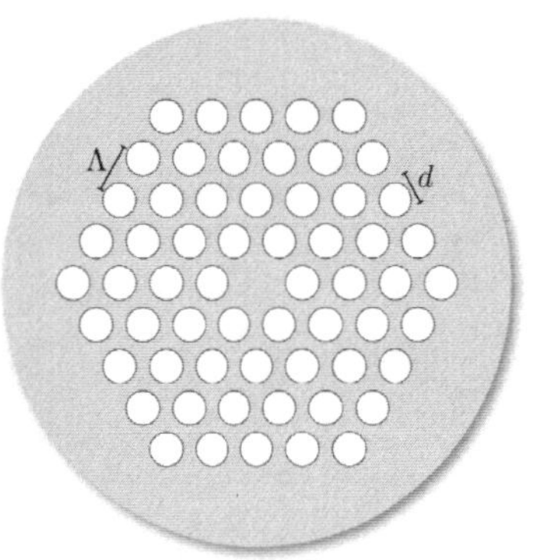

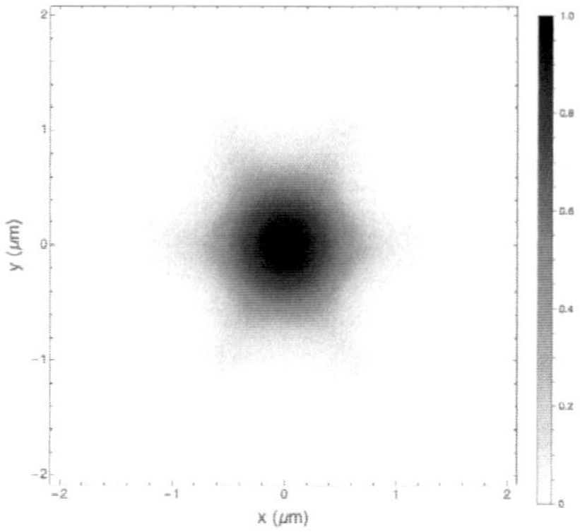

Figure 1. (Color online.) Left: Diagram of a typical PCF with hexagonal geometry. Gray represents the fiber material (silica) and the white circles are air holes. Right: The transversal mode in λ =800 nm, most of the energy goes inside the solid core, air-holes affect more longer wavelengths.

3. Dispersion of the photonic crystal fiber

One difficulty in the study of ultra-short pulses in real materials is the usual approximation of its dispersion relation with few terms of its Taylor expansion around a central frequency. The reason is that in this case, the width of the involved frequencies is comparable with the central frequency and, therefore, this approximation becomes invalid. Furthermore, this cannot be solved by taking more terms of the dispersion relation (see Amiranishvili et al. [2, 13]). Here we take a different approach, i.e., instead of the polynomial expansion we consider a rational one as it was done in Ref. [13], but in our case, we approximate the *phase index* $n_p(\omega)$ and our fit function is:

$$n_p(\lambda) = \frac{\lambda}{\lambda^2 - \lambda_r^2} \frac{\sum_{k=0}^{4} a_k \lambda^k}{\sum_{k=0}^{4} b_k \lambda^k}, \tag{7}$$

where $\lambda_r = 116.2414\,\text{nm}$ is a known resonance for silica and the ten coefficients a_k and b_k are the fitting parameters, which should have the correct units in order to obtain a dimensionless quantity n_p. We choose this model because the resonances of $n_p(\lambda)$ are known experimentally with high precision, which helps fixing the fit function in Eq. (7). Then we can obtain $\beta(\omega)$ for Eq. (4) through

$$\beta(\omega) = \frac{\omega n_p(\omega)}{c}, \quad \omega = \frac{2\pi c}{\lambda}. \tag{8}$$

The PCFs that generate the shortest optical pulses in the visible spectrum are the solid-core silica fibers with hexagonal geometry, see Fig. 1. For this work, we studied a commercial fiber from NKT Photonics: NL-1.5-590 [14], which is advertised to have zero dispersion wavelengths (ZDWs) around 590 nm and 1410 nm.

We obtain a numerical solution of $n_p(\lambda)$ through finite-element method using a FemSIM, a commercial software [15]. We obtain a very accurate result based on the material dispersion for air and silica (which are known with very high precision) and the two geometrical parameters of a PCF: the pitch Λ and the diameter d of the air holes (see Fig. 1). These two parameters were obtained through an electronic microscope photography of the fiber as Λ =1079 nm, d =763 nm. In the right-hand side of Fig. 1 we plot the amplitude of the electric field of the fundamental mode with $\lambda = 800\,\text{nm}$, where we can see that most of its energy travels inside the core, as needed in order to use the scalar approximation in Section 2. For longer wavelengths, the amount of energy traveling in the region with air holes increases, therefore, the dispersion is different from the material dispersion of silica in the infrared region. In Fig. 2 we plot the phase and group

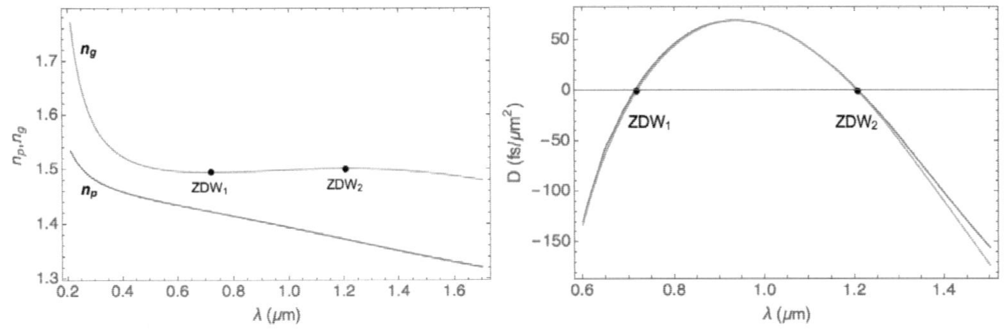

Figure 2. (Color online.) Left: Phase (blue) and group (orange) indexes obtained by our numerical solution using Eq. (7). Right: Dispersion function D from Eq. (9) (orange) compared with the one a function reconstructed from the data in Ref. [8] (blue). The ZDWs for our solution are marked in both plots.

indexes that we obtained numerically and the related function D, defined by

$$D(\lambda) = \frac{\partial^2 \beta}{\partial \lambda \partial \omega}, \tag{9}$$

(not to be confused with the full dispersion $\mathcal{D}$ in Eq. (2)). The ZDWs of our fit function are in 713 nm and 1207 nm, different form the advertised values. Between these two points the fiber presents anomalous GVD. The full dispersion data is not available from the company but it was also found by numerical means in Ref. [8], we compare it with our result in Fig. 2.

4. Higher order solitons

For a pulse in the anomalous GVD region of the spectrum, the nonlinearity can counter the dispersive effects, resulting in optical solitons or *solitary waves*. For the NLSE with second-order dispersion given in Eq. (1) there are localized solutions that either do not change (fundamental soliton) or change periodically (higher-order solitons) along the fiber. They are given by:

$$\mathcal{E}(\tau) = N\mathcal{E}_{\text{FS}} \operatorname{sech}(\tau/\tau_0) \exp(-i\omega_0\tau) \tag{10}$$

where $\mathcal{E}_{\text{FS}}$ is the amplitude of the fundamental soliton and N is known as the *soliton number*:

$$\mathcal{E}_{\text{FS}} = \sqrt{\frac{1}{\gamma L_{\text{D}}}}, \quad N^2 = \frac{L_{\text{D}}}{L_{\text{NL}}}, \quad L_{\text{D}} = \frac{\tau_0}{|\beta_2|}, \quad L_{\text{NL}} = \frac{1}{\gamma P_0}, \quad \tau_0 = \frac{\tau_{\text{FWHM}}}{2\log(1+\sqrt{2})}. \tag{11}$$

L_{D} and L_{NL} are the dispersion and nonlinear lengths and τ_{FWHM} is the time for full-width at half-maximum (FWHM), the most common way to measure time duration of optical pulses.

For $N = 1$ we have the fundamental soliton, whose shape does not change over propagation if we approximate dispersion given only by β_2 factor. Eq. (10) for $N = 1$ fixes a relation between duration τ_0 and amplitude $\sqrt{P_0}$. Keeping τ_0 fixed and increasing $\sqrt{P_0}$ leads to $N > 1$, which are the higher-order solitons, whose propagation is not constant but periodic. Usually, the integer part of N takes a relevant part on the actual form of its propagation, i.e., $\tilde{N}$ if we take $N = \tilde{N} + \epsilon$, with $\epsilon < 1/2$.

5. Ultra-short pulse propagation

Now we will use the model equations and the dispersion relation we have developed to explain through numerical simulations the dynamics of ultra-short higher-order solitons in a PCF. All

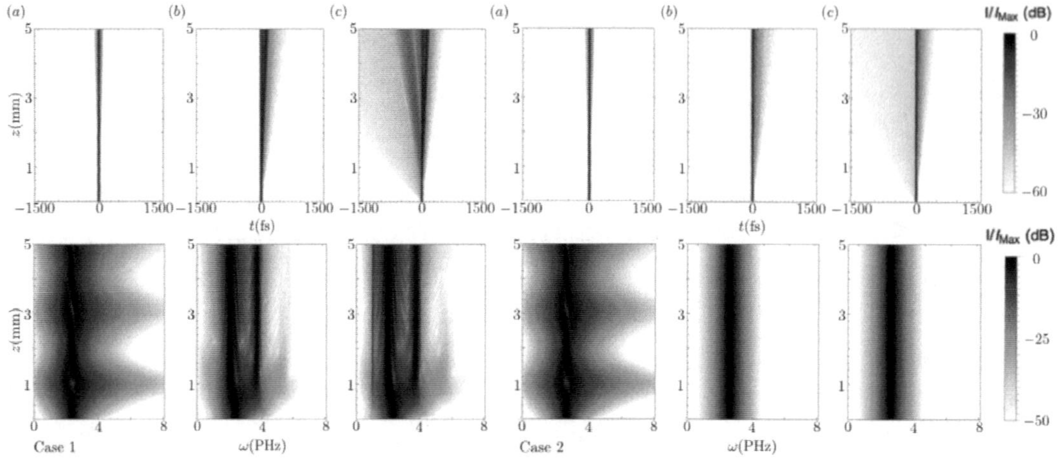

Figure 3. (Color online.) Left. Case 1, or base case, with λ =800 nm, $N = 2$ and $\tau_{\rm FWHM}$=7 fs. Right. Case 2, same as Case 1, except that now λ =720 nm, i.e., very close to the ZDW of the dispersion.

the simulations are done with a Dormand-Prince method, an adaptive version of the fourth-order Runge-Kutta method that contains an error monitoring function [16]. In addition, we choose the propagation step dz by checking norm conservation.

In our first studies, we determined that the SRS does not change the temporal shape, but for completeness we include it in our simulations. Furthermore, we found that the most important effects are given by higher-order terms in the dispersion. To study this, we will use three degrees of approximation in the dispersion: (a) second-order, (b) third-order and (c) rational approximation. For each of them, we will simulate four different pulses:

- Case 1. It is the base case, a higher-order soliton with λ =800 nm, $N = 2$ and $\tau_{\rm FWHM}$=7 fs, i.e., well inside the anomalous GVD of the dispersion. With second-order dispersion we see the usual dynamics from higher-order solitons and the dispersive effects do not enter. Already for the third-order dispersion some dispersive waves are shed around 500 nm. For the higher-order dispersion the pulse breaks further in time. See Fig. 3.
- Case 2. Similar to the base case except that λ =720 nm. Because we are very close to one ZDW, for the third-order and the full-dispersion there is no much dynamics except the usual broadening in time. Also, the effect of the higher-order dispersion is diminished and it does not allow steeper pulses. See Fig. 3.
- Case 3. Like the base case but with $N = 3$, the dynamics is much faster, as it is shown in shorter periods for the second-order dispersion and the faster break-up and shed of dispersive waves with the higher-order dispersions. See Fig. 4.
- Case 4. As the base case with $\tau_{\rm FWHM}$=15 fs, here the dynamics is slowed down, we have a slow broadening of the pulse. All of the dispersions give similar results except of few dispersive waves from the third- and higher-order dispersions. See Fig. 4.

6. Conclusions

In this work we studied the pulse dynamics of ultra-short higher-order solitons traveling through a PCF. We obtained that the third-order dispersion is essential to obtain the correct qualitative result in the spectrum. The inclusion of further terms modifies the pulse in time and accelerates its broadening. In most cases we found an initial steepening in one side of the soliton followed by the known soliton fission dynamics. On the case of higher soliton number, the pulses can

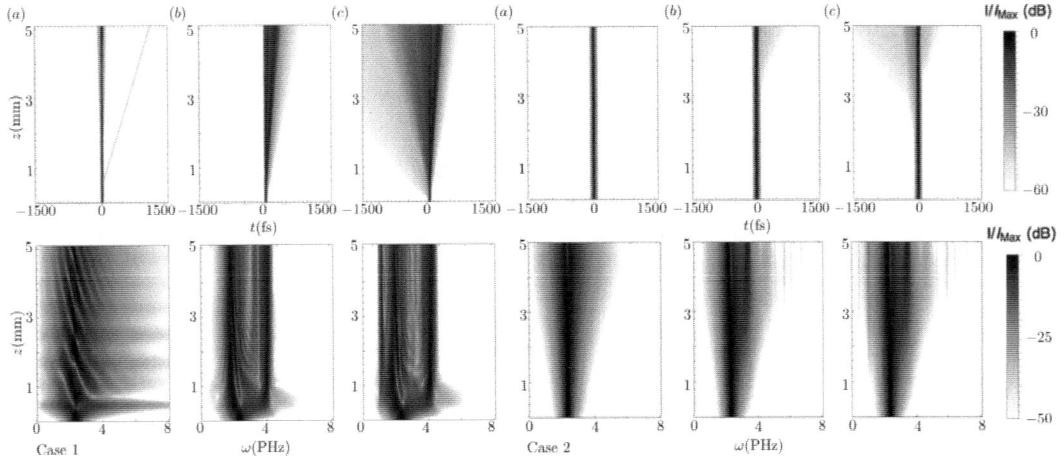

Figure 4. (Color online.) Left. Case 3, same as Case 1, but now the initial pulse has $N = 3$, i.e., larger energy. Right. Case 4, same as Case 1 with the exception of that $\tau_{\rm FWHM}$=15 fs.

be steeper but for a shorter propagation distance. The case of longer pulses starts with a disadvantage and even thought there is some steepening, it does not surpass other cases. Finally, we obtained pulses steeper than the initial ones in very short propagation distances, which can be useful in the search of an analogue of Hawking radiation [10, 11].

Acknowledgments

The author would like to thank Jonathan Drori for the information of the PCF. The author acknowledges the financial support of Conacyt (Mexico) through Project 152574.

References

[1] Agrawal G P 2013 *Nonlinear Fiber Optics* 5th ed (Oxford: Academic Press)
[2] Amiranashvili S, Bandelow U and Akhmediev N 2013 *Phys. Rev.* A **87** 013805
[3] Akhmediev N and Karlsson M 1995 *Phys. Rev.* A **51** 2602
[4] Husakou A V and Herrmann J 2001 *Phys. Rev. Lett.* **87** 203901
[5] Yin L, Lin Q and Agrawal G P 2007 *Opt. Lett.* **32** 391–393
[6] Dudley J M, Genty G and Coen S 2006 *Rev. Mod. Phys.* **78** 1135–1162
[7] Knight J C 2003 *Nature* **424** 847–851
[8] Karasawa N and Tada K 2012 Optical solitons from a photonic crystal fiber and their applications *Photonic Crystals. Introduction, Applications and Theory* ed Massaro A (InTech)
[9] Amiranashvili S, Bandelow U and Akhmediev N 2014 *Opt. Express* **22** 30251–30256
[10] Philbin T G, Kuklewicz C, Robertson S, Hill S, König F and Leonhardt U 2008 *Science* **319** 1367–1370
[11] Bermudez D, Drori J and Leonhardt U 2015 *Preprint* arXiv:1506.05086 [physics.optics]
[12] Lin Q and Agrawal G P 2006 *Opt. Lett.* **31** 3086–3088
[13] Amiranashvili S, Vladimirov A G and Bandelow U 2010 *Eur. Phys. J.* D **58** 219–226
[14] NKT Photonics 2015 *Nonlinear photonic crystal fibers for 800 nm pumping* URL www.nktphotonics.com
[15] Synopsys 2015 *FemSIM Product Overview* URL https://optics.synopsys.com
[16] Press W H, Teukolsky S A, Vetterling W T and Flannery B P 1997 *Numerical recipes in C* 2nd ed (Cambridge: Cambridge University Press)

Optical Lattice Design Assisted by Non-Hermitian Hamiltonians

B M Rodríguez-Lara

Instituto Nacional de Astrofísica, Óptica y Electrónica, Calle Luis Enrique Erro No. 1, Sta. Ma. Tonantzintla, Pue. CP 72840, México

E-mail: bmlara@inaoep.mx

Abstract. A brief introduction to non-Hermitian arrays of coupled waveguides is presented. The PT-symmetric dimer is revisited for the sake of clarity. It belongs to the class of photonic lattices with underlying $SO(2,1)$ symmetry that have been shown to provide all-optical conversion from phase to amplitude.

1. Introduction

In the last decade, arrays of coupled photonic waveguides have provided a robust and flexible platform for the optical simulation of standard quantum phenomena [1] thanks to the development of advanced fabrication techniques [2]. Reciprocally, the ability to mimic Hermitian Hamiltonians has inspired the development of passive photonic integrated circuits serving practical purposes [3, 4].

In recent years, the idea of non-Hermitian quantum mechanics [5–7] has also impacted the design of optical devices [8–16]. Here, a brief summary on the simplest non-Hermitian photonic element, which is the PT-symmetric dimer [17–20], is presented and used as a guiding example to uncover its relation to optical finite realizations of the Lorentz group [21]. Optical finite realizations of $SO(2,1)$ and $SO(3,1)$ may prove helpful in the design of all-optical photonic integrated circuits, such as the recently proposed PT-symmetric device providing conversion from amplitude (phase) to phase (amplitude) modulation [22].

2. Revisiting the PT-symmetric dimmer

A lossy two-waveguide coupler is described by the following non-Hermitian coupled-mode differential equation system [17],

$$-i\partial_z \begin{pmatrix} E_0(z) \\ E_1(z) \end{pmatrix} = \begin{pmatrix} n_0 & g \\ g & n_1 \end{pmatrix} \begin{pmatrix} E_0(z) \\ E_1(z) \end{pmatrix}, \tag{1}$$

where the complex field amplitudes at the j-th waveguide are given by E_j, the effective complex refractive indices by n_j, the effective real waveguide coupling by g, and the notation ∂_z is used for the derivative with respect to z. For the sake of simplicity, let us assume that the real part of the effective refractive indices are identical, $n_j = \alpha_R + i\alpha_j$ with $\alpha_x \in \mathbb{R}$, and write,

$$E_j(z) = e^{i\alpha_R z} e^{-\alpha_b z} \mathcal{E}_j(z), \tag{2}$$

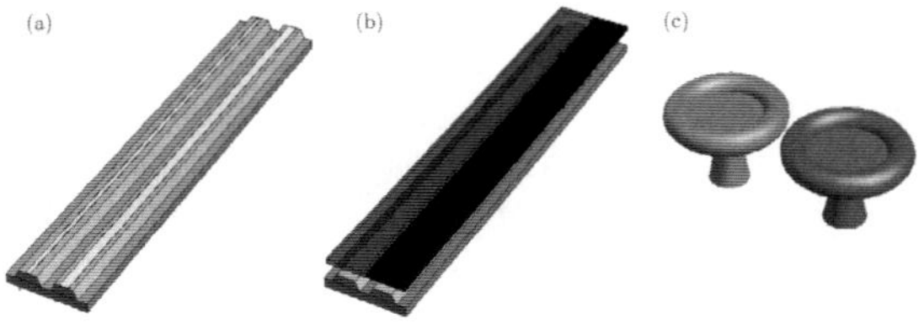

Figure 1. Experimental realization of the PT-Symmetric dimer in three different architectures: (a) lossy waveguides, (b) pumped waveguides, and (c) coupled whispering gallery microcavities.

that is, we consider the complex fields up to a common phase induced by the real part of the effective refractive index, α_R, and attenuation (amplification) induced by a positive (negative) bias,

$$\alpha_b = \frac{1}{2}\left(\alpha_0 + \alpha_1\right). \tag{3}$$

Under this assumption, the mode-coupling differential equation set becomes PT-symmetric,

$$-i\partial_z \begin{pmatrix} \mathcal{E}_0(z) \\ \mathcal{E}_1(z) \end{pmatrix} = \begin{pmatrix} i\alpha & g \\ g & -i\alpha \end{pmatrix} \begin{pmatrix} \mathcal{E}_0(z) \\ \mathcal{E}_1(z) \end{pmatrix}, \tag{4}$$

where the new real constant,

$$\alpha = \frac{1}{2}\left(\alpha_0 - \alpha_1\right), \tag{5}$$

can be interpreted as effective linear, $i\alpha$, or gain,$-i\alpha$, at each waveguide. This device is known in the literature as the PT-symmetric linear dimer and it has been experimentally demonstrated in lossy waveguides [18], pumped waveguides [19], and whispering-gallery mode microcavities [20].

Following standard techniques [23], it is straightforward to calculate the propagation through the PT-symmetric dimer,

$$\begin{pmatrix} \mathcal{E}_0(z) \\ \mathcal{E}_1(z) \end{pmatrix} = \begin{pmatrix} \cosh\Omega z - \frac{\alpha}{\Omega}\sinh\Omega z & \frac{ig}{\Omega}\sinh\Omega z \\ \frac{ig}{\Omega}\sinh\Omega z & \cosh\Omega z + \frac{\alpha}{\Omega}\sinh\Omega z \end{pmatrix} \begin{pmatrix} \mathcal{E}_0(0) \\ \mathcal{E}_1(0) \end{pmatrix}. \tag{6}$$

This uncovers three possible regimes: (i) a PT-symmetric phase for values $\alpha < g$ delivering a real propagation parameter $\Omega = \sqrt{g^2 - \alpha^2}$ where the dimer behaves like a periodic oscillator, Fig 2(a), (ii) a completely degenerate phase, $\alpha = g$ and $\Omega = 0$, where the dimer behaves like a directional amplifier with a power law, Fig 2(b), and (iii) a broken phase, $\alpha > g$ yielding an imaginary Ω, where the dimer becomes a directional amplifier with an exponential law [21], Fig 2(c).

3. A finite representation of the Lorentz group

Let us borrow Dirac notation for vectors and matrices [24] and rewrite the differential equation set for the PT-symmetric dimer,

$$i\partial_z|\mathcal{E}\rangle = 2\left(i\alpha J_0 + gJ_x\right)|\mathcal{E}\rangle, \tag{7}$$

where the operators $\left\{J_0, J_x, J_y\right\}$ form the $SU(2)$ group, which $j = 1/2$ representation delivers Eq. (4). This equation can also be written in terms of the $SO(2,1)$ group, which is equivalent to the 2+1D Lorentz group [21],

$$i\partial_z|\mathcal{E}\rangle = 2\left(\alpha K_0 + gK_x\right)|\mathcal{E}\rangle, \tag{8}$$

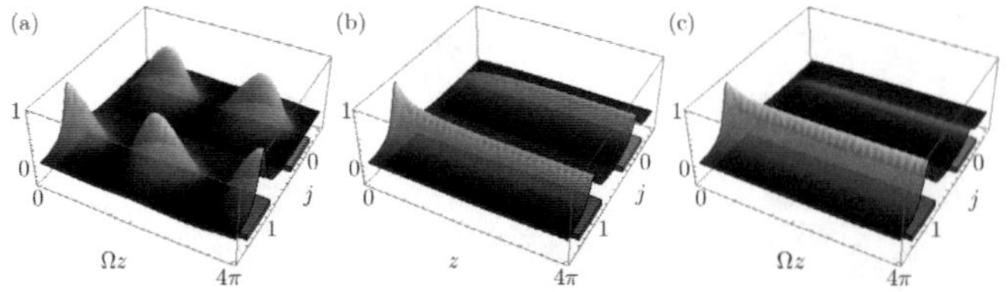

Figure 2. Visualization of the renormalized power propagation, $|\tilde{\mathcal{E}}_j(z)|^2 = |\mathcal{E}_j(z)|^2/\left(|\mathcal{E}_0(z)|^2 + |\mathcal{E}_1(z)|^2\right)$, through a $\mathcal{PT}$-symmetric dimer in the directional (a) oscillator, $\alpha = 0.2$ and $g = 1$, (b) power amplifier, $\alpha = 1$ and $g = 1$, and (c) exponential amplifier, $\alpha = 5$ and $g = 1$, regimes.

where we have just made imaginary some elements of $SU(2)$, $\left\{\hat{K}_0, \hat{K}_x, \hat{K}_y\right\} \equiv \left\{i\hat{J}_0, \hat{J}_x, i\hat{J}_y\right\}$. It is straightforward to note that any of the experimental architectures where the PT-symmetric dimer has been realized [17–20,25] may be tailored to produce optical finite representation of the Lorentz group with different Bargmann parameter, $j = 1/2, 1, 3/2, \ldots$. For example, starting from the PT-symmetric dimer, which is the $j = 1/2$ representation of SO(2,1), we can include an extra lossy waveguide, pumped waveguide, or pumped whispering mode cavity, and obtain a $j = 1$ representation of the $SO(2,1)$ group, and so on. Any device belonging to this class will show the directional oscillator and amplifier with power law and exponential laws behaviors described above [21].

4. All-optical PT-symmetric phase to amplitude conversion

The directional oscillator and amplifier behavior of PT-symmetric devices is well known and studied. Here we will bring forward a different application. The $j = 1$ optical representation of SO(2,1), which is a planar three-waveguide coupler described by the mode-coupling set [21],

$$-i\partial_z \begin{pmatrix} \mathcal{E}_0(z) \\ \mathcal{E}_1(z) \\ \mathcal{E}_2(z) \end{pmatrix} = \begin{pmatrix} i\alpha & g & 0 \\ g & 0 & g \\ 0 & g & -i\alpha \end{pmatrix} \begin{pmatrix} \mathcal{E}_0(z) \\ \mathcal{E}_1(z) \\ \mathcal{E}_2(z) \end{pmatrix}, \tag{9}$$

or in Dirac notation,

$$-i\partial_z|\mathcal{E}(z)\rangle = \mathbb{H}|\mathcal{E}(z)\rangle. \tag{10}$$

Introducing a scaled propagation distance, $\zeta = gz$, and a ratio between effective linear gain or loss with respect to the coupling paramenter, $\xi = g/\alpha$, yields a mode-coupling matrix described by the propagation parameter $\Omega = \sqrt{2 - \xi^2}$ and propagation ruled by the equation [21],

$$|\mathcal{E}(z)\rangle = \begin{cases} \left[\mathbb{1} + \frac{i}{\sin\Omega\zeta}\,\mathbb{H} + \frac{1}{\Omega^2}\left(\cos\Omega\zeta - 1\right)\mathbb{H}^2\right]|\mathcal{E}(0)\rangle, & \xi \neq \sqrt{2}, \\ \left[\mathbb{1} + i\zeta\mathbb{H} - \frac{1}{2}\zeta^2\mathbb{H}^2\right]|\mathcal{E}(0)\rangle, & \xi = \sqrt{2}. \end{cases} \tag{11}$$

Now, let us take a directional oscillator, $\xi < \sqrt{2}$, stop propagation at the scaled distance,

$$\zeta_f = \frac{1}{\Omega}\arccos\left(\xi^2 - 1\right), \tag{12}$$

and consider a two-waveguide input where the fields have the same power but different phases,

$$|\mathcal{E}(0)\rangle = \begin{pmatrix} e^{i\phi} \\ 1 \\ 0 \end{pmatrix}\mathcal{E}. \tag{13}$$

It is straightforward to calculate the output,

$$|\mathcal{E}(\zeta_f)\rangle = \begin{pmatrix} 0 \\ -\mathcal{E} \\ \left(2i\xi - e^{i\phi}\right)\mathcal{E} \end{pmatrix}, \tag{14}$$

and note that the power at the last waveguide varies with respect to the initial phase difference,

$$|\mathcal{E}_2(\zeta_f)|^2 = [1 + 4\xi(\xi - 1)\sin\phi]\,\mathcal{E}^2. \tag{15}$$

Thus, it is possible to provide phase to amplitude conversion in a finite optical representation of the 2+1D Lorentz group. The converse has also been demonstrated [22].

5. Conclusion

We have revisited the simplest non-Hermitian photonic device, which is the PT-symmetric dimer, in order to show its relation with the class of photonic lattices with underlying $SO(2,1)$ symmetry. We brought forward the possibility to use this class beyond the conventional directional oscillators and amplifiers behavior to convert phase (amplitude) modulation into amplitude (phase) modulation.

Acknowledgments
The author acknowledges fruitful discussion with J. Guerrero, R. El-Ganainy and A. J. Stoffel.

References

[1] Longhi S 2009 *Laser Photon. Rev.* **3** 243 – 261
[2] Szameit A and Nolte S 2010 *J. Phys. B: At. Mol. Opt. Phys.* **43** 163001
[3] El-Ganainy R, Eisfeld A, Levy M and Christodoulides D N 2013 *Appl. Phys. Lett.* **103** 161105
[4] Perez-Leija A, Keil R, Kay A, Moya-Cessa H, Nolte S, Kwek L C, Rodríguez-Lara B, Szameit A and Christodoulides D 2013 *Phys. Rev. A* **87** 012309
[5] Bender C M 1998 *Phys. Rev. Lett.* **80** 5243 – 5246
[6] Bender C M 2005 *Contemp. Phys.* **46** 277 – 292
[7] Bender C M 2007 *Rep. Prog. Phys.* **70** 947 – 1018
[8] Makris K G, El-Ganainy R, Christodoulides D N and Musslimani Z H 2008 *Phys. Rev. Lett.* **100** 103904
[9] Longhi S 2010 *Phys. Rev.* A **82** 031801
[10] Ramezani H, Kottos T, El-Ganainy R and Christodoulides D N 2010 *Phys. Rev.* A **82** 043803
[11] Makris K G, El-Ganainy R and Christodoulides D N 2010 *Phys. Rev.* A **81** 063807
[12] Longhi S and Valle G D 2012 *Phys. Rev.* A **85** 012112
[13] Lin Z, Ramezani H, Eichelkraut T, Kottos T, Cao H and Christodoulides D N 2011 *Phys. Rev. Lett.* **106** 213901
[14] Longhi S, Valle G D and Staliunas K 2011 *Phys. Rev.* A **84** 042119
[15] Regensburger A, Bersch C, Miri M A, Onishchukov G, Christodoulides D N and Peschel U 2012 *Nature* **488** 167 – 171
[16] Feng L, Xu Y L, Fegadolli W S, Lu M H, Oliveira J E B, Almeida V R, Chen Y F and Scherer A 2013 *Nature Mater.* **12** 108 – 113
[17] Somekh S, Garmire E, Yariv A, Garvin H L and Hunsperger R G 1973 *Appl. Phys. Lett.* **22** 46 – 47
[18] Guo A, Salamo G J, Duchesne D, Morandotti R, Volatier-Ravat M, Aimez V, Siviloglou G A and Christodoulides D N 2009 *Phys. Rev. Lett.* **103** 093902
[19] Rüter C E, Makris K G, El-Ganainy R, Christodoulides D N and Kip D 2010 *Nature Phys.* **6** 192 – 195
[20] Peng B, Ozdemir S K, Lei F C, Monifi F, Gianfreda M, Long G L, Fan S H, Nori F, Bender C M and Yang L 2014 *Nature Phys.* **10** 394 – 398
[21] Rodríguez-Lara B M and Guerrero J 2015 *Opt. Lett.* **40** 5682 –5685
[22] Zaragoza Gutirrez O I, Salinas Mendoza L F and Rodrguez-Lara B M 2016 *Opt. Express* **24** 3989 – 3998
[23] Villanueva Vergara L and Rodríguez-Lara B M 2015 *Opt. Express* **23** 22836–22846
[24] Rodríguez-Lara B M, Soto-Eguibar F and Christodoulides D N 2015 *Phys. Scr.* **90** 068014
[25] Schindler J, Lin Z, Lee J M, Ramezani H, Ellis F M and Kottos T 2012 *J. Phys A: Math. Theor.* **45** 444029

Quantum Fest 2015 IOP Publishing
Journal of Physics: Conference Series **698** (2016) 012019 doi:10.1088/1742-6596/698/1/012019

QUANTUM: A Wolfram *Mathematica* add-on for Dirac Bra-Ket Notation, Non-Commutative Algebra, and Simulation of Quantum Computing Circuits

J L Gómez Muñoz and F Delgado

Departamento de Física y Matemáticas, Escuela de Diseño, Ingeniería y Arquitectura, Tecnológico de Monterrey, Campus Estado de México, Atizapán, Estado de México, CP. 52926, México.

E-mail: `jose.luis.gomez@itesm.mx`, `fdelgado@itesm.mx`

Abstract. This paper introduces QUANTUM, a free library of commands of Wolfram *Mathematica* that can be used to perform calculations directly in Dirac braket and operator notation. Its development started several years ago, in order to study quantum random walks. Later, many other features were included, like operator and commutator algebra, simulation and graphing of quantum computing circuits, generation and solution of Heisenberg equations of motion, among others. To the best of our knowledge, QUANTUM remains a unique tool in its use of Dirac notation, because it is used both in the input and output of the calculations. This work depicts its usage and features in Quantum Computing and Quantum Hamilton Dynamics.

1. Introduction

QUANTUM is an add-on of *Mathematica* (versions 8 to 10) implementing calculations in Dirac notation [1], and allowing the end user to enter that notation using keyboard and palettes toolbars. As a simple example, the end user can easily write and obtain the following result:

$$\begin{aligned} \text{Input}: \quad & |\psi\rangle = \alpha\,|\phi_1\rangle + \beta\,|\phi_2\rangle\,; \\ & \text{Expand}[\,|\psi\rangle \cdot \langle\psi|\,] \\ \text{Output}: \quad & \alpha\alpha^*\,|\phi_1\rangle\,\langle\phi_1| + \alpha\beta^*\,|\phi_1\rangle\,\langle\phi_2| + \beta\alpha^*\,|\phi_2\rangle\,\langle\phi_1| + \beta\beta^*\,|\phi_2\rangle\,\langle\phi_2| \end{aligned} \tag{1}$$

The core of QUANTUM was constructed based on quantum information and quantum computing applications, therefore several of the advanced commands in the current version are restricted to discrete Hilbert spaces. The programming philosophy and implementation of QUANTUM are depicted in the second section. QUANTUM has three main modules; the third section describes the first one, Quantum'Notation', implementing bra-ket notation, non-commutative algebra of operators and commutators, and quantum measurements. The fourth section describes the second module, Quantum'Computing', including qubits and quantum gates for the simulation of algorithms and the automated drawing of quantum computing circuits. The third module in the fifth section, Quantum'QHD', implements the Quantized Hamilton Dynamics (QHD) approximation to the Heisenberg equations of motion. It can be used for the simulation of physical systems such as molecules. Last section is devoted to the conclusions.

2. Programming Philosophy and Implementation

Mathematica has a very consistent notation. For example, the standard command to graph a sine function from 0 to π is Plot[Sin[x],{x,0,Pi}], on the other hand, to integrate that function in the same interval the command is almost identical: Integrate[Sin[x],{x,0,Pi}]. We have implemented the same consistent behavior in the QUANTUM commands. As an example, the commands to obtain the bra-ket expression and the quantum computing circuit of a sequence of quantum computing gates are almost identical, as can be seen in Figure 1. Sometimes the Dirac notation, as it is used in books and papers of Quantum Mechanics and Quantum Computing, is not explicit and flexible enough for the calculations [2–4]. Therefore QUANTUM generalized this notation, as it is seen in the CNOT gate notation in Figure 1.

QuantumEvaluate[$C^{\{\hat{1}\}}[\mathcal{NOT}_{\hat{2}}]\cdot\mathcal{H}_{\hat{1}}$]	QuantumPlot[$C^{\{\hat{1}\}}[\mathcal{NOT}_{\hat{2}}]\cdot\mathcal{H}_{\hat{1}}$]
$\frac{\lvert 0_{\hat{1}},0_{\hat{2}}\rangle\cdot\langle 0_{\hat{1}},0_{\hat{2}}\rvert}{\sqrt{2}}+\frac{\lvert 1_{\hat{1}},1_{\hat{2}}\rangle\cdot\langle 0_{\hat{1}},0_{\hat{2}}\rvert}{\sqrt{2}}+\frac{\lvert 0_{\hat{1}},1_{\hat{2}}\rangle\cdot\langle 0_{\hat{1}},1_{\hat{2}}\rvert}{\sqrt{2}}+\frac{\lvert 1_{\hat{1}},0_{\hat{2}}\rangle\cdot\langle 0_{\hat{1}},1_{\hat{2}}\rvert}{\sqrt{2}}+\frac{\lvert 0_{\hat{1}},0_{\hat{2}}\rangle\cdot\langle 1_{\hat{1}},0_{\hat{2}}\rvert}{\sqrt{2}}-\frac{\lvert 1_{\hat{1}},1_{\hat{2}}\rangle\cdot\langle 1_{\hat{1}},0_{\hat{2}}\rvert}{\sqrt{2}}+\frac{\lvert 0_{\hat{1}},1_{\hat{2}}\rangle\cdot\langle 1_{\hat{1}},1_{\hat{2}}\rvert}{\sqrt{2}}-\frac{\lvert 1_{\hat{1}},0_{\hat{2}}\rangle\cdot\langle 1_{\hat{1}},1_{\hat{2}}\rvert}{\sqrt{2}}$	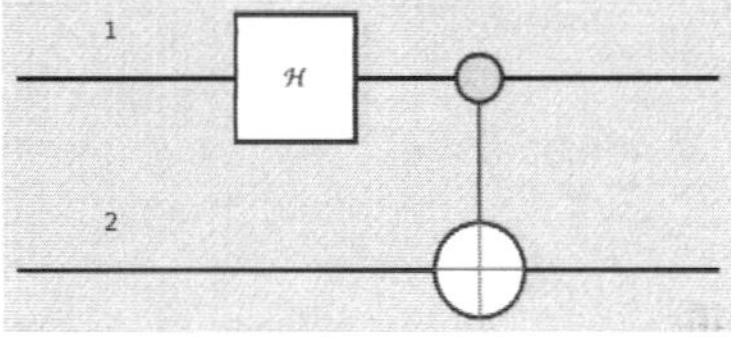

Figure 1. The same syntax is used to obtain a bra-ket expression and to graph the corresponding circuit. This consistent behavior was implemented across all QUANTUM commands.

An important aspect of Quantum'Notation' module compared with other packages [5–7] implementing quantum computations (based on matrix notation) is the use of an algebraic management (based on Dirac notation). The first approach could be computationally inefficient because states and operators become typically big after tensor products. Then, matrix products could be a slow technique compared with the algebraic, where only non-zero terms appear. There are several outstanding aspects there: a) the natural and easier way to introduce operators and kets, in particular for problems using a big number n of qudits (d-level quantum systems); b) the comparative speed performance in the calculations between algebraic management $O(d^n log(d^n))$ versus straight matrix management of calculations $O(d^{2n})$, at least in the most common problems in quantum information, which can be expressed by sparse matrices of order $O(log(d^n))$ (despite other packages could implement sparse matrix multiplication techniques, in QUANTUM this process is automatic); and c) the capabilities of *Mathematica* to do parallel computations which has reduced in half the QUANTUM processing time for large number of qudits in a benchmarking for problems in Adiabatic Quantum Computation [8].

To implement QUANTUM, several advanced *Mathematica* techniques were used, including the unprotection and redefinition of standard *Mathematica* commands like Expand and Simplify, instead of just be included in the general commands of the tool; the use of operators without built-in meaning, like CenterDot; and the low-level programming of *Mathematica* box constructs, like RowBox, SuperscriptBox and TagBox [9].

3. Quantum Notation Module

The Quantum'Notation' module implements the non-commutative algebra of operators, commutators, and the Dirac bra-ket notation. It includes commands for the simulation of quantum measurements, the calculation of partial trace, partial transpose, among others.

Figure 2 shows a hierarchy of some of the commands in this module as well as groups of commands developed mainly based on Kets, Operators and Operations there. Thus, main basic operations used in Dirac notation for discrete basis as qubits, qudits, operators, adjoint, scalar,

Journal of Physics: Conference Series **698** (2016) 012019 doi:10.1088/1742-6596/698/1/012019

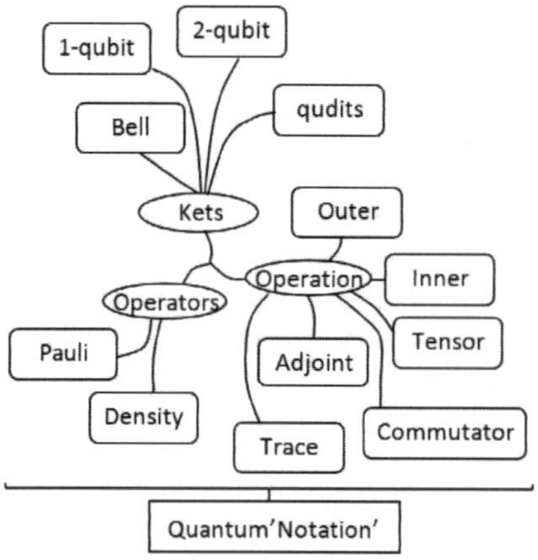

Figure 2. Main sections developed in Quantum'Notation' module integrating the basic aspects in Dirac notation: bra-ket's, operators and operations.

inner and tensor products, as well as other specific objects in the form of Bell states, Pauli matrices and Partial Trace operations are included. One of the palettes of this module ease the introduction of the notation containing the main elements in the module (Figure 3). Table 1 shows some of the main templates and their usage, which are entered either using the palettes or keyboard shortcuts. There, expressions after of symbol = represent traditional writing.

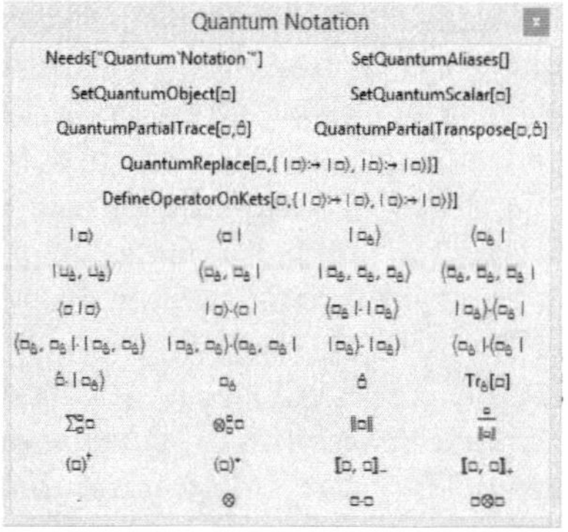

Figure 3. Quantum'Notation' palette showing the principal commands implemented in the module: bra-ket's and basic operations.

Thus, with these basic constructions, it is possible set and calculate states, operators, projections, measurements, probabilities and traces from quantum objects operating on a tensor product basis of discrete Hilbert spaces, covering lots of possibilities for quantum systems.

4. Quantum Computing Module

The module named Quantum'Computing' includes quantum computing gates and qubits. The bra-ket evaluation of the quantum computing gates is delayed, so that they can be used to generate circuits, as in Figure 1, and can be also used in algebra, for example to expand them in terms of Pauli operators. Bell states, Quantum Fourier Transform and measurements are also included. Figure 4 shows a hierarchy of some of the commands in this module. As in Quantum'Notation', some palettes (Quantum Computing Kets and Quantum Computing

Ket-Bra's		Operations	
Template	Example	Template	Example
$\left\vert \square_{\hat{\square}}, \square_{\hat{\square}}, \square_{\hat{\square}}, \right\rangle$	$\vert 0_{\hat{1}}, 0_{\hat{2}}, 1_{\hat{3}} \rangle$	$(\square)^{\dagger}$	$(\vert 1_{\hat{2}} \rangle)^{\dagger} = \langle 1_2 \vert$
$\left\vert \square_{\hat{\square}} \right\rangle \cdot \left\langle \square_{\hat{\square}} \right\vert$	$\vert 0_{\hat{1}} \rangle \cdot \langle 1_{\hat{1}} \vert$	$\left\vert \square_{\hat{\square}} \right\rangle \otimes \left\vert \square_{\hat{\square}}, \square_{\hat{\square}} \right\rangle$	$\vert 1_{\hat{3}} \rangle \otimes \vert 0_{\hat{1}}, 1_{\hat{2}}, \rangle = \vert 0_1 1_2 1_3 \rangle$
$\left\vert \square_{\hat{\square}} \right\rangle \otimes \left\vert \square_{\hat{\square}} \right\rangle$	$\vert 0_{\hat{1}} \rangle \otimes \langle 1_{\hat{2}} \vert$	$\mathrm{Tr}_{\hat{\square}}(\square)$	$\mathrm{Tr}_{\hat{1}}(\vert 0_{\hat{1}}, 1_{\hat{2}} \rangle \langle 0_{\hat{1}}, 0_{\hat{2}} \vert) = \vert 1_2 \rangle \langle 0_2 \vert$
$\square \cdot \left\vert \square_{\hat{\square}} \right\rangle$	$\mathcal{O} \cdot \vert 1_{\hat{2}} \rangle$	$\Vert \square \Vert$	$\Vert \, \vert 0_{\hat{1}} \rangle \, \Vert = \sqrt{\langle 0_1 \mid 0_1 \rangle}$

Table 1. Some objects included in Quantum'Notation' package and examples of their usage.

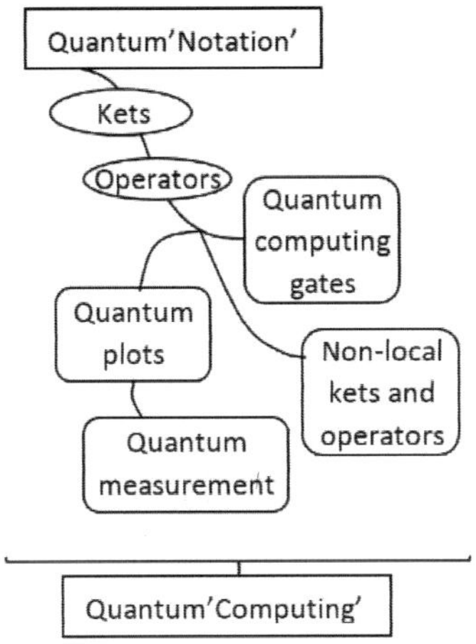

Figure 4. Main sections developed in Quantum'Notation' module integrating the basic aspects in Dirac notation: bra-ket's, operators and operations.

Gates) are included to ease the introduction of the templates for those elements and commands (Figure 5), but still each one of them can be entered with a keyboard shortcut. Table 2 reports a group of representative templates and examples for this module.

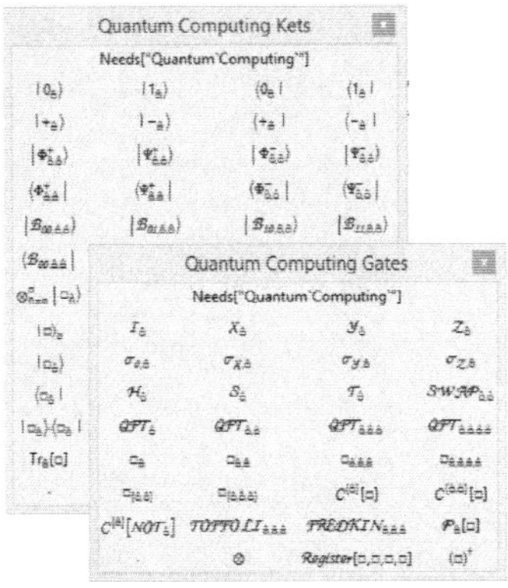

Figure 5. Main Quantum'Computing' palettes showing some commands implemented in the module: computing gates, measurements and plots.

Nevertheless Quantum'Computing' is based on computational basis from Quantum'Notation', the user can hold the calculations to not evolve in the final expressions on that basis, then

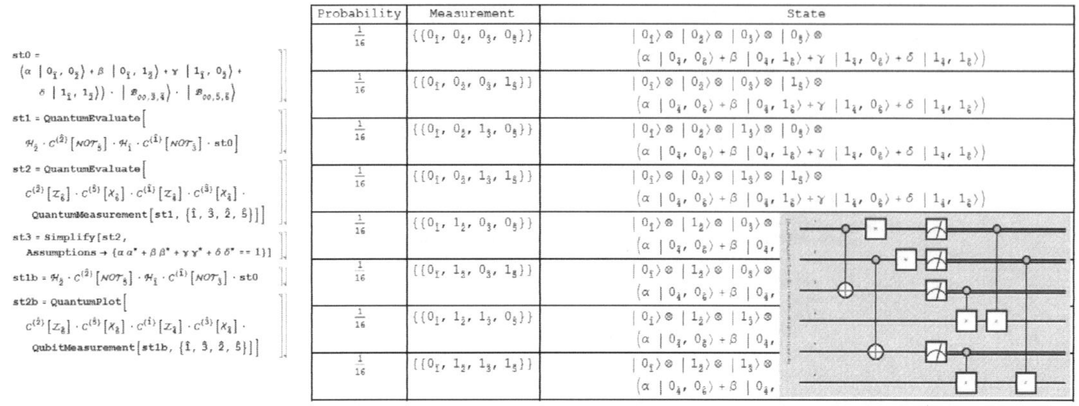

Figure 6. Code and measurement outcomes table in QUANTUM for a Quantum Teleportation algorithm of a 2-qubit state. The right-down inset is the quantum circuit plot obtained with the command QuantumPlot.

being capable to be processed by different commands to produce states, tables or plots. This library includes tensor product templates, norms, states sums for superposition, commutators, anti-commutators and quantum partial transpose as additional elements working together with those in Quantum'Notation'. On those elements, additional operators or gates related with circuit quantum computation are included. They are enriched with commands able to operate on them to generate measurements, probability tables and circuit plots (Figure 6). Table 2 reports a group of representative templates and examples for this module. Expressions after of symbol $\rightarrow$ are QUANTUM outputs.

Quantum'Computing' can perform both, simple and complex calculations in Quantum Information Processing or Quantum Computation. As a simple example, Figure 6 shows the code to realize the traditional Quantum Teleportation algorithm [10, 11] for a general 2-qubit state. Process begins defining the state on six qubits (st0) with state to teleport located in qubits 1 and 2. After, a state (st1) is obtained applying the common $\mathcal{H}_i$ and $C^i NOT_j$ gates in the algorithm. Finally, the teleportated state is obtained on the qubits 4 and 6 after of four measurements in the qubits 1, 2, 3 and 5, in the form of a probability table with the command QuantumMeasurement including further corrections using classical communication. Notice once again how almost the same code under alternative commands as QuantumPlot instead of QuantumEvaluate generates alternative outputs considered in the package (as the plot of respective quantum circuit). In similar ways, other more complex algorithms can be developed using the gates included or other defined by the users.

States		Gates	
Template	Example	Template	Example
$\left\|\beta_{ij,\hat{\square},\hat{\square}}\right\rangle$	$\left\|\beta_{10,\hat{1},\hat{2}}\right\rangle = \frac{\|0_1 0_2\rangle - \|1_1 1_2\rangle}{\sqrt{2}}$	$\sigma_{i,\hat{\square}}$	$\sigma_{1,\hat{2}} = \sigma_{x,2} = \mathcal{X}_2$
$\left\|\pm_{\hat{\square}}\right\rangle$	$\|+_{\hat{2}}\rangle = \frac{\|0_2\rangle + \|1_2\rangle}{\sqrt{2}}$	$\mathcal{H}_{\hat{\square}}$	$\mathcal{H}_{\hat{3}} = \frac{1}{\sqrt{2}}(\mathcal{X}_2 + \mathcal{Z}_2)$
$\bigotimes_{n=\hat{\square}}^{\hat{\square}} \left\|\square_{\hat{n}}\right\rangle$	$\bigotimes_{n=\hat{0}}^{\hat{2}} \left\|n+1_{\hat{n}}\right\rangle \rightarrow \|1_{\hat{0}}, 2_{\hat{1}}, 3_{\hat{2}}\rangle$	$\mathcal{C}^{\{\hat{\square}\}}[\mathcal{N}\mathcal{O}\mathcal{T}_{\hat{\square}}]$	$\mathcal{C}^{\{\hat{1}\}}[\mathcal{N}\mathcal{O}\mathcal{T}_{\hat{2}}] = C^1 NOT_2$
QuantumEvaluate[□]	QuantumEvaluate[$\mathcal{Z}_2$] $\rightarrow \|0_{\hat{2}}\rangle\langle 0_{\hat{2}}\| - \|1_{\hat{2}}\rangle\langle 1_{\hat{2}}\|$	$\mathcal{C}^{\{\hat{\square}\}}[\square]$	$\mathcal{C}^{\hat{1}}[\mathcal{H}_{\hat{3}}] = C^1\mathcal{H}_3$

Table 2. Some objects included in Quantum'Computing' package and examples of their usage.

5. Quantized Hamilton Dynamics Module

The evolution of the expected value of an observable A in the Heisenberg representation is given by the equation of motion (EOM):

$$i\hbar\frac{d}{dt}\langle A\rangle = \langle[A, H]\rangle \tag{2}$$

Heisenberg EOM
$\frac{d\langle p\rangle}{dt} = -\langle q\rangle - 3\,a\,\langle q^2\rangle$
$\frac{d\langle q\rangle}{dt} = \langle p\rangle$
$\frac{d\langle q^2\rangle}{dt} = 2\,\langle pq\rangle_s$
$\frac{d\langle pq\rangle_s}{dt} = \langle p^2\rangle - \langle q^2\rangle - 3\,a\,\langle q^3\rangle$
$\frac{d\langle p^2\rangle}{dt} = -2\,\langle pq\rangle_s - 6\,a\,\langle pq^2\rangle_s$
$\frac{d\langle q^3\rangle}{dt} = 3\,\langle pq^2\rangle_s$
It continues forever

Figure 7. A particular example of Heisenberg EOM. They are coupled and form an infinite hierarchy of equations.

Consider the averages of momentum, position and their products $\langle p\rangle$, $\langle q\rangle$, $\langle p^2\rangle$, $\langle q^2\rangle$, $\langle pq\rangle$, $\langle p^3\rangle$, $\langle p^2q\rangle$... The EOMs for their average values are coupled and, in general, form an infinite hierarchy of equations. As an example, the EOMs for the Hamiltonian $H = \frac{1}{2}(p^2 + q^2) + aq^3$ are shown in Figure 7. In order to obtain a finite hierarchy of equations, a closure procedure has to be applied. In the Quantized Hamilton Dynamics approximation [12], the expectation values of higher order expressions are approximated in terms of those other expectation values already included in the hierarchy, for instance, the approximation $\langle ABC\rangle \approx \langle AB\rangle\langle C\rangle + \langle AC\rangle\langle B\rangle + \langle BC\rangle\langle A\rangle - 2\langle A\rangle\langle B\rangle\langle C\rangle$ is one of the approximations used at second order, QHD-2. Figure 8 shows the finite QHD-2 hierarchy obtained in the example.

QHD - 2 Closure
$\frac{d\langle p\rangle}{dt} = -\langle q\rangle - 3\,a\,\langle q^2\rangle$
$\frac{d\langle q\rangle}{dt} = \langle p\rangle$
$\frac{d\langle q^2\rangle}{dt} = 2\,\langle pq\rangle_s$
$\frac{d\langle pq\rangle_s}{dt} =$ $\langle p^2\rangle + 6\,a\,\langle q\rangle^3 - \langle q^2\rangle - 9\,a\,\langle q\rangle\,\langle q^2\rangle$
$\frac{d\langle p^2\rangle}{dt} = 12\,a\,\langle p\rangle\,\langle q\rangle^2 - 6\,a\,\langle p\rangle\,\langle q^2\rangle -$ $2\,\langle pq\rangle_s - 12\,a\,\langle q\rangle\,\langle pq\rangle_s$

Figure 8. The QHD-2 closure procedure was applied to the EOM of the previous figure, resulting in hierarchy of only five equations.

The third module of QUANTUM has several commands that allow the automatic generation of EOM for arbitrary Hamiltonians, with or without closure (those commands were used to generate Figures 7 and 8). Closures can be applied at orders QHD-1 (which gives classical dynamics for the Hamiltonian), QHD-2 and QHD-3; then the module has commands to include the initial conditions, solve the system of equations, and generate plots of the evolution of the expected values in time, as well as phase plots. The website of QUANTUM [1] includes examples of the application of these commands to several physical systems, including the O-H bond of a water molecule.

6. Conclusions

In this paper we have presented QUANTUM, a Mathematica add-on for calculations in Dirac notation, non-commutative operator algebra, quantum computing and approximations to the Heisenberg equations of motion. It allows the end user to input the calculations directly in the usual notation for Quantum Mechanics, and to obtain results in the same format. This unique characteristic has appealed to students and researchers in more than 20 countries to the best of our knowledge. Actually some of the commands and features of QUANTUM were included by recommendation of those users. Future work is also based on those recommendations, and it includes modules for occupation number notation (Second Quantization and Quantum Field Theory), angular momentum, continuous Hilbert spaces, among others.

Aknowledgements
The support from Tecnológico de Monterrey to develop this research work is acknowledged.

References

[1] Gomez J L 2015 *Website of QUANTUM* (QUANTUM, Tecnológico de Monterrey) Retrieved from: http://homepage.cem.itesm.mx/jose.luis.gomez/quantum/
[2] Nielsen M A and Chuang I L 2000 *Quantum Computation and Quantum Information* (Cambridge University Press, Cambridge)
[3] Barnett S M 2009 *Quantum Information* (Oxford University Press, New York)
[4] Jaeger G 2007 *Quantum Information: an overview* (Sringer, New York)
[5] GitHub 2015 *Website of Julia Libraries for Quantum Science and Technology* (Julia, GitHub Inc.) Retrieved from: https://github.com/JuliaQuantum
[6] Wolfram 2015 *QuCalc: A Quantum Computation Package* (QuCalc, Universite de Montreal) Retrieved from: http://library.wolfram.com/infocenter/MathSource/657/
[7] Maplesoft 2015 *Website of Physics package* (Maplesoft, Waterloo Maple Inc.) Retrieved from: http://www.maplesoft.com/support/help/maple/view.aspx?path=Physics
[8] Díaz S 2010 *Diseño de y simulación de Hamiltonianos para Cómputo Cuántico Adiabático* (Master Thesis, Tecnológico de Monterrey, México)
[9] Wolfram 2015 *Wolfram Language and System: documentation center* Retrieved from: http://reference.wolfram.com/language/
[10] Bennett C H, Brassard G, Crépeau C, Jozsa R, Peres A, Wootters W K 1993 *Phys. Rev. Lett.* **70** 1895
[11] Delgado F 2015 *J. Phys.: Conf. Series* **624** 012006
[12] Prezhdo O V and Pereverzev Y V 2000 *J. Chem. Phys.* **113** (16) 6557

Quregisters, Symmetry Groups and Clifford Algebras

D Cervantes, G Morales-Luna

Computer Science Department, CINVESTAV-IPN, México City, Mexico

E-mail: gmorales@cs.cinvestav.mx

Abstract. Natural one-to-one and two-to-one homomorphisms from SO(3) into SU(2) are built conventionally, and the collection of qubits, is identified with a subgroup of SU(2). This construction is suitable to be extended to corresponding tensor powers. The notions of qubits, quregisters and qugates are translated into the language of symmetry groups. The corresponding elements to entangled states in the tensor product of Hilbert spaces reflect entanglement properties as well, and in this way a notion of entanglement is realised in the tensor product of symmetry groups.

1. Introduction

The quantum computations may be realised as geometric transformations of the three-dimensional real space [1, 2]. Several approaches have been published within this geometric approach [3, 4]. For instance, in [5] a relation is reviewed between local unitary symmetries and entanglement invariants of quregisters from the algebraic varieties point of view, and in [6] a construction is provided via symmetric matrices for pure quantum states where their reductions are maximally mixed. Through the well known identification of the symmetry group SU(2) with the unit sphere of the two-dimensional complex Hilbert space, namely, the collection of qubits, the sphere is provided with a group structure. The qu-registers, as tensor products of qubits, in the sense of Hilbert spaces, correspond then to group tensor products. We pose the problem to compute effectively this correspondence and to transfer the notion of entangled states into the corresponding elements of group tensor products.

2. The Clifford algebra of three-dimensional space

For $\mathbb{R}^3$ with the quadratic form $\mathbf{x} \mapsto \|\mathbf{x}\|^2 = |x_0|^2 + |x_1|^2 + |x_2|^2$ the Clifford algebra is

$$\mathrm{Cl}(\mathbb{R}^3, \|\cdot\|^2) = \left(\bigoplus_{n\geq 0} (\mathbb{R}^3)^{\otimes n} \right) / \left\langle (\mathbf{x}\mathbf{x} - \|\mathbf{x}\|^2 \mathbf{1})_{\mathbf{x}\in\mathbb{R}^3} \right\rangle .$$

Besides

$$\forall \mathbf{x}, \mathbf{y} \in \mathbb{R}^3 : \ \langle \mathbf{x}|\mathbf{y}\rangle = \frac{1}{2}(\mathbf{x}\,\mathbf{y} + \mathbf{y}\,\mathbf{x}) \quad , \quad \mathbf{x}\wedge\mathbf{y} = \frac{1}{2}(\mathbf{x}\,\mathbf{y} - \mathbf{y}\,\mathbf{x})$$

where $\wedge$ denotes the usual exterior product. For any $k \in \{0, 1, 2, 3\}$, let $\bigwedge^k \mathbb{R}^3$ be the linear space spanned by the k-vectors. The exterior algebra is $\bigwedge \mathbb{R}^3 = \bigoplus_{k=0}^{3} \bigwedge^k \mathbb{R}^3$, hence it is a real 2^3-dimensional linear space. If $\{\mathbf{s}_1, \mathbf{s}_2, \mathbf{s}_3\}$ is an orthonormal basis of the space of vectors

$\bigwedge^1 \mathbb{R}^3 \approx \mathbb{R}^3$ then: $\mathbf{s}_i^2 = \mathbf{1}$ & $(i \neq j \Rightarrow \mathbf{s}_i\mathbf{s}_j = -\mathbf{s}_j\mathbf{s}_i)$. It is naturally identified $\mathbf{s}_i \leftrightarrow \sigma_i$, for the Pauli matrices

$$\sigma_1 = \begin{pmatrix} 0 & 1 \\ 1 & 0 \end{pmatrix}, \sigma_2 = \begin{pmatrix} 0 & -\imath \\ \imath & 0 \end{pmatrix}, \sigma_3 = \begin{pmatrix} 1 & 0 \\ 0 & -1 \end{pmatrix} \in U(\mathbb{H}),$$

where $\mathbb{H} = \mathbb{C}^2$ is the complex 2-dimensional Hilbert space, and $\imath = \sqrt{-1}$. Since $\sigma_1\sigma_2\sigma_3 = \imath\,\mathrm{Id}_3$, it can be defined $\boldsymbol{\imath} = \mathbf{s}_1\mathbf{s}_2\mathbf{s}_3$. In the subspace of bivectors, let $\mathbf{t}_1 = \mathbf{s}_2\mathbf{s}_3$, $\mathbf{t}_2 = \mathbf{s}_3\mathbf{s}_1$ and $\mathbf{t}_3 = \mathbf{s}_1\mathbf{s}_2$. Then, if $\mathbf{t}_i = \mathbf{s}_j\mathbf{s}_k$, $\mathbf{t}_i^2 = -\mathbf{1}$ and, $\mathbf{t}_2\mathbf{t}_1 = \mathbf{t}_3$, $\mathbf{t}_1\mathbf{t}_3 = \mathbf{t}_2$, $\mathbf{t}_3\mathbf{t}_2 = \mathbf{t}_1$, namely, $\{\mathbf{t}_1, \mathbf{t}_2, \mathbf{t}_3\}$ generates a subalgebra in $\mathrm{Cl}(\mathbb{R}^3, \|\cdot\|^2)$ that is isomorphic to Hamilton's quaternions. This algebra is called the *even subalgebra* of $\mathrm{Cl}(\mathbb{R}^3, \|\cdot\|^2)$. Let $\mathcal{C}_3 = \mathrm{Cl}(\mathbb{R}^3, \|\cdot\|^2)$ and $\mathcal{C}_3^+$ be its even subalgebra.

3. Symmetry groups of small dimension linear spaces

For any $\theta \in [-\pi, +\pi]$, and unitary vector $\mathbf{r} \in \mathbb{R}^3$, let

$$\cos(\theta)\mathbf{1} + \sin(\theta)\,\boldsymbol{\imath}\,\mathbf{r} = \sum_{j\geq 0} \frac{1}{j!}(\boldsymbol{\imath}\mathbf{r}\,\theta)^j = e^{\boldsymbol{\imath}\,\mathbf{r}\theta},$$

with $e^{\boldsymbol{\imath}\,\mathbf{r}\theta}$ an unitary vector in the even subalgebra $\mathcal{C}_3^+$. Thus, it is properly an unitary quaternion, $e^{\boldsymbol{\imath}\,\mathbf{r}\theta} = a_0\mathbf{1} + a_1\mathbf{t}_1 + a_2\mathbf{t}_2 + a_3\mathbf{t}_3$ with $|a_0|^2 + |a_1|^2 + |a_2|^2 + |a_3|^2 = 1$.

Through the correspondence

$$\Phi : e^{\boldsymbol{\imath}\,\mathbf{r}\theta} \mapsto U = \begin{pmatrix} a_0 + \imath a_3 & a_1 + \imath a_2 \\ -a_1 + \imath a_2 & a_0 - \imath a_3 \end{pmatrix} = a_0\mathrm{Id}_{\mathbb{H}} + \imath a_2\sigma_1 + \imath a_1\sigma_2 + \imath a_3\sigma_3$$

SO(3) is identified with SU(2), the group of unitary transforms in $\mathbb{H}$ which preserve the orientation: $e^{-\boldsymbol{\imath}\,\mathbf{r}\frac{\theta}{2}}$ is a rotation of angle θ with $\mathbf{r}$ as axis of rotation.

A second identification of SO(3) with SU(2) is obtained by the composition of the following maps:

$$(\theta, \mathbf{r}) \overset{\Psi_0}{\mapsto} \mathbf{c} = \begin{bmatrix} c_0 \\ c_1 \end{bmatrix} = \begin{bmatrix} \cos(\theta) - \imath r_3 \sin(\theta) \\ r_1 \sin(\theta) + \imath r_2 \sin(\theta) \end{bmatrix} \overset{\Psi_1}{\mapsto} C = \begin{bmatrix} c_0 & -\overline{c_1} \\ c_1 & \overline{c_0} \end{bmatrix}. \tag{1}$$

Any rotation in SO(3) is seen as a *spinor*, and through the map Ψ_0 it corresponds to a qubit $\mathbf{c} \in S_1 = \{(c_0, c_1) \in \mathbb{H}|\ |c_0|^2 + |c_1|^2 = 1\}$. Via the map Ψ_1, any qubit is identified with an element of SU(2). Conversely, if $C = \begin{bmatrix} c_0 & d_0 \\ c_1 & d_1 \end{bmatrix} \in \mathrm{SU}(2)$ then $c_0d_1 - c_1d_0 = 1$. By assuming $(c_0, c_1) \in S_1$, the solutions of this equation are the points $(d_0, d_1) \in \mathbb{H}$ in the straight line passing through $(-\overline{c_1}, \overline{c_0})$ parallel to the straight-line orthogonal to $(c_0, -c_1)$. Since $(c_0, -c_1) \in S_1$, this line is tangent to S_1 at this point. The solution line is parameterised thus as $D(d) = (-\overline{c_1}, \overline{c_0}) + d(1, -\overline{c_0^{-1}c_1})$, with $d \in \mathbb{C}$. Also $(-\overline{c_1}, \overline{c_0}) \in S_1$, and the only solution (d_0, d_1) of $c_0d_1 - c_1d_0 = 1$ in S_1 is $(d_0, d_1) = (-\overline{c_1}, \overline{c_0})$. Thus $C = \Psi_1(c_0, c_1)$. Hence, Ψ_1 is a bijection $S_1 \to \mathrm{SU}(2)$. The operation in the group SU(2) translated into S_1 is

$$\star_1 : S_1 \times S_1 \to S_1\ ,\ \left(\begin{bmatrix} c_{00} \\ c_{10} \end{bmatrix}, \begin{bmatrix} c_{01} \\ c_{11} \end{bmatrix}\right) \mapsto \begin{bmatrix} c_{00} \\ c_{10} \end{bmatrix} \star_1 \begin{bmatrix} c_{01} \\ c_{11} \end{bmatrix} = \begin{bmatrix} c_{00}c_{01} - \overline{c_{10}}c_{11} \\ c_{10}c_{01} + \overline{c_{00}}c_{11} \end{bmatrix},$$

hence $(S_1, \star_1, \begin{bmatrix} 1 \\ 0 \end{bmatrix})$ is a group.

4. Tensor products

4.1. Tensor product of Hilbert spaces

For two complex Hilbert spaces H_0, H_1 their tensor product $H_0 \otimes H_1 = \mathrm{Lin}(H_1^*, H_0)$ satisfies a proper *Universal Property*:

> For any space H_2 and any continuous bilinear map $\phi : H_0 \times H_1 \to H_2$ there is a unique bounded linear map $\psi : H_0 \otimes H_1 \to H_2$ such that $\phi = \psi \circ \iota$, where $\iota : H_0 \times H_1 \to H_0 \otimes H_1$, $(\mathbf{x}, \mathbf{y}) \mapsto \mathbf{x} \otimes \mathbf{y}$, is the canonical embedding.

$H_0 \otimes H_1$ has as basis the Kroenecker product of corresponding basis in H_0 and H_1.

4.2. Tensor product of groups

Let V be a vector space and $\mathrm{GL}(V)$ its group of linear automorphisms. A *group representation* of a group G is a group homomorphism $\pi : G \to \mathrm{GL}(V)$. If V is a complex finite-dimensional linear space, of dimension n, then $\mathrm{GL}(V)$ is isomorphic to $\mathrm{GL}(n, \mathbb{C})$ and by composing the representation π with this isomorphism it is obtained a group homomorphism $G \to \mathrm{GL}(n, \mathbb{C})$.

Suppose that $\pi_0 : G_0 \to \mathrm{GL}(n_0, \mathbb{C})$ and $\pi_1 : G_1 \to \mathrm{GL}(n_1, \mathbb{C})$ are two finite dimensional representations of the groups G_0 and G_1. Let $(|i\rangle)_{i=0}^{n_0-1}$ and $(|j\rangle)_{j=0}^{n_1-1}$ be basis of $\mathbb{C}^{n_0}$ and $\mathbb{C}^{n_1}$ respectively. Then the tensor product $\mathbb{C}^{n_0} \otimes \mathbb{C}^{n_1} = \mathbb{C}^{n_0 n_1}$ has as basis $(|i\rangle \otimes |j\rangle)_{(i,j)\in\{0,\ldots,n_0\}\times\{0,\ldots,n_1\}}$. For any pair $(g_0, g_1) \in G_0 \times G_1$ let $P_{(g_0,g_1)} : \mathbb{C}^{n_0 n_1} \to \mathbb{C}^{n_0 n_1}$ be the linear map such that

$$\forall (i,j) \in \{0,\ldots,n_0\} \times \{0,\ldots,n_1\} : \; P_{(g_0,g_1)}(|i\rangle \otimes |j\rangle) = \pi_0(g_0)(|i\rangle) \otimes \pi_1(g_1)(|j\rangle).$$

The set of vectors $\left(P_{(g_0,g_1)}(|i\rangle \otimes |j\rangle)\right)_{(i,j)\in\{0,\ldots,n_0\}\times\{0,\ldots,n_1\}}$ is linearly independent, hence $P_{(g_0,g_1)} \in \mathrm{GL}(n_0 n_1, \mathbb{C})$. Consequently, the operator

$$\pi_0 \otimes \pi_1 : G_0 \times G_1 \to \mathrm{GL}(n_0 n_1, \mathbb{C}) \;,\; (g_0, g_1) \mapsto (\pi_0 \otimes \pi_1)(g_0, g_1) = P_{(g_0,g_1)}$$

is an embedding $G_0 \times G_1 \to \mathrm{GL}(n_0 n_1, \mathbb{C})$.

Let $F_{01} = (\pi_0 \otimes \pi_1)(G_0 \times G_1)$ be the image of $G_0 \times G_1$ under the embedding $\pi_0 \otimes \pi_1$ and $\langle F_{01} \rangle$ be the subgroup of $\mathrm{GL}(n_0 n_1, \mathbb{C})$ spanned by F_{01}. Then the group tensor product $G_0 \otimes G_1$ is represented by the group $\langle F_{01} \rangle < \mathrm{GL}(n_0 n_1, \mathbb{C})$. The group tensor product satisfies the following *Universal Property*:

> For any group J and any bilinear map $\phi : G_0 \times G_1 \to J$ there is a unique homomorphism $\psi : G_0 \otimes G_1 \to J$ such that $\phi = \psi \circ \iota$.

Thus the group tensor product is associative.

For any group G with a finite-dimensional representation, its tensor powers are defined recursively:

$$G^{\otimes 2} = G \otimes G \;\; \& \;\; \forall n > 2 : \; G^{\otimes n} = G^{\otimes(n-1)} \otimes G.$$

5. Embedding of $S_1^{\otimes n}$ into $\mathrm{SU}(2)^{\otimes n}$

5.1. General embedding

Let $\mathbb{H}_1 = \mathbb{H}$ and $\forall n > 1$: $\mathbb{H}_n = \mathbb{H}_{n-1} \otimes \mathbb{H}$. Clearly, $\dim \mathbb{H}_n = 2^n$. The unit sphere S_{2^n-1} of $\mathbb{H}_n$ is the set of n-quregisters.

The map Ψ_1 defined at relation (1) is a representation of the group $(S_1, \star_1, \begin{bmatrix} 1 \\ 0 \end{bmatrix})$, isomorphic to $\mathrm{SU}(2) < \mathrm{GL}(2, \mathbb{C})$. Then, the group tensor power $S_1^{\otimes n}$ is isomorphic to the group $\mathrm{SU}(2)^{\otimes n} = \langle F_1^{\otimes n} \rangle < \mathrm{GL}(2^n, \mathbb{C})$ spanned by $F_1^{\otimes n} = (\Psi_1(S_1))^{\otimes n}$ in $\mathrm{GL}(2^n, \mathbb{C})$. Namely, the elements in $\mathrm{SU}(2)^{\otimes n}$ can be written as finite length words of the form $T_0^{\varepsilon_0} \cdots T_{k-1}^{\varepsilon_{k-1}}$, with $T_i \in F_1^{\otimes n}$, $\varepsilon_i \in \{-1, +1\}$ and concatenation assumed as map composition, being two such words equivalent if they represent the same element in $\mathrm{GL}(2^n, \mathbb{C})$. In this context, the length of the minimal length word representing an element $T \in \mathrm{SU}(2)^{\otimes n}$ can be seen as a degree of separability of the map $T : \mathbb{H}_n \to \mathbb{H}_n$.

Let G be any of SO(3) or SU(2) and let ϕ be an embedding $S_1 \to G$. By acting componentwise, there is a natural embedding $\phi^n : S_1^n \to G^n$. Then there are maps α_n, due to the Universal Property of the Hilbert Space Tensor Product, and β_n, due to the Universal Property of the Group Tensor Product, that make the following diagram commutative:

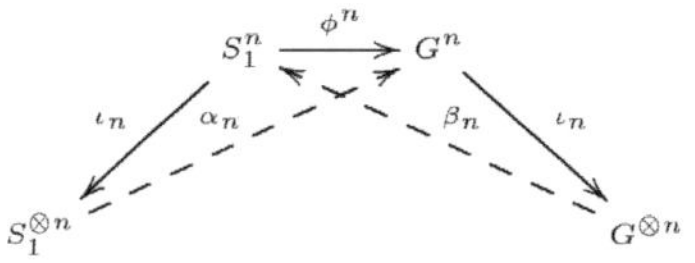

(2)

Again by the Universal Properties there are two maps ψ_n and ω_n commuting the following diagram:

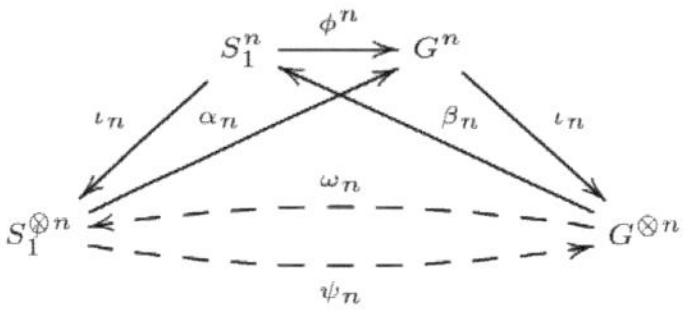

(3)

Indeed if the elements in the group G are represented by matrices, then the map ψ_n is transforming vector tensor product into matrix tensor products. Hence, $G^{\otimes n}$ can be identified with $S_1^{\otimes n}$.

Besides, there exists a bijection $\psi_n : S_{2^n-1} \to \mathrm{SU}(2)^{\otimes n}$, where S_{2^n-1} is the unit sphere of $\mathbb{H}_n$ and through this bijection, the collection S_{2^n-1} of n-quregisters is realised as the n-fold group tensor product of $S_1 \approx \mathrm{SU}(2)$.

5.2. *Examples of the embedding for lower dimensions*

Let us consider the following pairs of index and sign matrices:

$n = 1$

$$I_1 = \begin{bmatrix} 0 & 1 \\ 1 & 0 \end{bmatrix} \quad \Sigma_1 = \begin{bmatrix} 1 & -1 \\ 1 & 1 \end{bmatrix}$$

$n = 2$

$$I_2 = \begin{bmatrix} 0 & 1 & 2 & 3 \\ 1 & 0 & 3 & 2 \\ 2 & 3 & 0 & 1 \\ 3 & 2 & 1 & 0 \end{bmatrix} \quad \Sigma_2 = \begin{bmatrix} 1 & -1 & -1 & 1 \\ 1 & 1 & -1 & -1 \\ 1 & -1 & 1 & -1 \\ 1 & 1 & 1 & 1 \end{bmatrix}$$

$n = 3$

$$I_3 = \begin{bmatrix} 0 & 1 & 2 & 3 & 4 & 5 & 6 & 7 \\ 1 & 0 & 3 & 2 & 5 & 4 & 7 & 6 \\ 2 & 3 & 0 & 1 & 6 & 7 & 4 & 5 \\ 3 & 2 & 1 & 0 & 7 & 6 & 5 & 4 \\ 4 & 5 & 6 & 7 & 0 & 1 & 2 & 3 \\ 5 & 4 & 7 & 6 & 1 & 0 & 3 & 2 \\ 6 & 7 & 4 & 5 & 2 & 3 & 0 & 1 \\ 7 & 6 & 5 & 4 & 3 & 2 & 1 & 0 \end{bmatrix} \quad \Sigma_3 = \begin{bmatrix} 1 & -1 & -1 & 1 & -1 & 1 & 1 & -1 \\ 1 & 1 & -1 & -1 & -1 & -1 & 1 & 1 \\ 1 & -1 & 1 & -1 & -1 & 1 & -1 & 1 \\ 1 & 1 & 1 & 1 & -1 & -1 & -1 & -1 \\ 1 & -1 & -1 & 1 & 1 & 1 & -1 & -1 \\ 1 & 1 & -1 & -1 & 1 & -1 & -1 & -1 \\ 1 & -1 & 1 & -1 & 1 & -1 & 1 & -1 \\ 1 & 1 & 1 & 1 & 1 & 1 & 1 & 1 \end{bmatrix}$$

Let $\rho_n = \exp\left(\imath \frac{2\pi}{2^n}\right) = \exp\left(\imath \frac{\pi}{2^{n-1}}\right)$ be the 2^n-th primitive root of unit. Then, the corresponding bijections $\psi_n : S_{2^n-1} \to \mathrm{SU}(2)^{\otimes n}$ are

$$\begin{array}{llllll} \psi_1 : S_1 & \to & \mathrm{SU}(2) & , & [x_i]_{i=0}^{2^1-1} & \mapsto \left[\sigma_{1ij} \rho_1^j x_{I_{1ij}}\right]_{i,j=0}^{2^1-1} \\ \psi_2 : S_1^{\otimes 2} & \to & \mathrm{SU}(2)^{\otimes 2} & , & [x_i]_{i=0}^{2^2-1} & \mapsto \left[\sigma_{2ij} \rho_2^j x_{I_{2ij}}\right]_{i,j=0}^{2^2-1} \\ \psi_3 : S_1^{\otimes 3} & \to & \mathrm{SU}(2)^{\otimes 3} & , & [x_i]_{i=0}^{2^3-1} & \mapsto \left[\sigma_{3ij} \rho_3^j x_{I_{3ij}}\right]_{i,j=0}^{2^3-1} \end{array}$$

where I_{kij} and σ_{kij} are that ij-th entries of matrices I_k and Σ_k, respectively.

It can be seen in a direct way that each identification ψ_n coincides with the linear operator tensor product when restricted to separable states.

6. Maximally non-separable elements

Let $Q = \{0, 1\}$ be the set of classical bits and let $\{|0\rangle, |1\rangle\}$ denote the canonical basis of $\mathbb{H}_1$.

For any $n \geq 1$ and any $\varepsilon = \varepsilon_{n-1} \cdots \varepsilon_1 \varepsilon_0 \in Q^n$ let $|\varepsilon\rangle = |\varepsilon_{n-1}\rangle \otimes \cdots \otimes |\varepsilon_1\rangle \otimes |\varepsilon_0\rangle$. Then $(|\varepsilon\rangle)_{\varepsilon \in Q^n}$ is the canonical basis of $\mathbb{H}_n$.

For any $n \geq 2$ and any $\varepsilon = \varepsilon_{n-1} \cdots \varepsilon_1 \varepsilon_0 \in Q^n$ let

$$\mathbf{b}_\varepsilon = \frac{1}{\sqrt{2}} \left(|0\varepsilon_{n-2} \cdots \varepsilon_1 \varepsilon_0\rangle + (-1)^{\varepsilon_{n-1}} |1\overline{\varepsilon_{n-2}} \cdots \overline{\varepsilon_1}\, \overline{\varepsilon_0}\rangle \right).$$

Then $(\mathbf{b}_\varepsilon)_{\varepsilon \in Q^n}$ is the Bell basis of $\mathbb{H}_n$, and it consists of maximally entangled states.

The corresponding image of linear automoprhisms

$$(\psi_n(\mathbf{b}_\varepsilon))_{\varepsilon \in Q^n} \subset \mathrm{SU}(2)^{\otimes n} < \mathrm{GL}(2^n, \mathbb{C})$$

is a collection of maximally entangled elements in $\mathrm{SU}(2)^{\otimes n}$.

In this way, the notion of entanglement is transported directly into the context of geometric transformations.

Since the even algebra $\mathcal{C}_3^+$ of the Clifford algebra $\mathrm{Cl}(\mathbb{R}^3, \|\cdot\|^2)$ is isomorphic to Hamilton's quaternions, the calculation procedures given in [7, 8] may be used within the context of the current paper.

7. Conclusion

The main purpose of the current paper is the translation of the most basic notions of Quantum Computing into symmetry groups. Through this translation, qubits become orientation preserving unitary transforms, and quregisters become tensor products of orientation preserving unitary transforms. Thus, the collection of qubits acquire a group structure and the collection of quregisters, which are tensor products of qubits, acquire as well a group structure which is, in a consistent way, the group tensor product of the corresponding quit group structure. By the way, using group representations, the involved group tensor products can efficiently be represented, e.g., in a manageable way.

References

[1] Havel T F and Doran J L 2001 Geometric algebra in quantum information processing *CoRR* arXiv:quant-ph/0004031v3

[2] Lomonaco S J and Brandt H E 2002 *Quantum Computation and Information: AMS Special Session Quantum Computation and Information*, January 19-21, 2000, Washington. Number pt. 3 in Contemporary mathematics - American Mathematical Society. American Mathematical Society

[3] Labunets V, Rundblad E, and Astola J 2002 *Applications of Geometric Algebra in Computer Science and Engineering* (Is the brain a Clifford algebra quantum computer?) ed L Dorst, C Doran, and J Lasenby (Birkhäuser Boston) 285–295

[4] Cafaro C and Mancini S 2011 A geometric algebra perspective on quantum computational gates and universality in quantum computing *AACA* **21(3)** 493–519

[5] Johansson M 2014 Local unitary symmetries and entanglement invariants *J Phys. A: Math. Theor.* **47(46)** 465301

[6] Feng K, Jin L, Xing C, and Yuan C 2015 Multipartite entangled states, symmetric matrices and error-correcting codes *CoRR* arXiv:quant-ph/1511.07992

[7] Girard P R 2004 *Quaternions, algèbre de Clifford et physique relativiste* (Romandes PPU)

[8] Girard P R 2007 *Quaternions, Clifford Algebras and Relativistic Physics* (Birkhäuser)

Journal of Physics: Conference Series **698** (2016) 012021 doi:10.1088/1742-6596/698/1/012021

On the Eigenvalue Problem in Multipartite Quantum Systems

M Enríquez[1,2]

[1]UPIITA, Instituto Politécnico Nacional, Av. Instituto Politécnico Nacional 2580 CP 07340 México D.F., México
[2]Departamento de Física, Cinvestav, AP 14-740, 07000 México DF, Mexico

E-mail: menriquezf@fis.cinvestav.mx

Abstract. We show that the Hamiltonian of a multiqudit system can be diagonalized through a sequence of unitary transformations written in terms of Hubbard operators. As an application this formalism is applied to particular cases of two and three-qubit systems.

1. Introduction

The eigenvalue problem plays a relevant role in quantum mechanics. Indeed, the set of eigenvalues of a linear operator A yields the numerical results that one would obtain when the observable represented by A is measured [1]. In the case where the system of study is integrated by several parts some algebraic subtleties should be taken into account. For instance, consider a system of n subsystems with d levels each, the pure states are represented by vectors in the Hilbert space $\mathcal{H}_d^{\otimes n}$, here $\otimes$ denotes the Kronecker (tensor or direct) product [2, 3], and $\mathcal{H}_d$ stands for the Hilbert space of a single d-level system (*qudit*). An operator A_k acting on the k-th subsystem is promoted to act on the entire space in the following way

$$\mathbf{A}_k \leftrightarrow \underbrace{\mathbb{I}_d \otimes \mathbb{I}_d \otimes \cdots \otimes \mathbb{I}_d}_{(k-1)-times} \otimes A_k \underbrace{\otimes \mathbb{I}_d \otimes \cdots \mathbb{I}_d}_{(n-k)-times}, \tag{1}$$

where $\mathbb{I}_d$ is the identity operator in $\mathcal{H}_d$. It is not difficult to realize that the matrix representation of (1) has at most $d^{n+1} < d^{2n}$ entries different from zero. This fact suggests to look for a convenient framework to handle with the involved cumbersome calculations. To approach the tensor product algebra in a more convenient way we proposed the use of Hubbard operators [2–4]. Indeed, these are also refered as X-operators and satisfy the following properties

i) $X^{i,j}X^{k,m} = \delta_{jk}X^{i,m}$ (multiplication rule)

ii) $\sum_k X^{k,k} = \mathbb{I}$ (completeness)

iii) $(X^{i,j})^\dagger = X^{j,i}$ (non-hermiticity)

Because of property ii) any linear operator A acting on the whole system can be expressed in terms of the X-operators. For instance, the Hamiltonian of the system in this representation reads

$$H = \sum_{i_{n-1},\cdots,i_0} h_{i_{n-1},\ldots,i_0;j_{n-1},\ldots,j_0} X^{i_{n-1},j_{n-1}} \otimes \cdots \otimes X^{i_1,j_1} \otimes X^{i_0,j_0}, \tag{2}$$

Of particular interest is the diagonalization of Hamiltonians with nearest-neighbour spin interaction. Among them are the bipartite three dimensional anisotropic Ising model [5], the XY Heisenberg model for spin-1/2, which can be exactly diagonalized using the Jordan-Wigner transformation [6], etc. For higher spins the diagonalization of some Heisenberg models have been accomplished numerically [7]. In the present work we use the tensor product properties of the Hubbard operators to diagonalize some spin interacting Hamiltonians. Indeed, it has been shown in [2,4] that the tensor products in equation (2) can be written as an X-operator depending on just two indices. Such a representation allows us to generalize the Unitary Transformation Method (UTM) introduced by Ovchinnikov [8] to diagonalize a Hermitian operator in an easy manner.

This contribution is organized as follows. In Section 2 we present a short review of the UTM for a single d-level system. The formalism of Hubbard operators for systems of several qudits is discussed in Section 3. It is shown that cumbersome calculations of the tensor product can be turned into simple relations of indices. Moreover, we generalize the UTM for hermitian operators acting on a multiqudit space. In Sec. 4 this formalism is applied to specific cases of two and three-qubit Hamiltonians. Finally, some remarks are given in Sec. 5.

2. A review of the unitary transformation method

Let $B = \{|p\rangle : p = 0, 1, \ldots, d-1\}$ be an orthonormal basis of $\mathcal{H}_d$. Thus, a vector $|\psi\rangle \in \mathcal{H}$ can be written as a normalized linear combination of the basis elements

$$|\psi\rangle = \sum_{p=0}^{d-1} c_p|p\rangle, \quad c_p \in \mathbb{C}, \quad \sum_{p=0}^{d-1} |c_p|^2 = 1. \tag{3}$$

On the other hand, the simplest representation of Hubbard operators acting on $\mathcal{H}_d$ is constructed as the following outer products

$$X_d^{p,q} := |p\rangle\langle q|, \quad p, q = 0, 1, \ldots, d-1, \tag{4}$$

where the sub-index d makes reference to the dimension of the space. It is not difficult to prove that the operator so defined fulfill the properties i) – iii). Furthermore, the matrix representation of (4) has 1 in the entry $(i+1, j+1)$ and zero elsewhere. The action of the operator $X_d^{p,q}$ on any basis vector follows immediately

$$X_d^{p,q}|r\rangle = \delta_{q,r}|p\rangle. \tag{5}$$

Thus, the Hamiltonian H of the system can be written as a linear combination of Hubbard operators

$$H = \sum_{p=0}^{d-1} \epsilon_p X_d^{p,p} + \sum_{p\neq q}^{d-1} V_{p,q} X_d^{p,q}, \quad \epsilon_d \in \mathbb{R}, \quad V_{p,q} \in \mathbb{C}, \quad V_{q,p} = V_{p,q}^*, \tag{6}$$

the conditions on the coefficients follow from the fact $H^\dagger = H$. We restrict ourselves to the case where all coefficients are real, hence $V_{q,p} = V_{p,q}$. In order to solve the eigenvalue equation

$$H|E_p\rangle = E_p|E_p\rangle, \quad p = 0, 1, \ldots, d-1, \tag{7}$$

we look for an unitary operator U so that the transformed Hamiltonian

$$H' = UHU^\dagger = \sum_{p=0}^{d-1} E_p X_d^{p,p}. \tag{8}$$

is indeed diagonal. Note that in this representation the states $|p\rangle$ are eigenstates of (8) with eigenvalue E_p. Remark that degeneracy is not been taking into account. Moreover, the states fulfilling (7) are obtained as $|E_p\rangle = U^\dagger|p\rangle$, for $p = 0, 1, \ldots, d-1$. In order to find U we first consider the $d(d-1)/2$ unitary operators defined as

$$U_{k,\ell}(\alpha_{k,\ell}) = \exp[\alpha_{k,\ell}(X_d^{k,\ell} - X_d^{\ell,k})], \quad k < \ell, \quad \alpha_{k,\ell} \in \mathbb{R}. \tag{9}$$

For fixed indices k and ℓ and using the Baker-Campell-Hausdorff formula it is easy to show that the operator (9) changes H to

$$H^{(1)} = U_{k,\ell}(\alpha_{k,\ell})HU_{k,\ell}^\dagger(\alpha_{k,\ell}) = \sum_{p=0} \epsilon_p^{(1)} X_n^{p,p} + \sum_{p\neq q} V_{p,q}^{(1)} X_n^{p,q}, \tag{10}$$

where

$$\begin{aligned}
\epsilon_{k_l}^{(1)} &= \tfrac{1}{2}(\epsilon_k + \epsilon_\ell) + \tfrac{1}{2}(\epsilon_k - \epsilon_\ell)\cos 2\alpha_{k,\ell} + V_{k,\ell}\sin 2\alpha_{k,\ell}, \\
\epsilon_\ell^{(1)} &= \tfrac{1}{2}(\epsilon_k + \epsilon_\ell) - \tfrac{1}{2}(\epsilon_k - \epsilon_\ell)\cos 2\alpha_{k,\ell} - V_{k,\ell}\sin 2\alpha_{k,\ell}, \\
V_{k,\ell}^{(1)} &= -\tfrac{1}{2}(\epsilon_k - \epsilon_\ell)\sin 2\alpha_{k,\ell} + V_{k,\ell}\cos 2\alpha_{k,\ell}, \\
V_{k,p}^{(1)} &= V_{k,p}\cos\alpha_{k,\ell} + V_{\ell,p}\sin\alpha_{k,\ell}, \quad V_{\ell,p}^{(1)} = V_{p,\ell}\cos\alpha_{k,\ell} - V_{p,k}\sin\alpha_{k,\ell}, \quad p \neq k,\ell, \\
\epsilon_p^{(1)} &= \epsilon_p, \quad V_{p,q}^{(1)} = V_{p,q}, \quad p,q \neq k,\ell.
\end{aligned} \tag{11}$$

The former procedure can be repeated iteratively to get $H^{(r)} = VHV^\dagger$ where the transformation $V = V(\alpha_{k_r,\ell_r}, \ldots, \alpha_{k_1,\ell_1}) = U_{k_r,\ell_r}\cdots U_{k_1,\ell_1}$, with $r \leq n(n-1)/2$. To erase the off-diagonal elements we require $V_{k_1,\ell_1}^{(r)} = V_{k_2,\ell_2}^{(r)} = \ldots = V_{k_r,\ell_r}^{(r)} = 0$. These conditions constitue a system of r non-linear equations in the parameters $\alpha_{k_1,\ell_1}, \ldots, \alpha_{k_r,\ell_r}$, whose solution allows to find the operator V. Clearly, the eigenvalues of both $H^{(r)}$ and H are the numbers $\epsilon_p^{(r)}$. Moreover, the eigenvectors of H are computed as $|E_p\rangle = V^\dagger|p\rangle$. According to Ovchinnikov [8] we should solve $d(d-1)/2$ non-linear equations, however it will be shown that in some cases the number of equations and parameters can be reduced.

3. Multiquidit systems

The Hilbert space $\mathcal{H}_d^{\otimes n}$ of a system of n qudits is constructed as follows

$$\mathcal{H}_d^{\otimes n} = \text{Span}\{|i_{n-1}\rangle \otimes \cdots \otimes |i_1\rangle \otimes |i_0\rangle, \; i_\ell = 0, 1, \ldots, d-1\}. \tag{12}$$

According to Ref. [4] the basis elements can be labeled using a single-index notation identifying $|i_{n-1}\rangle \otimes \cdots \otimes |i_1\rangle \otimes |i_0\rangle \equiv |i_{n-1}\cdots i_0\rangle = |i\rangle$, where $i = (i_{n-1}\cdots i_0)_d$. The last expression indicates that the numbers i_ℓ are the coefficients of the expansion of i in base d and they are related with the physical state of the $(n-\ell)$-th subsystem. Moreover, any vector $|\psi\rangle \in \mathcal{H}_d^{\otimes n}$ is a linear combination of the form

$$|\psi\rangle = \sum_{i=0}^{d^n-1} c_i|i\rangle = \sum_{i=0}^{d^n-1} c_i|i_{n-1}\cdots i_0\rangle, \quad i = (i_{n-1}\cdots i_0)_d. \tag{13}$$

On the other hand, it was shown in [4] that any linear operator A acting on the Hilbert space of a multipartite system can be written in terms of Hubbard operators using a single-index for both rows and columns, for instance the Hamiltonian (2) reads

$$H = \sum_{i,j=0}^{d^n-1} h_{i_{n-1},\ldots,i_0;j_{n-1},\ldots,j_0} X_{d^n}^{i,j}, \quad i = (i_{n-1}\cdots i_0)_d, \quad j = (j_{n-1}\cdots j_0)_d. \tag{14}$$

This representation is convenient to deal with some algebraic calculations. For instance, the action of H on a linear combination (13) follows immediately from equation (5). Note that if a n-qudit Hamiltonian is written in this form then can be diagonalized by the UTM. This observation generalizes the method for multiqudit systems. In the following Section two particular cases are studied.

4. Applications

An operator defined as follows

$$H_2 = \epsilon_n X_d^{n,n} + \epsilon_m X_d^{m,m} + V_{n,m}\{X_d^{n,m} + X_d^{m,n}\}, \quad m > n, \quad \epsilon_m \geq \epsilon_n. \tag{15}$$

is known as a *two-level form*. In particular, when $d = 2$ the well-known two-level Hamiltonian is recovered. We now consider the transformation (10) on H_2 with $k = n$ and $\ell = m$. To take the operator (15) to a diagonal form we require $V_{n,m}^{(1)} = 0$, which yields the condition

$$(\epsilon_n - \epsilon_m)\sin(2\alpha_{n,m}) = 2V_{n,m}\cos(2\alpha_{n,m}), \tag{16}$$

hence it is obtained

$$\sin(2\alpha_{n,m}) = \frac{-2V_{n,m}}{[(\epsilon_n-\epsilon_m)^2 + 4(V_{n,m})^2]^{1/2}}, \quad \cos(2\alpha_{n,m}) = \frac{\epsilon_m - \epsilon_n}{[(\epsilon_n-\epsilon_m)^2 + 4(V_{n,m})^2]^{1/2}}. \tag{17}$$

Substituting the last equations in the expressions for $\epsilon_n^{(1)}$ and $\epsilon_m^{(1)}$ in (11) the two eigenvalues of H_2 are immediately computed

$$\begin{aligned} E_n &= \epsilon_n^{(1)} = \tfrac{1}{2}(\epsilon_n + \epsilon_m) - \tfrac{1}{2}\left[(\epsilon_n - \epsilon_m)^2 + 4(V_{n,m})^2\right]^{1/2}, \\ E_m &= \epsilon_m^{(1)} = \tfrac{1}{2}(\epsilon_n + \epsilon_m) + \tfrac{1}{2}\left[(\epsilon_n - \epsilon_m)^2 + 4(V_{n,m})^2\right]^{1/2}. \end{aligned} \tag{18}$$

The eigenvectors are calculated as $|E_p\rangle = U_{n,m}^\dagger(\alpha_{n,m})|p\rangle$, for $p = n, m$. Hence, we obtain

$$|E_n\rangle = \cos\alpha_{n,m}|n\rangle + \sin\alpha_{n,m}|m\rangle, \quad |E_m\rangle = \cos\alpha_{n,m}|n\rangle - \sin\alpha_{n,m}|m\rangle, \tag{19}$$

where

$$\cos\alpha_{n,m} = \left[\frac{E_m - \epsilon_n}{E_m - E_n}\right]^{1/2}. \tag{20}$$

As an example consider the Hamiltonian for the two-qubit anisotropic Ising model proposed by Delgado [5]

$$H_2 = -\sum_{i=1}^{2} \lambda_i \sigma_i \otimes \sigma_i + b_3(1 \otimes \sigma_3) + c_3(\sigma_3 \otimes 1), \tag{21}$$

where σ_i are the Pauli matrices and λ_i, b_3 and c_3 are real parameters. Using equation (14) and identifying $\lambda_1 = \lambda$, $\lambda_2 = \chi$ and $\lambda_3 = \beta$ this operator can be written as $H_2 = h_1 + h_2$, where

$$\begin{aligned} h_1 &= (\chi - \lambda)\{X_4^{0,3} + X_4^{3,0}\} + (b_3 + c_3 - \beta)X_4^{0,0} - (b_3 + c_3 + \beta)X_4^{3,3} \\ h_2 &= (\chi + \lambda)\{X_4^{1,2} + X_4^{2,1}\} + (-b_3 + c_3 + \beta)X_4^{1,1} - (b_3 - c_3 + \beta)X_4^{2,2}. \end{aligned} \tag{22}$$

are two-level forms. In order to diagonalize H_2 we apply a sequence of two transformations (9) as follows $H_2^{(2)} = U_{1,2}U_{0,3}H_2U_{0,3}^\dagger U_{1,2}^\dagger$. Note that the action of $U_{0,3}$ ($U_{1,2}$) leaves $h_2(h_1)$ invariant. So, the conditions $V_{0,3}^{(2)} = V_{1,2}^{(2)} = 0$ yield the parameters $\alpha_{0,3}$ and $\alpha_{1,2}$. Moreover, the eigenvalues are computed using equations (19)

$$\begin{aligned} E_0 &= -\beta - \left[(b_3 + c_3)^2 + (\chi - \lambda)^2\right]^{1/2}, \quad E_1 = \beta - \left[(b_3 - c_3)^2 + (\chi + \lambda)^2\right]^{1/2}, \\ E_2 &= \beta + \left[(b_3 - c_3)^2 + (\chi + \lambda)^2\right]^{1/2}, \quad E_3 = -\beta + \left[(b_3 + c_3)^2 + (\chi - \lambda)^2\right]^{1/2}. \end{aligned} \tag{23}$$

The corresponding eigenvectors in the single-index and computational basis read

$$\begin{aligned} |E_0\rangle &= \cos\alpha_{0,3}|0\rangle + \sin\alpha_{0,3}|3\rangle = \cos\alpha_{0,3}|00\rangle + \sin\alpha_{0,3}|11\rangle, \\ |E_1\rangle &= \cos\alpha_{1,2}|1\rangle + \sin\alpha_{1,2}|2\rangle = \cos\alpha_{1,2}|01\rangle + \sin\alpha_{1,2}|10\rangle, \\ |E_2\rangle &= \cos\alpha_{1,2}|1\rangle - \sin\alpha_{1,2}|2\rangle = \cos\alpha_{1,2}|01\rangle - \sin\alpha_{1,2}|10\rangle, \\ |E_3\rangle &= \cos\alpha_{0,3}|0\rangle - \sin\alpha_{0,3}|3\rangle = \cos\alpha_{0,3}|00\rangle - \sin\alpha_{0,3}|11\rangle, \end{aligned} \tag{24}$$

where

$$\cos\alpha_{0,3} = \left[\frac{E_3 - \epsilon_0}{E_3 - E_0}\right]^{1/2}, \quad \text{and} \quad \cos\alpha_{1,2} = \left[\frac{E_2 - \epsilon_1}{E_2 - E_1}\right]^{1/2}. \tag{25}$$

We now consider a particular case of a *four-level form* given by

$$\begin{aligned} H_4 &= \epsilon_p X_d^{p,p} + \epsilon_q \left(X_d^{q,q} + X_d^{r,r} + X_d^{s,s}\right) \\ &+ \left\{\gamma\left(X_d^{p,q} + X_d^{p,r} + X_s^{p,s}\right) + \delta\left(X_d^{q,r} + X_d^{q,s} + X_d^{r,s}\right) + h.c.\right\}, \quad \gamma, \delta \in \mathbb{R}. \end{aligned} \tag{26}$$

In order to diagonalize H_4 we take a sequence of transformations (9). The entries of the transformed operator $H_4^{(1)} = U_{q,s}H_4U_{q,s}^\dagger$ read

$$\begin{aligned} \epsilon_q^{(1)} &= \epsilon_q + \delta\sin(2\alpha_{q,s}), \quad \epsilon_s^{(1)} = \epsilon_q - \delta\sin(2\alpha_{q,s}), \quad V_{q,s}^{(1)} = \delta\cos(2\alpha_{q,s}), \\ V_{p,q}^{(1)} &= \gamma\cos\alpha_{q,s} + \gamma\sin\alpha_{q,s}, \quad V_{q,r}^{(1)} = \delta\cos\alpha_{q,s} + \delta\sin\alpha_{q,s}, \\ V_{p,s}^{(1)} &= \gamma\cos\alpha_{q,s} - \gamma\sin\alpha_{q,s}, \quad V_{r,s}^{(1)} = \delta\cos\alpha_{q,s} - \delta\sin\alpha_{q,s}. \end{aligned} \tag{27}$$

Remark that $V_{\ell,k}^{(1)} = V_{k,\ell}^{(1)}$ and the other entries remain the same. As a second step we make the transformation $H_4^{(2)} = U_{q,r}H_4^{(1)}U_{q,r}^\dagger$, where

$$\begin{aligned} \epsilon_q^{(2)} &= \tfrac{1}{2}(\epsilon_q^{(1)} + \epsilon_r^{(1)}) + \tfrac{1}{2}(\epsilon_q^{(1)} - \epsilon_r^{(1)})\cos(2\alpha_{q,r}) + V_{q,r}^{(1)}\sin(2\alpha_{q,r}), \\ \epsilon_r^{(2)} &= \tfrac{1}{2}(\epsilon_q^{(1)} + \epsilon_r^{(1)}) - \tfrac{1}{2}(\epsilon_q^{(1)} - \epsilon_r^{(1)})\cos(2\alpha_{q,r}) - V_{q,r}^{(1)}\sin(2\alpha_{q,r}) \\ V_{q,r}^{(2)} &= -\tfrac{1}{2}(\epsilon_q^{(1)} - \epsilon_r^{(1)})\sin(2\alpha_{q,r}) + V_{q,r}^{(1)}\cos(2\alpha_{q,r}), \\ V_{q,s}^{(2)} &= V_{q,s}^{(1)}\cos\alpha_{q,r} + V_{r,s}^{(1)}\sin\alpha_{q,r}, \quad V_{q,p}^{(2)} = V_{q,p}^{(1)}\cos\alpha_{q,r} + V_{r,s}^{(1)}\sin\alpha_{r,p}, \\ V_{p,r}^{(2)} &= V_{p,r}^{(1)}\cos\alpha_{q,r} - V_{p,q}^{(1)}\sin\alpha_{q,r}, \quad V_{s,r}^{(2)} = V_{s,r}^{(1)}\cos\alpha_{q,r} - V_{s,q}^{(1)}\sin\alpha_{r,p}. \end{aligned} \tag{28}$$

To erase the off-diagonal elements we require $V_{q,s}^{(2)} = V_{q,r}^{(2)} = 0$. The first condition fulfils if $V_{q,s}^{(1)} = V_{r,s}^{(1)} = 0$ and hence $\alpha_{q,s} = \pi/4$, the latter yields $\cos^2(\alpha_{q,r}) = 1/9$ and $H_4^{(2)}$ reduces to a two-form

$$H_4^{(2)} = \epsilon_p X_d^{p,p} + (\epsilon_q + 2\delta) X_d^{q,q} + (\epsilon_q - \delta)(X_d^{r,r} + X_d^{s,s}) + \sqrt{3}\gamma\{X_d^{p,q} + X_d^{q,p}\}. \tag{29}$$

This can be diagonalized using the previous method. The eigenvalues of H_4 read

$$\begin{aligned} E_p &= \tfrac{1}{2}(\epsilon_p + \epsilon_q + 2\delta) - \tfrac{1}{2}\left[(\epsilon_p - \epsilon_q - 2\delta)^2 + 12\gamma^2\right]^{1/2}, \\ E_q &= \tfrac{1}{2}(\epsilon_p + \epsilon_q + 2\delta) - \tfrac{1}{2}\left[(\epsilon_p - \epsilon_q - 2\delta)^2 + 12\gamma^2\right]^{1/2}, \quad E_r = E_s = \epsilon_q - \delta. \end{aligned} \tag{30}$$

The normalized eigenvectors are obtained as $|E_k\rangle = U_{q,s}^\dagger U_{q,r}^\dagger U_{p,q}^\dagger |k\rangle$, for $k = p, q, r, s$. We find

$$\begin{aligned} |E_p\rangle &= \cos\alpha_{p,q}|p\rangle + \frac{1}{\sqrt{3}}\sin\alpha_{p,q}\left(|q\rangle + |r\rangle + |s\rangle\right), \\ |E_q\rangle &= -\sin\alpha_{p,q}|p\rangle + \frac{1}{\sqrt{3}}\cos\alpha_{p,q}\left(|q\rangle + |r\rangle + |s\rangle\right), \\ |E_r\rangle &= -\frac{1}{\sqrt{6}}|q\rangle + \sqrt{\frac{2}{3}}|r\rangle + \frac{1}{\sqrt{6}}|s\rangle, \\ |E_s\rangle &= \frac{1}{\sqrt{2}}\left(|s\rangle - |q\rangle\right), \quad \text{where } \cos\alpha_{p,q} = \left[\frac{E_q - \epsilon_p}{E_q - E_p}\right]^{1/2}. \end{aligned} \tag{31}$$

The four-level form (26) is useful in the study of a three-qubit system with pairwise interaction given by (21). In such a case it can be shown that the corresponding Hamiltonian is written as the sum of two four-forms (26) with $d = 2^3$, and its eigenvalues and eigenvectors are obtained using the method formerly presented [9]. More general problems can be approached using this formalism. For instance, one can consider the eigenvalue problem of the hierarchy of n-qubit Hamiltonians studied in [10, 11]. In the case of higher spin, the Hamiltonian (21) can be generalized by changing $\sigma_i \to J_i$, where J_i is the i-th generator of $SU(2)$ with arbitrary j. In such a case the corresponding eigenvalues and eigenvectors can be found by diagonalizing two 2^{2j}-level forms. These results will be reported elsewhere.

5. Conclusions

The unitary transformation method was originally proposed to diagonalize a single d-level system Hamiltonian. However, we have shown that the method can be extended to the case of n qudits by means the tensor product algebra of Hubbard operators. The developed formalism was applied to particular cases with $d = 2$ and $n = 2, 3$, still the general case can be treated once the Hamiltonian is written as a sum of appropriate k-level forms.

Acknowledgments

The author is in indebted with the organizing committee members of Quantum Fest 2015 for their kind invitation. The financial support of CONACyT is acknowledged.

References

[1] Mielnik B and Rosas-Ortiz O 2007/Rev. 2009 *Quantum Mechanical Laws* in Fundamentals of Physics, ed J L Morán-López (Encyclopedia of Life Support Systems EOLSS), Developed under the Auspices of the UNESCO, Eolss Publishers, Oxford, United Kingdom
[2] Enríquez M and Rosas-Ortiz O 2013 *Ann. Phys.* **339** 218
[3] Enríquez M and Rosas-Ortiz O 2014 *J. Phys.: Conf. Series* **538** 012007

[4] Enríquez M 2015 *J. Phys.: Conf. Series* **624** 012019
[5] Delgado F 2015 *Int. J. Quantum Inf.* **13** (7) 1550055
[6] Niemeijer T 1967 *Physica* **36** 377
[7] Lin H Q 1990 *Phys. Rev.* B **42** 6561
[8] Ovchinnikov S G and Val'kov VV 2004 *Hubbard Operators in the Theory of Strongly Correlated Electrons* (London, Imperial College Press)
[9] Enríquez M 2016 Entanglement in the three-qubit anisotropic Ising model in terms of the minimal decomposition entropy, *in preparation*
[10] Steeb W-H and Hardy Y 2011 *Z. Naturforsch.* **66a** 615
[11] Steeb W-H 2015 *Preprint arXiv:1509.07955*

Journal of Physics: Conference Series **698** (2016) 012022 doi:10.1088/1742-6596/698/1/012022

Additional Quantum Properties of Entangled Bipartite Qubit Systems Coupled to Photon Baths

C Quintana
Physics Department, Cinvestav, A.P. 14-740, 07000, México D.F, Mexico
E-mail: `lquintana@fis.cinvestav.mx`

Abstract. The time evolution of an entangled bi-partite qubit interacting with two independent photon baths in isolated cavities is not unitary. It is shown that the bi-partite qubit oscillates between pure and mixed states, and that the initial entanglement is lost and recovered in time by time as a consequence of its interaction with the baths.

1. Introduction

It is well know that composite quantum systems are not separable in general. Indeed, if quantum correlations between the parts are present, phenomena as the 'spooky action at a distance' described by Einstein [1] are activated. Thus, even if the parts are separated, the *fate* of a given subsystem depends on the behaviour of the other parts and vice versa. Entanglement is the clearest manifestation of quantum correlations. In the simplest case the quantum state of a bipartite system $\mathcal{S} = \mathcal{S}_A + \mathcal{S}_B$ is not separable in the direct product $|\varphi_A\rangle \otimes |\varphi_B\rangle$, with $|\varphi_{A,B}\rangle$ an eigenvector of M associated to $\mathcal{S}_{A,B}$, if the subsystems $\mathcal{S}_A$ and $\mathcal{S}_B$ are entangled in a given observable M. This lack of separability distinguishes the quantum properties of the entire system $\mathcal{S}$ and finds diverse applications in contemporary physics [2–4]. In particular, if $\mathcal{S}$ is an entangled bi-partite qubit system interacting with two independent photon baths in separated cavities, the related concurrence evolves in time by oscillating between a given initial amount and zero [5–7]. This last because the whole system (two qubits + two photon baths) is closed, isolated of any interaction with other possible systems, so that the quantum correlations are transferred from the qubits to the bath and vice versa. If one pays attention only to the bi-partite qubit state, some information about the transitions between the states of the whole system is lost [7]. Interestingly, such an information is encoded in the determinant of the reduced density operators of each of the qubits involved [5–7], so that this can be recovered in terms of the concurrence. In this contribution we analyze some additional properties of the time-evolution of the entangled bi-partite qubit system described above. To be precise, we shall see that the state evolves in time by oscillating from an initial pure entangled state to a disentangled pure state, and being in different degrees of probabilistic mixture in between.

2. Time-evolution of the bi-partite qubit state

The dynamics of two entangled qubits, each one interacting with a photon bath in an isolated cavity, has been analyzed in [5–7]. The tetra-partite quantum system (two qubits + two photon baths) is in a pure state that evolves in time unitarily. After the partial trace over the fields, the reduced bi-partite qubit state can be diagonalized as follows

$$\rho_q = \text{diag}(\lambda_1, \lambda_2, \lambda_3, \lambda_4), \quad (1)$$

with

$$\lambda_1 = \Omega_{11}, \quad \lambda_2 = \frac{(\Omega_{22}+\Omega_{33})+\sqrt{(\Omega_{22}+\Omega_{33})^2+4(\Omega_{23}^2-\Omega_{22}\Omega_{33})}}{2}, \\ \lambda_3 = \Omega_{22}+\Omega_{33}-\lambda_2, \quad \lambda_4 = \Omega_{44}, \quad (2)$$

and

$$\begin{aligned} \Omega_{11} &= \cos^2\alpha\cos^2(\gamma t\sqrt{r+1})\sin^2(\gamma t\sqrt{s}) + \sin^2\alpha\cos^2(\gamma t\sqrt{s+1})\sin^2(\gamma t\sqrt{r}), \\ \Omega_{22} &= \cos^2\alpha\cos^2(\gamma t\sqrt{r+1})\cos^2(\gamma t\sqrt{s}) + \sin^2\alpha\sin^2(\gamma t\sqrt{s+1})\sin^2(\gamma t\sqrt{r}), \\ \Omega_{33} &= \cos^2\alpha\sin^2(\gamma t\sqrt{r+1})\sin^2(\gamma t\sqrt{s}) + \sin^2\alpha\cos^2(\gamma t\sqrt{s+1})\cos^2(\gamma t\sqrt{r}), \\ \Omega_{44} &= \cos^2\alpha\sin^2(\gamma t\sqrt{r+1})\cos^2(\gamma t\sqrt{s}) + \sin^2\alpha\sin^2(\gamma t\sqrt{s+1})\cos^2(\gamma t\sqrt{r}), \\ \Omega_{23} &= \Omega_{32} = \cos\alpha\sin\alpha\cos(\gamma t\sqrt{r+1})\cos(\gamma t\sqrt{r})\cos(\gamma t\sqrt{s+1})\cos(\gamma t\sqrt{s}). \end{aligned}$$

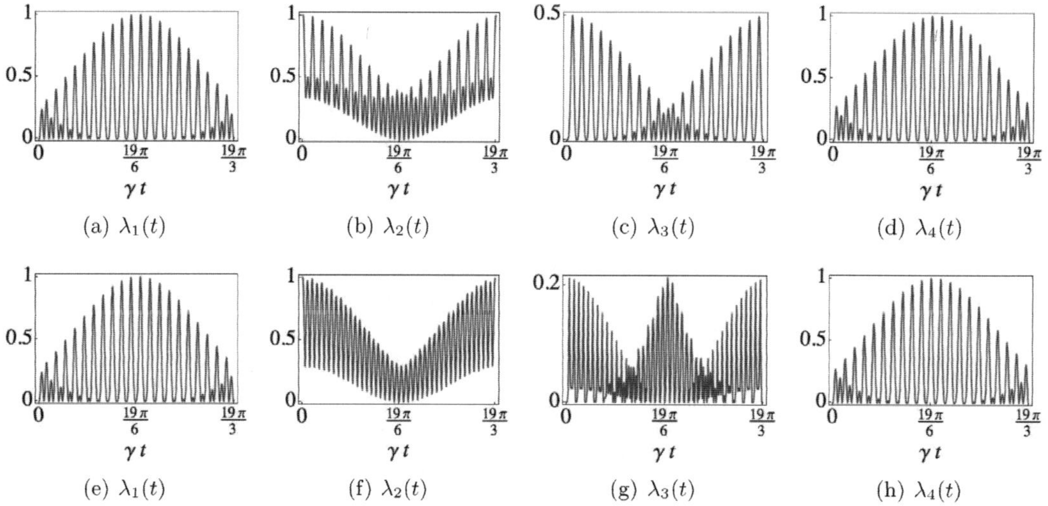

Figure 1. Eigenvalues $\lambda_k(t)$ of the density matrix ρ_q as a function of time. First row ($\alpha = \frac{\pi}{4}$), the bipartite qubit system is initially in the Bell state $|\beta_3\rangle = \frac{1}{\sqrt{2}}(|+,-\rangle + |-,+\rangle)$. Second row ($\alpha = \frac{9\pi}{20}$), the system is in the initial configuration $|-,+\rangle$ with probability $\mathcal{P} \approx 0.976$.

Here, the bath in cavity $A(B)$ has $r(s)$ photons and $\gamma_A = \gamma_B = \gamma$ is the coupling constant. The parameter $\alpha \in [0, \frac{\pi}{2}]$ determines the related set of eigenvectors

$$|\phi_{\lambda_1}\rangle = |+,+\rangle, \quad |\phi_{\lambda_2}\rangle = \cos\alpha|+,-\rangle + \sin\alpha|-,+\rangle, \quad (3)$$

$$|\phi_{\lambda_3}\rangle = \cos\alpha|+,-\rangle - \sin\alpha|-,+\rangle, \quad |\phi_{\lambda_4}\rangle = |-,-\rangle. \quad (4)$$

It is straightforward to verify that $\text{Tr}(\rho_q) = 1$ and $\text{Tr}(\rho_q^2) < 1$. Then, the diagonal expression (1) verifies that ρ_q is the probabilistic mixture of two disentangled pure states $|\phi_{\lambda_1,\lambda_4}\rangle$ and two partially entangled pure states $|\phi_{\lambda_2,\lambda_3}\rangle$. In Figure 1 we have depicted the behaviour of the eigenvalues (2) as a function of time when the ket $|\phi_{\lambda_2}\rangle$ is given as initial condition for $r = s = 10$

and two different values of α. Notice that the time-evolution is periodic with period $\tau \approx \frac{19}{3}\pi$. At $t = 0\,(=\tau)$ and $t = \frac{\tau}{2}$ the system is in the partially entangled state $|\phi_{\lambda_2}\rangle$ and in the disentangled state $|\phi_{\lambda_4}\rangle$ respectively. For other values of $t \in [0, \tau]$ the system is in a mixed state.

We would like to emphasize that the initial condition $\rho_q(t=0) = |\beta_3\rangle\langle\beta_3|$ implies $n = r+s+1$ photons, that is $r+s$ photons in the cavities plus the one associated with the excitation of either qubit A or B. This amount of photons is preserved because the whole system is closed. As the system is in the state $|\phi_4\rangle = |-,-\rangle$ at $t = \frac{\tau}{2}$, disentanglement means that all the n photons are in the cavities, so that the fields configuration is in either $|r, s+1\rangle$ or $|r+1, s\rangle$. Therefore, although the states of the entire system $|-,-,r,s+1\rangle$ and $|-,-,r+1,s\rangle$ are the result of unitary time-evolution, the transition $|\beta_3\rangle \to |-,-\rangle$ is not unitary because the system $\mathcal{S} = \mathcal{S}_A + \mathcal{S}_B$ is not closed, this interchanges a photon with the environment (the fields in the cavities) in time by time. To get a deeper insight on the matter let us consider the corresponding Von Neumann entropy

$$S(\rho_q) = -\sum_{k=1}^{4} \lambda_k \log \lambda_k, \tag{5}$$

this is depicted in Figures 2(a) and 2(c) for the values of α used in Figure 1. Consistently, at the times $t = 0, \frac{\tau}{2}$ and τ, the entropy (5) is equal to zero (i.e, the system is in a pure state). For other values of t the entropy oscillates between 0 and 1 (i.e., the system is in a mixed state). Quite interestingly, $S(\rho_q)$ is almost 1 in the vicinity of the above indicated values of t. This last means that the transitions from pure to mixed state are taken in very short intervals of time.

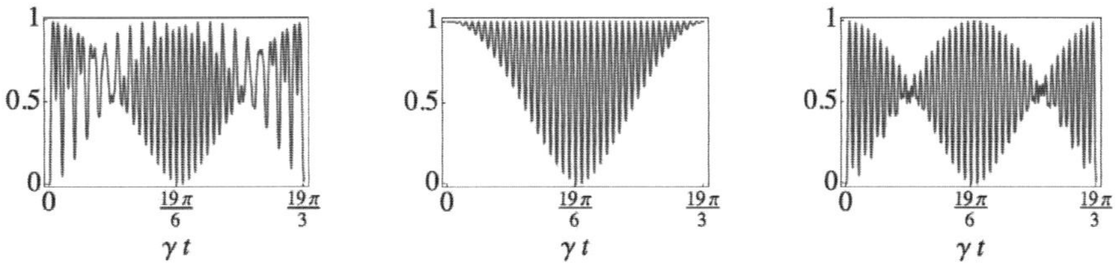

(a) $S(\rho_q)$, $\alpha = \frac{\pi}{4}$ (b) $E(\rho_q)$, $\alpha = \frac{\pi}{4}$ (c) $S(\rho_q)$, $\alpha = \frac{9\pi}{20}$

Figure 2. Time-evolution of the Von Neumann entropy (a, c) associated with the first and second rows of Figure 1 respectively. In (c) it is shown the time-dependence of the entropy of entanglement associated with the first row of Figure 1.

As a measure of the entanglement between the qubits we can use the Entropy of entanglement

$$E(\rho) = S(\rho_1) = S(\rho_2), \tag{6}$$

this last is possible only for $\alpha = \pi/4$, that is for a maximally entangled initial state $|\phi_{\lambda_2}\rangle = |\beta_3\rangle = \frac{1}{\sqrt{2}}(|+,-\rangle + |-,+\rangle)$. The time-evolution of (6) can be appreciated in Figure 2(b). As expected, $E(\rho_q)$ is equal to 1 when the system is maximally entangled (i.e. at $t = 0, \tau$) and it is equal to 0 when the system is disentangled (i.e. at $t = \frac{\tau}{2}$). For other values of t the entropy $E(\rho_q)$ oscillates between 1 and values that are close to zero as t is closer to $\frac{\tau}{2}$. However, although $E(\rho_q)$ takes the value 1 at different times (meaning that the qubit is in a maximally mixed state) the entropy $S(\rho_q)$ of the bi-partite qubit system is neither 1 nor 0 in such times, so that the bi-partite system is in a partially mixed state. Thus, the condition $E(\rho_q) = 1$ gives information of the entanglement just when the bi-partite qubit system is in a pure state. If this last is in a (partially) mixed state, other measures of entanglement are necessary (see e.g. [6, 7] and references quoted therein).

Coming back to a general bipartite system $\mathcal{S} = \mathcal{S}_I + \mathcal{S}_{II}$, other important correlations appear when the number of degrees of freedom is larger in $\mathcal{S}_{II}$ than in $\mathcal{S}_I$. Then one says that the system S_I is embedded in the bath (reservoir or environment) S_{II}, and the subject is the form in which $\mathcal{S}_{II}$ affects the evolution of $\mathcal{S}_I$. That is, $\mathcal{S}_I$ is an open system that interacts with the environment $\mathcal{S}_{II}$ in such a manner that the whole $\mathcal{S}$ is isolated (closed). In this case, the quantum state ρ of $\mathcal{S}$ is pure ($\rho^2 = \rho$) while the state $\rho_{I,II}$ of $\mathcal{S}_{I,II}$ is mixed ($\rho^2_{I,II} \neq \rho_{I,II}$). This produces different dynamical laws for ρ and $\rho_{I,II}$. Namely,

$$\dot{\rho} = \mathcal{L}_0\{\rho\}, \quad \dot{\rho}_{I,II} = \mathcal{L}_0\{\rho_{I,II}\} + i\Gamma\{\rho_{I,II}\}, \tag{7}$$

where

$$\mathcal{L}_0\{\rho\} = -i[H, \rho] \tag{8}$$

is the Liouville operator acted on ρ. The (super)operator Γ represents non-Hermitian corrections associated to the coupling of the system $\mathcal{S}_I$ with the environment $\mathcal{S}_{II}$. The equation obeyed by ρ is named after Liouville and the one fulfilled by $\rho_{I,II}$ is known as the *master equation* (see e.g., [8]). Considering $\mathcal{S}_I = \mathcal{S}$ and $\mathcal{S}_{II}$ as the system of the two photon baths we see that the study of transitions like $|\beta_3\rangle \rightarrow |-,-\rangle$, described above, requires the identification of the superoperator Γ that connects the master equation (7) with the dynamics of the reduced density matrix $\rho_q(t)$ defined in (1). Results in this direction are in progress and will be reported elsewhere.

3. Concluding remarks

We have shown that an entangled bi-partite qubit subjected to the action of independent photon baths in isolated cavities evolves in time by oscillating between pure and mixed states. The initial entanglement is lost and recovered in time by time because the bi-partite qubit is an open system, although the whole tetra-partite one (two qubits + two phton baths) is closed.

Acknowledgments

The author gratefully acknowledges the funding received through a CONACyT Scholarship.

4. References

[1] Born M (Ed.) 1971 *The Born-Einstein-Letters* (Macmillan, London) p 158
[2] Haroche S 1998 Entanglement, decoherence and quantum/classical boundary *Phys. Today* **51** 36
[3] Nielsen M A and Chuang L L 2000 *Quantum Computation and Quantum Information* (Cambridge University Press, Cambridge)
[4] Bouwmeester B, Ekert A and Zeilinger A (Eds.) 2000 *The physics of quantum information: Quantum cryptography, quantum teleportation, quantum computation* (Springer, New York)
[5] Quintana C 2014 *Quantum entanglement: time-evolution of bipartite systems* (MSc Thesis (in Spanish), Physics Department, Cinvestav, México)
[6] Enríquez M, Quintana C and Rosas-Ortiz O 2014 Time-evolution of entangled bipartite atomic systems in quantized radiation fields *J. Phys.: Conf. Ser.* **512** 012022
[7] Quintana C and Rosas-Ortiz O 2015 Note on the quantum correlations of two qubits coupled to photon baths *J. Phys.: Conf. Ser.* **624** 012022
[8] Ernst R R, Bodenhausen G and Wokaun A 1988 *Principles of Nuclear Magnetic Resonance in One and Two Dimensions* (Oxford University Press, New York)

Correlation Dimension in Two-Dimensional Disordered Systems with Rashba Spin-Orbit Coupling

E M Hernández, J A Otero and F Delgado

Departamento de Física y Matemáticas, Escuela de Diseño, Ingeniería y Arquitectura, Tecnológico de Monterrey, Campus Estado de México, Atizapán, Estado de México, CP. 52926, México.

E-mail: emcooper@itesm.mx, j.a.otero@itesm.mx, fdelgado@itesm.mx

Abstract. The metal-insulator phase transition that arises in the Integer Quantum Hall Effect has been characterized through the multifractal nature of extended states near the center of the Lowest Landau Level. In this work, we obtain numerical solutions for the one-electron Hamiltonian with disorder, where the correlation dimension of extended states in the first two Landau Levels is obtained, by taking into account the Rashba spin-orbit coupling in the Hamiltonian. Although, spin-orbit coupling at moderate field intensities has been determined experimentally, there is no theoretical evidence for the nature of the transition in this case. The correlation dimension of extended states for the resolved spin levels is obtained, and within the statistical error, it is found that the Rashba Hamiltonian in presence of disorder, belongs to the same universality class of spin unresolved systems.

1. Introduction

After the experimental confirmation of the Integer Quantum Hall Effect (IQHE) by Klaus Von Klitzing [1], the main concern was to find a localization theory for two dimensional electron systems (2DES) in presence of disorder. The general idea was that the origin of localized states, responsible for the plateau regions in the Hall resistance ρ_{xy}, was due to the presence of fixed impurities in the semiconductors used to host the 2DES. Unfortunately, when impurities are included in the physical model, an exact analytic solution becomes extremely difficult. With numerical simulations it is possible to find a solution to the one-electron Hamiltonian with disorder and understand the electronic transport in the 2DES, as a phase transition between localized and extended states. The obtained results from the numerical models proved to be successful, since they were able to predict localization, and find the critical exponents [2, 3] of the localization-delocalization (LD) transition, which were experimentally confirmed in the late 80's [4], and early 90's [5, 6], from temperature and sample size scaling experiments.

There is a major interest to study 2DES with zero field spin splitting, where it is possible to control the spin-orbit coupling (SOC) parameter by tuning the gate voltage. The source of the splitting energy can be attributed to the coupling between the electron's spin magnetic moment with a magnetic field induced by the interfacial electric field. The splitting energies, corresponding to the Rashba SOC, have been observed in several experiments [7–10], by studying the beating patterns in the Shubnikov de Hass (SdH) oscillations at relatively small magnetic fields. The SOC inside hererostructures, can be tuned by applying different gate voltages; which is very appealing for spintronic applications [11, 12]. From the theoretical point of view, the dependence of the critical exponents in the LD transition and multifractal spectra on the coupling parameter, has not been addressed. In this work, we study the nature of the phase transition on 2DES with Rashba SOC, in presence of disorder. We found extended and localized states within each spin resolved level for different SOC strengths. Even though, the Rashba SOC term in the Hamiltonian, couples adjacent Landau levels; we are able to obtain for different SOC parameters, the correlation dimension of extended states near the center of each spin split level. We apply multifractal analysis to the extended states [13–18], in order to determine the critical behavior of states near the center of each resolved level; and found that, for different SOC strengths, the correlation dimension does not change whithin the statistical error. This work is organized as follows: in section 2 we offer a brief description of the Rashba SOC term and the one-electron Hamiltonian in presence of disorder. In section 3, we introduce the numerical model for disorder and the resulting matrix elements of the Hamiltonian when we take into account the SOC term, in section 4 we present the results obtained from the numerical simulations for several SOC strengths and finally; in section 5, the conclusions are presented.

2. Problem statement

In 2DES formed in inversion layers, when a uniform magnetic field is applied in a direction perpendicular to the interface, the kinetic energy of the electrons is fixed into a series of Landau levels. The two dimensionality and freezing of the kinetic energy due to the magnetic field, play a fundamental role in the electronic transport properties of the system when disorder and electron interactions, are brought into the picture. In the Landau gauge where $\vec{A} = (0, xB)$, the two dimensional Schrödinger equation for the one-electron Hamiltonian, in absence of disorder, can easily be solved. The magnetic field gives rise to Landau levels $E_n = (n + \frac{1}{2})\hbar\omega_c$, fixing the kinetic energy at these values, and giving rise to highly degenerate levels with $N_\phi = A/2\pi l^2$ states, where A is the system size of area $A = L_x L_y$, $\omega_c = eB/m$ is the cyclotron frequency and $l = \sqrt{\hbar/eB}$ is the magnetic length or cyclotron radius.

The Hamiltonian must take into account the interactions between electrons and the impurities within the heterostructure. When disorder is brought into the picture, the problem becomes more complex because many body interactions, between the electron and impurities, must be taken into account. Early numerical models were focused on the microscopic characteristics of disorder; for example, scattering from neutral impurities can be considered with short range interactions [19] in which the cyclotron radius is larger than the range of these potentials. A more general model for disorder was introduced by Huckestein and Kramer [20] where the potential is assumed to have a normal probability distribution:

$$P[V] = Ne^{-\frac{1}{2V_0^2}\int d^2r[V(\vec{r})]^2} \tag{1}$$

where V_0 represents the disorder strength. On highly doped heterostructures, where the mean free path is small enough, the electron-electron interactions can be neglected or treated as a perturbation of the disorder. In this situation, the physical behavior of the system can be described by the one particle Hamiltonian:

$$H = \frac{(\vec{p} - e\vec{A})^2}{2m} + V(\vec{r}) \tag{2}$$

where the first term, is the kinetic energy of the electron that is being fixed by the applied magnetic field, and the second term represents the disorder potential with the probability distribution given by equation (1). The effect of disorder is to break the degeneracy of each Landau level, which is observed as a broadening in the density of states (DOS). Equation (2) has been solved numericaly for magnetic fields where $\hbar\omega_c \gg V_0$, which is the lowest Landau level (LLL) approximation. By averaging the results over different realizations of disorder, it is possible to explain the phase transition that takes place between adjacent plateaus in the Hall resistance, through a multifractal analysis of the extended states near the center of the LLL.

In this work, we also consider a SOC term in the Hamiltonian. A surface electron will see in it's rest frame, an effective magnetic field, due to it's relative motion to the electric field at the interface of the inversion layer. On a free 2DES, the interaction energy between the spin of the electron and it's orbital motion expressed in terms of the Pauli matrices; known as the Rashba Hamiltonian, is given by [21]:

$$H_R = \alpha(\vec{p} \times \vec{\sigma}) \cdot \hat{z} \tag{3}$$

where α is linearly proportional to the electric field and provides the spin-coupling parameter which can be tuned through the gate voltage. From the experimental point of view, the zero field splitting energy is determined through extrapolations of the low field behavior of the SdH oscillations. Therefore, one way to study the SOC, is to solve the Rashba Hamiltonian when a external magnetic field is being applied. In this situation, the electron's spin will interact with the external magnetic field as well, giving rise to a Zeeman Hamiltonian which is proportional to the field's intensity. Experiments where the SOC term is dominant, are performed on magnetic field intensities in which the Zeeman splitting can be neglected. In the presence of an external magnetic field, the kinetic momentum changes to $\vec{\Pi} = \vec{p} + q\vec{A}$ and the Hamiltonian transforms to:

$$H = \frac{1}{2m}\vec{\Pi}^2 + \alpha(\vec{\Pi} \times \vec{\sigma}) \cdot \hat{z} + \frac{g}{2}\mu_B B \sigma_z \tag{4}$$

where μ_B is the Bohr's magneton $\mu_B = |e|/2m$. Due to the external magnetic field, we can use creation and annihilation operators to find the solution to equation (4) in the Landau representation [21]:

$$E_0 = \frac{^-\hbar\omega}{2} \text{ , and } E_n^{\pm} = \hbar\omega_c n \pm \sqrt{2n\alpha^2 m\hbar\omega_c + \frac{1}{4}(\hbar\omega_c + g\mu_B B)^2} \tag{5}$$

By considering the SOC of the Rashba type in equation (2), we obtain the one-electron Hamiltonian:

$$H = \frac{1}{2m}\vec{\Pi}^2 + \alpha(\vec{\Pi} \times \vec{\sigma}) \cdot \hat{z} + V(\vec{r}) \tag{6}$$

which for moderate magnetic field intensities, the Zeeman term is much smaller than the SOC term, $g \ll \alpha^2 m\hbar$.

3. Numerical model

In our simulations we use a Gaussian White Noise Potential (GWNP) as the source of disorder, by working equation (1) in momentum space:

$$P[V] = \frac{1}{\sqrt{2\pi s}} e^{-\frac{|v(\vec{q})|^2}{2s^2}}, \text{ with } \langle v(\vec{q})v(\vec{q}\,')\rangle = s\delta_{\vec{q}+\vec{q}\,'}, \tag{7}$$

where $V_0 = s$. In this case, the potential $V(\vec{r})$ can be calculated as the Fourier transform of $v(\vec{q})$:

$$V(\vec{r}) = \frac{1}{A}\sum_{\vec{q}} v(\vec{q}) e^{\imath\vec{q}\cdot\vec{r}} \tag{8}$$

Journal of Physics: Conference Series **698** (2016) 012023 doi:10.1088/1742-6596/698/1/012023

The key to find the solutions of equation (2), is to use the states in absence of disorder as a basis to obtain the matrix elements of $V(\vec{r})$. This is possible by using a linear combination of the Landau states, in order to build a basis with toroidal boundary conditions (periodic functions in x and y):

$$\phi_{n,m}(x,y) = \frac{1}{L_y l\sqrt{\pi}} \sum_{s=-\infty}^{\infty} e^{\frac{\imath x_{m,s} y}{l^2}} e^{-\frac{1}{2l^2}\bar{x}^2} \chi\left(\frac{\bar{x}}{l}\right) \tag{9}$$

where $x_{m,s} = 2m\pi l^2/L_y + sL_x$, $\bar{x} = x - x_{m,s}$ and a system of units where $\hbar\omega_c = 1$ is being used. The functions $\chi_n(\bar{x}/l) = \mathrm{H}_n(\bar{x}/l)/\sqrt{2^n n!})$, where $\mathrm{H}_n(\bar{x}/l)$ are the Hermite polynomials of degree n. With this basis, and generating uncorrelated complex random variables $v(\vec{q})$, the matrix elements of disorder can be obtained. In the LLL approximation, where $V_0 \ll {}^-h\omega$, each level can be studied independently. In any LL, the resulting matrix elements of the term $e^{\imath\vec{q}\cdot\vec{r}}$, appearing in equation (8), are given by:

$$\langle m,n|e^{\imath\vec{q}\cdot\vec{r}}|m',n\rangle = \delta_{x_{m',0}-x_{m,0}+q_y l^2,0}\mathrm{L}_n\left(\frac{1}{2}\vec{q}^2 l^2\right)e^{-\frac{1}{4}\vec{q}^2 l^2}e^{\imath x_{m'}q_x}e^{\frac{1}{2}q_x q_y l^2} \tag{10}$$

with $q_x = \frac{2\pi n_x}{L_x}$ ($q_y = \frac{2\pi n_y}{L_y}$), where n_x (n_y) are integer numbers, and the functions L_n are the Laguerre polynomials of order n. Using the Rashba Hamiltonian in terms of creation and annihilation operators, the matrix elements of the SOC term in the Landau basis, are given by

$$\langle m,n,\sigma|H_R|m',n',\sigma'\rangle = -\frac{1}{\sqrt{2}}\tilde{\alpha}\delta_{m,m'}(\sqrt{n'}\langle\sigma|\sigma^-|\sigma'\rangle\delta_{n,n'-1} + \sqrt{n'+1}\langle\sigma|\sigma^+|\sigma'\rangle\delta_{n,n'+1}) \tag{11}$$

The symbol σ represents the index for a state with spin up (+) or spin down (−) orientation and the operators σ^+ (σ^-) are the creation and annihilation operators in terms of the Pauli matrices σ_x and σ_y. In the system of units where $\hbar\omega_c = 1$, the SOC parameter α can expressed in terms of a dimensionless parameter $\tilde{\alpha} = \alpha/\omega_c l$. Even though, we are using the LLL approximation, it is evident from equation (11) that the Rashba Hamiltonian is coupling adjacent LL's. Finally, the matrix elements of the one-electron Hamiltonian given by equation (6), can be obtained by applying equations (10) and (11) to the LLL ($n = 0$) and the next LL ($n = 1$).

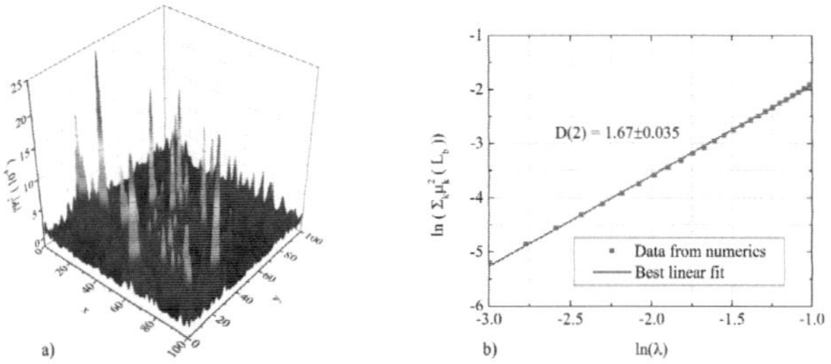

Figure 1. a) Probability density, $|\psi|^2$ (10^{-4}), for an extended state in a system of $L = 100l$, and b) disorder averaged $D(2)$ for states with the smallest IPN in a system of $L = 50l$, which is consistent with that of [14] within the statistical error.

4. Results and discussion

In the first part of this section, we will describe the results obtained by solving the Hamiltonian in equation (2). In the LLL approximation, we only consider the matrix elements of disorder in the LLL, by substituting $n = 0$ in equation (10). The system is solved by diagonalization of the resulting matrix for N_ϕ states over a square system of size $L = L_x = L_y$. All lengths and energies are expressed as multiples of l and $\hbar\omega_c$, respectively. In Figure 1, we show the most extended state; that is, the state with the smallest inverse participation number (IPN) [22], which is close to the center of the LLL. We apply the multifractal analysis of Mandelbrot [23] to obtain the correlation dimension $D(2)$ of extended states by studying the scale invariance for the second moment of the box probability μ_k, defined as

$$P^2 = \sum_k \mu_k^2(L_b) \sim \lambda^{D(2)} \tag{12}$$

where $\lambda = L_b/L$ and $D(2)$ is the corresponding scaling exponent. For localized states, $D(2) = 0$, since all the probability density falls within one box; however, extended states exhibit a fractal behavior, where the scaling exponent has been calculated for several models of disorder. Aoki and Ando [13] found a value of $D(2) = 1.57 \pm 0.03$ using delta scatterers, Huckestein [14] obtained a value of $D(2) = 1.62 \pm 0.02$ by studying the scaling of equation (12) in a lattice Hamiltonian. In Figure 1, we also show the scaling exponent found by using the GWNP introduced in section 3, for a system of $L = 50l$ over 205 realizations of the random potential.

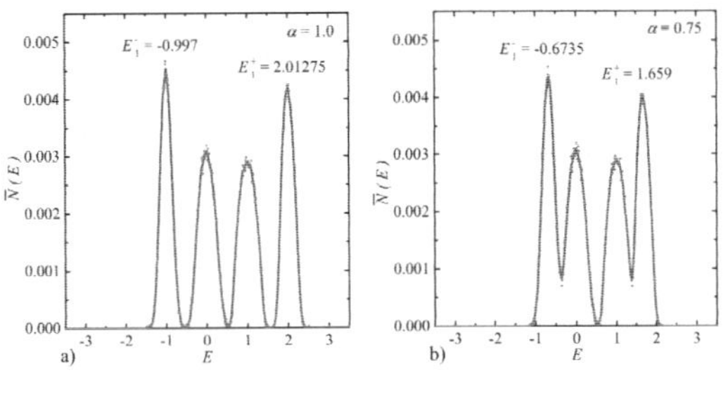

Figure 2. a) Spin resolved DOS for the LLL and first LL with $\widetilde{\alpha} = 1.0$ and b) spin resolved DOS for $\widetilde{\alpha} = 0.75$ for a system of $35l$. The energies shown are obtained numerically and can be verified from equation (5).

In this part of the results, we will discuss the multifractal nature of extended states for spin resolved LLs. We solve the system by diagonalizing the Hamiltonian of equation (6) with the matrix elements obtained in section 3 for different values of $\widetilde{\alpha}$. In Figure 2, is shown the disorder averaged DOS for the lowest and first LLs, taken from 200 different realizations of the random potential on a system of $L = 35l$. Since the Rashba Hamiltonian is coupling adjacent levels, and considering two spin orientations for each level according to equation (11), we need to diagonalize a $4N_\phi \times 4N_\phi$ matrix. The center of each spin resolved level corresponds to the values predicted by equation (5), which are also shown in Figure 2 in our system of units.

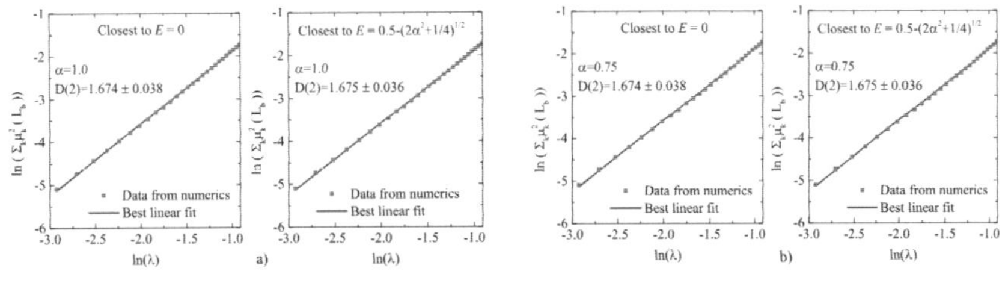

Figure 3. Average $D(2)$ for extended states with the smallest IPN, closest to the center of the LLL and spin down oriented levels for a) $\widetilde{\alpha} = 1.0$ and b) $\widetilde{\alpha} = 0.75$.

From the numerical solutions, we can study the dependence of the correlation dimension $D(2)$ on the SOC parameter for different spin oriented levels. In Figure 3, we show the value of $D(2)$ for extended states with the smallest IPN. The numerical value was obtained by averaging the scaling exponent $D(2)$, over 125 realizations of the random potential on a system of $35l$. As can be observed from Figure 3, within the statistical error, the correlation dimension is practically the same as the value shown in Figure 1 for the spin degenerate case.

5. Conclusions

In this work, it has been shown that extended states are also present near the center of each spin resolved Landau level. By including the Rashba spin-orbit coupling term into the one-electron Hamiltonian with disorder; it has been shown, that extended states belonging to each spin split Landau level are scale invariant, and show multifractality in a similar way to the spin unresolved Hamiltonian. As in the spin degenerate case, we found that the correlation dimension $D(2)$ of extended states near the center of each resolved level, shows universality relative to disorder strength; but more importantly, it is also universal relative to the strength of the spin-orbit coupling parameter. These findings seem to suggest, that disordered systems under moderate magnetic fields, with a Rashba spin-orbit coupling interaction of any strength, belong to the same universality class of the localization-delocalization transition observed in the Integer Quantum Hall Effect.

References

[1] Von Klitzing K, Dorda G and Pepper M 1980 *Phys. Rev. Lett.* **45** 494
[2] Chalker J T and Coddington P D 1988 *J. Phys. C. Solid State* **21** 2665
[3] Huckestein B and Kramer B 1990 *Phys. Rev. Lett.* **64** 1437
[4] Wei H P, Tsui D C, Paalanen M A and Pruisken A M M 1988 *Phys. Rev. Lett.* **61** 1294
[5] Wei H P, Hwang S W, Tsui D C and Pruisken A M M 1990 *Surf. Sci.* **229** 34
[6] Koch S, Haug R J, Von Klitzing K and Ploog K 1991 *Phys. Rev.* B. **43** 6828
[7] Nitta J, Akazaki T and Takayanagi H 1997 *Phys. Rev. Lett.* **78** 1335
[8] Engels G, Lange J, Schapers Th and Luth H 1997 *Phys. Rev.* B. **55** R1958
[9] Sato Y, Kita T, Gozu S and Yamada S 2001 *J. App. Phys.* **89** 8017
[10] Helda J P, Van Wees B J, Kuipers J J and Klapwijk T M 1988 *Phys. Rev.* B. **57** 11911
[11] Sinova J, Valenzuela S, Wunderlich J, Back C H and Jungwirth T 2015 *Rev. Mod. Phys.* **87** 1213
[12] Sinova J, Culcer D, Niu Q, Sinitsyn N A, Jungwirth T and MacDonald A H 2004 *Phys. Rev. Lett.* **92** 126603
[13] Aoki H 1986 *Phys. Rev.* B. **33** 7310
[14] Huckestein B and Schweitzer L 1994 *Phys. Rev. Lett.* **72** 713
[15] Janßen M, Viehweger O, Fastenrath U and Hajdu J 1994 *Introduction to the Theory of the Integer Quantum Hall Effect* (Wiley, Weinheim)
[16] Pook W and Janßen M 1991 *Z. Phys.* B. **82** 295
[17] Huckestein B 1995 *Rev. Mod. Phys.* **67** 357
[18] Huckestein B and Schweitzer L 1992 *Helv. Phys. Acta.* **65** 317
[19] Ando T 1982 *J. Phys. Soc. Jpn.* **52** 1740
[20] Huckestein B and Kramer B 1989 *Solid State Commun.* **71** 445
[21] Schliemann J, Egues J C and Loss D 2003 *Phys. Rev. Lett.* **90** 146801
[22] Hikami S 1986 *Prog. Theor. Phys.* **76** 1210
[23] Mandelbrot B B 1983 *The Fractal Geometry of Nature* (Freeman, San Francisco)

Journal of Physics: Conference Series **698** (2016) 012024 doi:10.1088/1742-6596/698/1/012024

Generation of Control by SU(2) Reduction for the Anisotropic Ising Model

F Delgado

Departamento de Física y Matemáticas, Escuela de Ciencias e Ingeniería, Tecnológico de Monterrey, Campus Estado de México, Atizapán, Estado de México, CP. 52926, México.

E-mail: fdelgado@itesm.mx

Abstract. Control of entanglement is fundamental in Quantum Information and Quantum Computation towards scalable spin-based quantum devices. For magnetic systems, Ising interaction with driven magnetic fields modifies entanglement properties of matter based quantum systems. This work presents a procedure for dynamics reduction on $SU(2)$ subsystems using a non-local description. Some applications for Quantum Information are discussed.

1. Introduction

Quantum control exploits the fine management of physical variables to improve the capacity and speed of the information processing [1,2]. The magnetic driven Ising interaction [3] generates that control [4] for bipartite qubits, chains and lattices [5–7], involving temperature, field strengths and geometry in the analysis [8–10]. Control for a single or a couple of spin elements is still at the heart of a scalable spin-based quantum computer because the controlled state exchange lets to obtain universal quantum operations [11,12] in terms of DiVincenzo criteria [13] for quantum states reliability. This paper discusses how the bipartite dynamics of Ising model in $SU(4)$ is reduced in two $SU(2)$ subsystems on a non-local basis. Section 2 sets the Hamiltonian, notation and $SU(2)$ reduction. Section 3 discusses some applications: Quantum Error Correction, Evolution Loops, Exchange Operations, $SU(2)$ optimal control and gate design based on Unitary Factorization. Section 4 states the conclusions and future work.

2. $SU(2)$ decomposition for the anisotropic Ising Hamiltonian in a non local basis

Ising interaction has been analyzed for several systems and configurations (XX, XY, XYZ, etc.) [14–16] with structured control effects [17,18] due to the geometry or the physical properties [19–21]. We use a Hamiltonian with a driven magnetic field in $h = 1,2,3$ (x,y,z) direction:

$$H_h = \sum_{k=1}^{3} J_k \sigma_{1_k} \sigma_{2_k} - B_{1_h} \sigma_{1_h} - B_{2_h} \sigma_{2_h} \qquad (1)$$

including the previous models. Due to the Bell basis is privileged and to comprise the three field cases, we adopt the notation [22]: $|\beta_{\mu\nu}\rangle \equiv |\beta_{AB}\rangle\,; \mu,\nu \in \{-,+\}; A,B \in \{0,1\}; A = \frac{1+\mu}{2}, B = \frac{1+\nu}{2}$. Capital scripts $A, B, \ldots$ are $0,1$ for the computational basis; greek scripts for $\pm$ or ± 1, because they evolve to ±1 in algebraic expressions; latin scripts $h, i, j, ...$ for coordinates. Thus

$$E_{h_{\mu\nu}} = \mu J_h + \nu R_{h_{-\mu}} \tag{2}$$

are the eigenvalues for the states $\left|\phi^h_{\mu\nu}\right\rangle$ [22]. $U(t) \in SU(4)$ because they add zero. There

$$R_{h\pm} \equiv \sqrt{B^2_{h\pm} + J^2_{\{h\}_\mp}}, \quad J_{\{h\}\pm} \equiv J_i \pm J_j, \quad B_{h\pm} \equiv B_{1_h} \pm B_{2_h}, \tag{3}$$

h, i, j is a cyclic permutation of $1, 2, 3$; then $\{h\}$ is equivalent to the pair i, j. The parameters

$$b_{h\pm} = \frac{B_{h\pm}}{R_{h\pm}}, \quad j_{h\pm} = \frac{J_{\{h\}\mp}}{R_{h\pm}} \in [-1, 1], \quad \Delta_{h\mu}^{\nu} = \frac{t}{2}(E_{h\mu+} + \nu E_{h\mu-}), \tag{4}$$
$$e_{h\alpha}^{\beta} = \cos\Delta_{h\alpha}^{-} + i\beta j_{h_{-\alpha}} \sin\Delta_{h\alpha}^{-}, \quad d_{h\alpha} = b_{h_{-\alpha}} \sin\Delta_{h\alpha}^{-},$$

reduce the evolution matrix expressions for the time independent Hamiltonian. Reader should be aware about double scripts to avoid misconceptions. As instance, $R_{h_{-\mu}}$ in 2 is R_{3_+} for $\mu = -1$ when driven field is in the z direction. While, B_{1_h}, B_{2_h} are the driven fields in h direction on particles 1,2 respectively. Then, in the Bell basis, evolution operators and bipartite states are

$$U_h(t) = \sum_{\mu,\nu,\gamma,\delta} U_{h\mu\nu,\gamma\delta} \left|\beta_{\mu\nu}\right\rangle \left\langle\beta_{\gamma\delta}\right|, \quad |\psi\rangle = \sum_{\mu,\nu\in\{-,+\}} \mathcal{B}_{\mu\nu} \left|\beta_{\mu\nu}\right\rangle, \tag{5}$$

and $U_h(t)$ becomes split on the direct-sum of subspaces $\mathcal{H}_h^{\otimes 2} = \mathcal{H}_{h,1} \oplus \mathcal{H}_{h,2}$ as function of h [23]:

$$\mathcal{H}_{h,1} \equiv \text{span}(\{|\beta_{--}\rangle, |\beta_{s^{0,a}s^{0,b}}\rangle\}), \quad \mathcal{H}_{h,2} \equiv \text{span}(\{|\beta_{s^{1,a}s^{h,b}}\rangle, |\beta_{+s^{1,h}}\rangle\}), \tag{6}$$

with $a = \frac{1}{2}(h-2)(h-3), b = \frac{1}{2}(h-1)(h-2), s^{c,d} = (-1)^{(c+d)\text{mod}2}$. Then, $U_h(t)$ becomes the direct sum $U_h(t) = s_{h1} \oplus s_{h2}$ with s_{hj} as a $U(1) \times SU(2) \subset U(2)$ block on each $\mathcal{H}_{h,j}, j = 1, 2$. $|\psi\rangle = \sum_j \alpha_j |\psi_j\rangle$ and $|\psi_j\rangle \in \mathcal{H}_{hj}$ evolving in each subspace. For the time independent case

$$s_{h_j} = e^{i\Delta_{h\alpha}^{+}} \begin{pmatrix} e_{h\alpha}^{\beta *} & -qi^h d_{h\alpha} \\ qi^{*h} d_{h\alpha} & e_{h\alpha}^{\beta} \end{pmatrix} = e^{i\Delta_{h\alpha}^{+}} \left(\cos\Delta_{h\alpha}^{-} \mathbf{I}_{h_j} - i\sin\Delta_{h\alpha}^{-} \mathbf{n}\cdot\mathbf{S}_{h_j}\right), \tag{7}$$

where $\mathbf{n} = (qb_{h_{-\alpha}} \sin\frac{h\pi}{2}, qb_{h_{-\alpha}} \cos\frac{h\pi}{2}, \beta j_{h_{-\alpha}}), \alpha = (-1)^{h+j+1}, \beta = (-1)^{j(h+l_j-k_j+1)}, q = \beta(-1)^{h+1}$, h is the direction of magnetic field and $j = 1, 2$ a position label for each block in the whole evolution matrix. k_j, l_j are the labels for its rows. As instance, $k_2 = 3, l_2 = 4$ are the rows of the second block s_{21}, with $j = 2$ in $U_{h=1}(t)$. Pauli matrices $\mathbf{S}_{h_j} = (\mathbf{X}_{h_j}, \mathbf{Y}_{h_j}, \mathbf{Z}_{h_j})$ and 2×2 identity $\mathbf{I}_{h_j}$ are straight forms in the matrix block to avoid confusion with the traditional operators in the computational basis. $U_h(t)$ forms abelian subgroups for fix $j_{h\pm}, b_{h\pm}$ [22]. Thus, inverses $U_h^\dagger(t)$ are obtained as another $U_h(t')$ for the same $j_{h\pm}, b_{h\pm}$ in the system, so $\mathbf{I}_{hj}$ can be achieved in finite time with the same fields. Properties and applications of the decomposition arise from those known for $SU(2)$ systems. A direct calculation gives the dynamics of the concurrence for the Bell states $|\beta_{\mu\nu}\rangle$ evolution

$$\mathcal{C}^h_{\mu\nu} = 1 - 4j^2_{h_{-f^h_{\mu\nu}}} b^2_{h_{-f^h_{\mu\nu}}} \sin^4\Delta^{-}_{h\, f^h_{\mu\nu}}, \quad f^h_{\mu\nu} = \mu\delta_{1h} + \mu\nu\delta_{2h} + \nu\delta_{3h}, \tag{8}$$

showing a simple behavior on only one Rabi frequency (contrasting with separable states [18]), an inherited feature from the separation of $\mathcal{H}^{\otimes 2}$ in two subspaces. There are not a maximal entangled evolution path with finite constant fields, more than trivial $j_{h f^h_{\mu\nu}} = 0$ or $b_{h f^h_{\mu\nu}} = 0$. Intermediate separable states are possible only for the tuning $4j^2_{h f^h_{\mu\nu}} b^2_{h f^h_{\mu\nu}} = 1$.

Journal of Physics: Conference Series **698** (2016) 012024 doi:10.1088/1742-6596/698/1/012024

3. Properties and potential applications of $SU(2)$ block reduction

3.1. Quantum error correction

Expression for a desired s_{hj} in terms of $\mathbf{S}_{h_j}$ on each subspace sets a discrete language for quantum errors on it and their related syndromes. As instance, for s_{12}, the subspace is span by $|\beta_{+-}\rangle, |\beta_{++}\rangle$, then, $\mathbf{X}_{1_2}$ is a bit flip in the second qubit; $\mathbf{Z}_{1_2}$ is a phase flip on both qubits; and $\mathbf{Y}_{1_2} = -i\mathbf{Z}_{1_2}\mathbf{X}_{1_2}$ is the combination of both errors. Traditional quantum error correction codes can be applied on each sector [24] and addressed on a specific evolution noting if $\mathcal{D} \equiv \delta\mathbf{p} \cdot \nabla_{\mathbf{p}}$, then, a tiny imprecision in the prescription parameters $\delta\mathbf{p}\ (t, \Delta_{h\alpha}^{\pm}, j_{h\pm}, b_{h\pm})$ to reproduce s_{hj}^{0}, induces an error $\delta s_{hj} \approx \mathcal{D}s_{hj}^{0} + \frac{1}{2}\mathcal{D}^2 s_{hj}^{0}$, thus, for a desired final state $\left|\psi_{fj}\right\rangle$ and its fidelity $\mathcal{F}$:

$$\delta\left|\psi_{fj}\right\rangle \approx \delta s_{hj} s_{hj}^{0\dagger}\left|\psi_{fj}\right\rangle, \quad 1-\mathcal{F}^2 \approx \left\langle s_{hj}^{0}\mathcal{D}s_{hj}^{0\dagger}\mathcal{D}s_{hj}^{0}s_{hj}^{0\dagger}\right\rangle - \left|\left\langle \mathcal{D}s_{hj}^{0}s_{hj}^{0\dagger}\right\rangle\right|^2, \tag{9}$$

where s_{hj} unitary properties are used. The quadratic dependence on $\delta\mathbf{p}$ shows a mild impact.

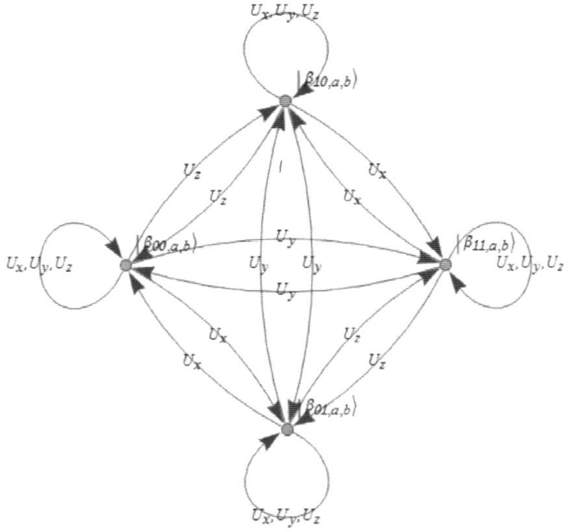

Figure 1. Bell states transformations through Evolution Loops and Exchange Operations.

3.2. Diagonal and antidiagonal basic forms

If the dynamics in $t = T$ reduces to the forms (diagonal and antidiagonal cases) [22]:

$$s_{hj} = \pm\mathbf{I}_{hj} \tag{10}$$

$$s_{hj} = \pm\mathbf{X}_{hj} \quad \text{or}: \quad s_{hj} = \pm i\mathbf{Y}_{hj} \tag{11}$$

we could achieve evolution loops [25–27] and exchange operations [18] in $\mathcal{H}^{\otimes 2}$ combining them, recovering or switching Bell states. Last operations let manipulate the Bell states in a planned way by applying a sequence of magnetic field pulses in adequate directions. Figure 1 summarizes the achievable loops and transitions. There are several ways to s_{hj} fulfills (10) or (11). Using only one field pulse, a direct analysis shows $s_{hj} = (-1)^{m_\alpha}\mathbf{I}_{hj}$ at $t = T$ if $n_\alpha, m_\alpha \in \mathbf{Z}$:

$$T = \frac{m_\alpha - n_\alpha}{\alpha J_h}\pi > 0 \quad , \quad B_{h-\alpha}^{2} = \left(\frac{J_h n_\alpha}{m_\alpha - n_\alpha}\right)^2 - J_{\{h\}_\alpha}{}^{2}. \tag{12}$$

To get the evolution loop $U_h(T) = (-1)^{m_\alpha}\mathbf{I_4}$, (12) should be fulfilled in both blocks $\pm\alpha$. It is possible if $m_\alpha - n_\alpha = n_{-\alpha} - m_{-\alpha}$ and $m_{\pm\alpha}$ have the same parity. If (11) is required in one block and (10) in another, $U_h(T)$ becomes an exchange operation transforming two Bell states between them, while the remaining become unchanged. Nevertheless, there are not solutions for block antidiagonalization (11) in one pulse compatible with (12) without restrict J_h and use non finite fields. But a two blocks product (7) for consecutive pulses $s'_{hj}s_{hj}$ fulfill it (α, β, q are the same). Complementary diagonalization prescriptions (10) in two pulses for block α are [23]:

$$\Delta_{h\alpha}^- + \mathrm{sign}(J_{\{h\}_\alpha}J'_{\{h\}_\alpha})\Delta'^-_{h\alpha} = n_\alpha\pi, \quad \frac{B_{h-\alpha}}{J_{\{h\}_\alpha}} = \frac{B'_{h-\alpha}}{J'_{\{h\}_\alpha}}, \quad \Delta_{h\alpha}^+ + \Delta'^+_{h\alpha} = (m_\alpha + n_\alpha)\pi, \tag{13}$$

$m_\alpha, n_\alpha \in \mathbf{Z}$, giving: $s'_{hj}s_{hj} = (-1)^{m_\alpha}\mathbf{I}_{hj}$. Antidiagonalization (11) for block $-\alpha$ gives [23]:

$$\begin{aligned} \Delta_{h-\alpha}^- &= \tfrac{2n_{-\alpha}+1}{2}\pi, \quad \Delta'^-_{h-\alpha} = \tfrac{2n'_{-\alpha}+1}{2}\pi, \quad \frac{B_{h\alpha}}{J_{\{h\}_{-\alpha}}} = -\frac{J'_{\{h\}_{-\alpha}}}{B'_{h\alpha}} \\ \Delta_{h-\alpha}^+ + \Delta'^+_{h-\alpha} &= \tfrac{\pi}{2}M_{h,q,-\alpha,\beta,n_{-\alpha},n'_{-\alpha},s_{-\alpha}} \\ &\equiv -\tfrac{\pi}{2}(h + \mathrm{sign}(q\beta b'_{h_\alpha}j_{h_\alpha}) + 2(n_{-\alpha} + n'_{-\alpha} - s_{-\alpha} + 1)), \end{aligned} \tag{14}$$

with: $s_{-\alpha}, n_{-\alpha}, n'_{-\alpha} \in \mathbf{Z}$, then $s'_{hj}s_{hj} = (-1)^{s_{-\alpha}}i^{h\mathrm{mod}2}\mathbf{S}_{hj}^{1+h\mathrm{mod}2}$ ($1+h\mathrm{mod}2$ component of $\mathbf{S}_{hj}$) as (11). $s_{-\alpha}$ and h introduce a phase when the Bell states are exchanged, while $s_{-\alpha}$ is not an integer for all h values due to it depends on the restrictions for $\Delta_{h-\alpha}^+ + \Delta'^+_{h-\alpha}$ in another sector.

3.3. Evolution loops and exchange operations

Evolution loops and exchange operations are involved in quantum characterization, teleportation, state discrimination and repreparation [18]. As it was stated, evolution loops are reached in one field pulse using (12) for both blocks $\pm\alpha$: $U_h(T) = (-1)^{m_\alpha}\mathbf{I_4}$. Exchange operations are possible in two pulses depending on: $B_{h\pm\alpha}, B'_{h\pm\alpha}, t, t'$. Due to (14), the antidiagonal-antidiagonal case implies an extra condition that (13), limiting J_i. Thus, we analyze only the diagonal-antidiagonal case where only two Bell states are exchanged while other remain unchanged. Combining (13) and (14), solutions are reached as a program to get $B_{h-\alpha}, B'_{h\pm\alpha}, t, t'$ in terms of $B_{h\alpha}$. We set $\pm\alpha$ for the diagonal/antidiagonal block. Decoupling $B_{h\alpha}$:

$$|\xi| = \frac{-AB \pm \sqrt{A^2 + B^2 - 1}}{B^2 - 1}, \quad A = \frac{(2n_{-\alpha}+1)J_h}{2(m_\alpha + n_\alpha)|J_{\{h\}_{-\alpha}}|}, \quad B = \frac{(2n'_{-\alpha}+1)J'_h}{2(m_\alpha + n_\alpha)|J'_{\{h\}_{-\alpha}}|}, \tag{15}$$

where $\xi \equiv \frac{B_{h\alpha}}{J_{\{h\}_{-\alpha}}}$. It has solutions if $A^2 + B^2 \geq 1$ and positivity is fulfilled. It is possible for finite and anisotropic strengths selecting $n_{-\alpha}, n'_{-\alpha}, n_\alpha, m_\alpha$ properly. Solutions are mainly around of opposite signs for A and B, limiting physical cases. $n_{-\alpha}, n'_{-\alpha} \geq 0$ and $m_\alpha + n_\alpha, J_h, J'_h$ determine the selection of A, B. Figure 2 shows the regions with solutions for $|\xi|$ in the plane AB. Note there are solutions in the four quadrants for finite J_h, J'_h with an adequate selection of $n_{-\alpha}, n'_{-\alpha}, n_\alpha, m_\alpha$. Dashed regions have not solutions for (15). Remaining prescriptions are

$$\begin{aligned} &\frac{B_{h-\alpha}^2}{J_{\{h\}_\alpha}^2} \equiv \chi^2 = \left(\frac{2n_\alpha\sqrt{\xi^2+1}}{S_\alpha(2n_{-\alpha}+1) + P_\alpha S'_\alpha(2n'_{-\alpha}+1)|\xi|}\right)^2 - 1, \quad \frac{B'_{h\alpha}}{J'_{\{h\}_{-\alpha}}} = -\xi^{-1}, \quad \frac{B'_{h-\alpha}}{J'_{\{h\}_\alpha}} = \chi, \\ &\frac{|J_{\{h\}_{-\alpha}}|t}{(2n_{-\alpha}+1)} = \frac{|J'_{\{h\}_{-\alpha}}|t'}{(2n'_{-\alpha}+1)|\xi|} = \frac{\pi}{2\sqrt{\xi^2+1}}, \quad P_\alpha = \mathrm{sign}(J'_{\{h\}_\alpha}J_{\{h\}_\alpha}), \quad S_\alpha = \left|\frac{J_{\{h\}_\alpha}}{J_{\{h\}_{-\alpha}}}\right|, \quad S'_\alpha = \left|\frac{J'_{\{h\}_\alpha}}{J'_{\{h\}_{-\alpha}}}\right|, \end{aligned} \tag{16}$$

in addition, block phases should be synchronized: $2(m_\alpha + n_\alpha) = -M_{h,q,-\alpha,\beta,n_{-\alpha},n'_{-\alpha},s_{-\alpha}}$. Then, $m_\alpha + n_\alpha$ fixes $s_{-\alpha}$ value in $s'_{hj}s_{hj}$. Note our analysis has preserved the non decisive possibility of the strengths could change during each pulse, a few common, but not impossible situation.

Journal of Physics: Conference Series **698** (2016) 012024 doi:10.1088/1742-6596/698/1/012024

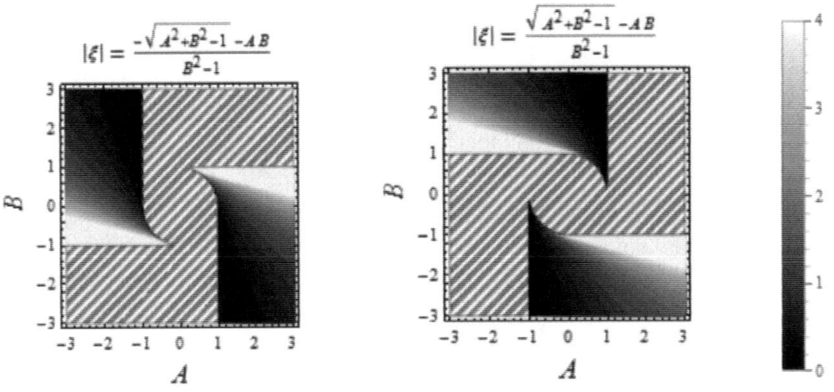

(a) (b)

Figure 2. Solution for $|\xi|$ values in gray scale for a) Negative and b) Positive signs.

3.4. Remarkable natural quantum gates

An analysis for exchange operations shows the evolution (for $T = t + t'$) as one of the gates

$$\begin{array}{lll} U_1(T): & \mathcal{A}_{1,1} \equiv i\mathbf{Y}_{1_1} \oplus \mathbf{I}_{1_2} = X_1 C^1(iY_2) X_1, & \mathcal{A}_{1,2} \equiv \mathbf{I}_{1_1} \oplus i\mathbf{Y}_{1_2} = C^1(iY_2) \\ U_2(T): & \mathcal{A}_{2,1} \equiv i\mathbf{X}_{2_1} \oplus \mathbf{I}_{2_2} = X_1 C^{1|2}(iX_1X_2) X_1, & \mathcal{A}_{2,2} \equiv \mathbf{I}_{2_1} \oplus i\mathbf{X}_{2_2} = C^{1|2}(iX_1X_2) \\ U_3(T): & \mathcal{A}_{3,1} \equiv \mathbf{I}_{3_1} \oplus i\mathbf{Y}_{3_2} = X_2 C^2(iY_1) X_2, & \mathcal{A}_{3,2} \equiv i\mathbf{Y}_{3_1} \oplus \mathbf{I}_{3_2} = C^2(iY_1) \end{array} \tag{17}$$

remembering C^aNOT_b gates on the $|\beta_{\alpha\beta}\rangle$ scripts (extending our notation under: $0 \leftrightarrow -, 1 \leftrightarrow +$). Realizing $\mathcal{A}_{1,2}$ is $C^1(iY_2)$ gate (understood for the Bell basis), an analysis shows the second expressions in (17) with the traditional X_i, Y_i in the 4×4 matrices. $C^{1|2}(G)$ is a controlled gate G where $1|2$ means $(A+B)\text{mod}2$ (using the equivalence with computational scripts $|\beta_{AB}\rangle = |\beta_{\alpha\beta}\rangle$). These operations give alternative gates to those used with computational basis in the circuit model of quantum computation, being an adaptation to the Bell basis and deserving the same treatment (by example, gate $\mathcal{A}_{1,2}$ was used to implement a Teleportation algorithm [28]).

3.5. Optimal cost quantum control and gate design by Unitary Factorization

Compatible trends of the reduction are devised. Optimal control for two level quantum systems is developed for energy cost [29] and minimal time [30] and the Hamiltonian is fully compatible with the current: (1) can be decomposed as $H_h = \sum_j H_{j,h}$, with each $H_{j,h}$ acting on each $\mathcal{H}_{hj}$,

$$H_{j,h} = -s_0 J_h \mathbf{I}_{2,h_j} + s_1 J_{\{h\}_{s_0}} \mathbf{Z}_{hj} + s_2 B_{h-s_0} \mathbf{S}^q_{hj}, \tag{18}$$

where $s_0 = (-1)^{h+j+1}, s_1 = s_0^p, s_2 = (-1)^p s_0^{p+q}, p = 1 + \frac{1}{2}(h-1)(h-2)$ and $q = h\text{mod}2 \in \{1,2\}$.

Another application is the Unitary Factorization gate design [31]. Diagonal-antidiagonal forms on (5) can reproduce unitary matrices $M^{(i,j)^\dagger}$ as factors [32]. Then, a unitary gate $U \in SU(4)$ can be written as ($\rightarrow$ states a forward product stacking factors from left to right):

$$U = \prod_{\substack{1\leq j<n \\ n>i\geq j}}^{\rightarrow} M^{(i,j)\dagger} \tag{19}$$

4. Conclusions

Spin-based quantum computing (superconducting integrated systems, superconducting flux qubits, straight nuclear magnetic resonance or quantum dots) exploits Ising interactions [33] where several technological issues have been solved around of stability and decoherence to set stable isolated qubits. There, contemporary control physics reports times and magnetic fields achievable in the experimental setup around of $t \sim 10^{-9} - 10^{-6}s$ and $B \sim 1 - 10T$ operating in regions around of $r \ll 5nm$ [34]. Despite, in nowadays, deep control of physical parameters in quantum magnetic systems is still limited to set a programmable artificial spin network under full control [33]. Current work develops control models including several freedom degrees. Operations constructed under $SU(2)$ decomposition state elementary physical operations to reproduce a planned evolution on a Bell states based grammar, able to be scaled to simulate complex computational problems on them. Moreover, combining these operations (5) for magnetic fields in different directions, we can increase the control possibilities. Creation of universal procedures by factorization as arbitrary gates is an alternative design based completely on $SU(2)$ reduction. Future work should be addressed considering finite temperature effects and deeper error correction procedures for the contemporary experimental limitations. The extension of this formalism to multiqubit case only requires to group arbitrarily the Hamiltonian eigenstates by pairs, but the search of universal basis (as Bell basis was here) is still open.

Acknowledgements

The support from Tecnológico de Monterrey to develop this research work is acknowledged.

References

[1] Jozsa R and Linden N 2002 *Proc. Royal Soc. A: Mathematical, Physical and Engineering Sciences* **459** 2011
[2] Bennett C H, DiVincenzo D P, Smolin J A and Wootters W K 1996 *Phys. Rev.* A **54** 3824
[3] Brush S G 1967 *Rev. Mod. Phys.* **39** 883
[4] Nielsen M A 1998 *Ph. D. Thesis: Quantum information theory* (Albuquerque: University of New Mexico)
[5] Terzis A F and Paspalakis E 2004 *Phys. Lett.* A **333** 438
[6] Stelmachovic P and Buzek V 2004 *Phys. Rev.* A **70** 032313
[7] Novotny J, Stefanak M, Kiss T and Jex, I 2005 *J. Phys.* A: **38** 9087
[8] Arnesen M C, Bose S and Vedral V 2001 *Phys. Rev. Lett.* **87** 017901
[9] Wang X 2001 *Phys. Rev.* A **64** 012313
[10] Khaneja N, Brockett R and Glaser S J 2001 *Phys. Rev.* A **63** 032308
[11] Recher P and Loss D 2002 *Proc. of SPINTRONICS 2001: Int. Conf. on Novel Aspects of Spin-Polarized Transport and Spin Dynamics* vol 15 (Springer, New York) p 49
[12] Koppens F H L, Buizert C, Tielrooij K J, Vink I T, Nowack K C, Meunier T, Kouwenhoven L P and Vandersypen L M K 2006 *Nature* **442** 766
[13] DiVincenzo D P 1997 *Mesoscopic Electron Transport, NATO Advanced Study Institute* vol 345 E (Kluwer, Dordrecht) p 657
[14] Berman G P, Doolen G D, López G V and Tsifrinovich V I 1998 *Generalized Quantum Control-Not Gate in Two-Spin Ising System* (quant-ph/9802013v1)
[15] Wang X 2001 *Phys. Lett.* A **281** 101
[16] Dalessandro D 2007 *Introduction to Quantum Control and Dynamics* (Boca Raton: Chapman Hall Applied Mathematics Nonlinear Science)
[17] Xi Z and Jin G 2007 *Int. J. Quant. Info.* **5** 857
[18] Delgado F 2010 *Phys. Rev.* A **81**
[19] Kamta G L and Starace A F 2002 *Phys. Rev. Lett.* **88** 107901
[20] Sun Y, Chen Y and Chen H 2003 *Phys. Rev.* A **68** 044301
[21] Zhou L, Song H S, Guo CY Q and Li C 2003 *Phys. Rev.* A **68** 024301
[22] Delgado F 2015 *Int. J. Quant. Info.* **13** 1550055
[23] Delgado F 2014 *Generation of non-local evolution loops and exchange operations for quantum control in three dimensional anisotropic Ising model* (quant-ph/1410.5515)
[24] Kaye P, Laflamme R and Mosca M 2007 *An Introduction to Quantum Computing* (Oxford University Press, New York, 2007)

[25] Mielnik B 1986 *J. Math. Phys.* **27** 2290
[26] Fernandez D J 1994 *Int. J. Theor. Phys.* **33** 2037
[27] Delgado F and Mielnik B 1997 *J. Phys.* A **31** 309
[28] Delgado F 2015 *J. Phys.: Conf. Series* **624** 012006
[29] D'Alessandro D and Dahleh M 2001 *IEEE Transactions on Automatic Control* **46** 866
[30] Boscain U and Mason P 2006 *J. Math. Phys.* **47** 062101
[31] Li C K, Jones R and Yin X 2013 *Int. J. Quant. Inf.* **11** 1350015
[32] Delgado F 2014 *Factorization of unitary matrices induced by 3D anisotropic Ising interaction* (quant-ph/1410.5891)
[33] Johnson M et al 2011 *Nature* **473** 194
[34] Kloeffel Ch and Loss D 2013 *Annu. Rev. Condens. Matter Phys.* **4** 51

Factorization of the Quantum Fractional Oscillator

F Olivar-Romero and O Rosas-Ortiz
Physics Department, Cinvestav, A.P. 14-740, 07000, México D.F, Mexico
E-mail: folivar@fis.cinvestav.mx, orosas@fis.cinvestav.mx

Abstract. The applicability of the factorization method is extended to the case of quantum fractional-differential Hamiltonians. In contrast with the conventional factorization, it is shown that the 'factorization energy' is now a fractional-differential operator rather than a constant. As a first example, the energies and wave-functions of a fractional version of the quantum oscillator are determined. Interestingly, the energy eigenvalues are expressed as power-laws of the momentum in terms of the non-integer differential order of the eigenvalue equation.

1. Introduction

In fractional calculus the orders of integration and derivation are real numbers rather than natural ones. Thereby, besides the conventional derivatives of order $n \in \mathbb{N}$, one is able to evaluate derivatives of orders as arbitrary as 0.5 or $\sqrt{2}$. The fractional derivatives are non-commutative and non-associative in general. Moreover, applied on a product of functions, they usually give rise to infinite series with derivatives and integrals of diverse non-integer orders (see e.g. [1–3]). Although such properties seem 'counterintuitive', they find immediate application in describing systems with nonlocal properties of power-law type and are useful in different branches of physics and engineering (see e.g. [4–7]). Quite recently, the Feynman path integral approach was extended to the fractional case [8, 9], the main idea was to replace the involved Brownian trajectories by Lévy flights. As a consequence, a space-fractional version of the well known Schrödinger equation has come to light [10]. Besides the cases of the quantum oscillator [10] (see also [11]) and the hydrogen-like potential [10], the space-fractional Schrödinger equation has been solved for some piece-constant potentials [11, 12], and has been extended to the time-fractional case [13]. Interesting discussions about the physical implementation of the quantum fractional oscillator can be found in [14, 15]. The physical realization of other fractional quantum systems is, as far as we know, an open problem.

With the present communication we extend the applicability of the factorization method [16–18] to the case of fractional-differential Hamiltonians. Without loss of generality, we pay attention to the one-dimensional oscillator and show that the 'factorization energy', which is a constant in the conventional factorization, must be replaced by a fractional-differential operator in the extended fractional formulation. This last permits the application of algebraic methods to determine the energies and wave-functions of the quantum fractional oscillator. Remarkably, it is found that the energy eigenvalues are expressed as power-laws of the momentum in terms of the non-integer differential order of the eigenvalue equation. As indicated above, our method is not restricted to the oscillator-like interactions, it is also useful in solving the fractional eigenvalue problem of other one-dimensional potentials.

2. Problem and solution

For one-dimensional systems the time-independent, space-fractional Schrödinger equation [10] is of the form

$$\left[-\mathcal{D}_\alpha \hbar^\alpha \frac{d^\alpha}{dx^\alpha} + V(x)\right]\psi(x) = E\psi(x), \tag{1}$$

where α is the Lévy index ($1 < \alpha \leq 2$), $\frac{d^\alpha}{dx^\alpha}$ is the Riesz fractional derivative of order α, and $\mathcal{D}_\alpha$ is a scale constant. The conventional Schrödinger equation is recovered if $\alpha = 2$. From now on we use proper units such that $\mathcal{D}_\alpha \hbar^\alpha = 1$.

In [10], the fractional oscillator potential is defined as $|x|^\beta$, with $1 < \beta \leq 2$. Here, for the sake of clarity, this parameter is fixed as $\beta = 2$. The arbitrariness of β makes no substantial difference in the method and can be retrieved in a further step. Therefore, the fractional-differential equation to be solved is given by

$$H_\alpha \psi(x) \equiv \left[-\frac{d^\alpha}{dx^\alpha} + x^2\right]\psi(x) = E\psi(x), \quad 1 < \alpha \leq 2. \tag{2}$$

As this will be clear in the sequel, it is useful to express (2) in momentum-representation

$$\left[|k|^\alpha - \frac{d^2}{dk^2}\right]\phi(k) = E\phi(k). \tag{3}$$

2.1. The fractional-factorization method

We look for a pair of operators A_α and B_α such that

$$H_\alpha = B_\alpha A_\alpha + \epsilon_\alpha, \tag{4}$$

where ϵ_α can be either a number (as in the conventional factorization) or a fractional-differential operator. The explicit form of the factors can be as general as in the conventional factorization whenever the identity (4) is fulfilled. Here we present just one of their simpler expressions. Namely,

$$A_\alpha = \frac{d^{\alpha/2}}{dx^{\alpha/2}} + x \qquad \text{and} \qquad B_\alpha = -\frac{d^{\alpha/2}}{dx^{\alpha/2}} + x \tag{5}$$

give

$$B_\alpha A_\alpha = -\frac{d^\alpha}{dx^\alpha} - \frac{\alpha}{2}\frac{d^{\alpha/2-1}}{dx^{\alpha/2-1}} + x^2. \tag{6}$$

Comparing (6) with (2) we see that (4) is fulfilled if

$$\epsilon_\alpha = \frac{\alpha}{2}\frac{d^{\alpha/2-1}}{dx^{\alpha/2-1}}. \tag{7}$$

That is, for $\alpha \neq 2$, the factorization energy ϵ_α is a fractional-differential operator of order $\frac{\alpha}{2} - 1$ rather than a constant. Notice that $\alpha = 2$ leads to the conventional result $\epsilon = 1$, with $A_2 = a$ and $B_2 = a^\dagger$ the boson operators of the (mathematical) quantum oscillator $V_{osc}(x) = x^2$.

2.2. Spectrum and wavefunctions

To identify the kernel of A_α we have to solve the fractional-differential equation

$$A_\alpha \psi_0^{(\alpha)}(x) = \left[\frac{d^{\alpha/2}}{dx^{\alpha/2}} + x\right]\psi_0^{(\alpha)}(x) = 0. \tag{8}$$

Equivalently, one can solve this last equation in position-representation

$$i\left[k^{\alpha/2}\text{sgn}(k)+\frac{d}{dk}\right]\phi_0^{(\alpha)}(k)=0. \quad (9)$$

Substituting the solution

$$\phi_0^{(\alpha)}(k)=\exp\left(-\frac{2|k|^{\alpha/2+1}}{\alpha+2}\right) \quad (10)$$

into (3) we obtain an expression for the related energy

$$E_0(k,\alpha)=1\left(\frac{\alpha}{2|k|^{1-\alpha/2}}\right). \quad (11)$$

Thus, the wave-function $\phi_0^{(\alpha)}(k)$ and energy $E_0(k,\alpha)$ have a power-law dependence on the momentum k that is determined by the Lévy index α. In particular, the solutions for the ground state $E_0=1$ of the (mathematical) quantum oscillator are recovered if $\alpha=2$. Similar results have been reported for the 'Cauchy oscillator' by using the so called Strang splitting method in [11]. In our case, (10) and (11) are just a consequence of the factorization (4). Indeed, introducing (8) in (4) and using (2), one obtains the equation

$$\left[-\frac{d^\alpha}{dx^\alpha}+x^2-\frac{\alpha}{2}\frac{d^{\alpha/2-1}}{dx^{\alpha/2-1}}\right]\psi_0^{(\alpha)}(x)=0. \quad (12)$$

In momentum-representation this last acquires the form

$$\left[|k|^\alpha-\frac{d^2}{dk^2}-\frac{\alpha}{2}|k|^{\alpha/2-1}\right]\phi_0^{(\alpha)}=0. \quad (13)$$

Clearly, the function (10) is solution of (13), so that our approach is self-consistent.

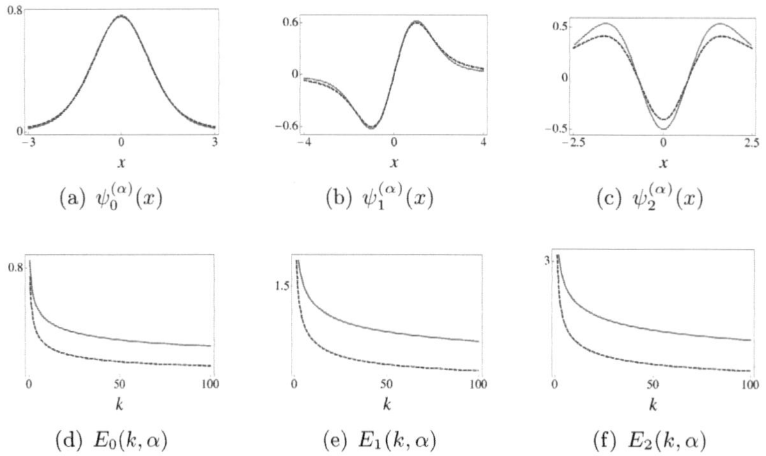

Figure 1. Wave-functions (a–c) and energies (d–f) of the first three states of the quantum fractional oscillator. In all cases the dashed-blue curve corresponds to $\alpha=1.2$ while the solid-red one is for $\alpha=1.5$. The solutions of the conventional quantum oscillator are recovered if $\alpha=2$.

The behaviour of the (numerically calculated) Fourier transform $\psi_0^{(\alpha)}(x)$ of (10) and the energy (11) are depicted in Figure 1(a) and (d) for two different values of the Lévy index. Note that the variations of $\phi_0^{(\alpha)}(x)$ with respect to the parameter α are almost negligible (the power-law dependence on k is in the argument of an exponential function). The situation for the energy eigenvalue $E_0(k,\alpha)$ is quite different because the power-law dependence on k is in the denominator of (11).

As in the conventional factorization, the proper application of the operators A_α and B_α give rise to the other solutions of the problem. For instance, to get the first excited state $\phi_1^{(\alpha)}(k)$ we apply B_α on $\phi_0^{(\alpha)}(k)$. After some calculations one gets

$$\phi_1^{(\alpha)}(k) = -2i|k|^{\alpha/2}\text{sgn}(k)\exp\left(-\frac{2|k|^{\alpha/2+1}}{\alpha+2}\right), \tag{14}$$

and using (3) we arrive at the expression for the energy

$$E_1(k,\alpha) = 3\left(\frac{\alpha}{2|k|^{1-\alpha/2}}\right) - \frac{\alpha}{2|k|^2}\left(\frac{\alpha}{2}-1\right). \tag{15}$$

A similar procedure is used to construct the other excited sates. In particular, the second excited state

$$\phi_2^{(\alpha)}(k) = (\alpha|k|^{\alpha/2-1} - 4|k|^{\alpha})\exp\left(-\frac{2|k|^{\alpha/2+1}}{\alpha+2}\right), \tag{16}$$

belongs to the energy

$$E_2(k,\alpha) = 5\left(\frac{\alpha}{2|k|^{1-\alpha/2}}\right)\left(\frac{\alpha-1-2|k|^{3\alpha/2-1}}{\frac{\alpha}{2}-2|k|^{\alpha/2+1}}\right) + \frac{\alpha}{2|k|^2}\left(\frac{\alpha}{2}-1\right)\left[\frac{2-\frac{\alpha}{2}+|k|^{\alpha/2+1}}{\frac{\alpha}{2}-2|k|^{\alpha/2+1}}\right] \tag{17}$$

The behaviour of these solutions can be appreciated in Figure 1. In all cases the conventional results of the quantum oscillator are recovered if $\alpha = 2$. Note that the dependence of the wave-functions on the Lévy index α starts to be significative for the excited states.

3. Discussion and further applications

With some further modifications, the method can be immediately applied to solve the eigenvalue problem of the fractional potential $|x|^\beta$, $1 < \beta \leq 2$, discussed in [10]. In such a case, the solutions depend on the parameters α and β, and behave quite similar to the ones presented in this communication. Noticeably, one can use more general fractional-differential operators A_δ and B_γ, with Lévy indices δ and γ, such that the identity $H_\alpha = A_\delta B_\gamma + \epsilon_{\delta,\gamma}$ is true. This last includes the case in which $\alpha = 2$ from the very beginning. That is, the case in which we look for the fractional-factorization of the conventional Hamiltonian in quantum mechanics. As expected, even in this last case, the fractional-factorization program gives rise to new algebraic properties of the factorizing operators [19]. This last opens new possibilities for the trends in supersymmetric quantum mechanics as this last is based fundamentally in the factorization method. Other possible applications of the fractional-factorization presented in this paper include the paraxial theory of optical beam propagation [20,21]. Further progress in these subjects will be presented elsewhere.

4. Conclusions

We have introduced an algebraic technique to solve the eigenvalue problem of the Laskin time-independent, space-fractional Schrödinger equation [10]. This is based on a modification of the well known factorization method and requires that the factorization energy, which is a constant in the conventional factorization, be expressed as a differential operator of non-integer order. Although we have specialized our discussion in the case of the (mathematical) oscillator potential $V_{osc}(x) = x^2$, the generalities of the method can be glimpsed from the expressions derived in Section 2.1. Thus, with the appropriate refinements, the fractional-factorization can be implemented in practically all the cases where the conventional factorization is known to work. Of special interest, we have shown that the energies of the quantum fractional oscillator studied here depend on the momentum in terms of a power-law that is determined by the non-integer order of the fractional eigenvalue equation. Some insights of such a behaviour can be obtained from the classical fractional version of the same potential for which a dissipative-like term is included in the motion equation [19]. Definitive conclusions on the matter are in progress.

Acknowledgments
The comments and suggestions by the anonymous referees are acknowledged. FOR acknowledges the funding received through a CONACyT Scholarship.

5. References

[1] Miller K S and Ross B 1993 *An introduction to the fractional calculus and fractional differential equations* (John Wiley and Sons, New York)
[2] Kilbas A A, Srivastava H M and Trujillo J J 2006 *Theory and Applications of Fractional Differential Equation* (Elsevier, Amsterdam)
[3] Duarte M 2011 *Fractional Calculus for Scientists and Engineers* (Springer, New York)
[4] Diethelm K 2010 *The Analysis of Fractional Differential Equations: An Application-Oriented Exposition Using Differential Operators of Caputo Type* (Springer, New York)
[5] Sabatier J, Agrawal O P and Tenreiro J A (Eds.) 2007 *Advances in Fractional Calculus: Theoretical Developments and Applications in Physics and Engineering* (Springer, New York)
[6] Hilfer R (Ed.) 2000 *Applications of fractional Calculus in Physics* (World Scientific Singapore)
[7] Tarasov V E 2011 *Fractional Dynamics: Applications of Fractional Calculus to Dynamics of Particles, Fields and Media* (Springer, New York)
[8] Laskin N 2000 Fractional Quantum Mechanics and Levy Path Integrals *Phys. Lett.* A **268** 298
[9] Laskin N 2000 Fractional Quantum Mechanics *Phys. Rev.* E **62** 3135
[10] Laskin N 2002 Fractional Schröndinger equation *Phys. Rev.* E **66** 056108
[11] Zaba M and Garbaczewski P 2014 Solving fractional Schrödinger-type spectral problems: Cauchy oscillator and Cauchy well *J. Math. Phys.* **55** 092103
[12] Guo X and Xu M 2006 Some physical applications of fractional Schrödinger equation *J. Math. Phys.* **47** 082104
[13] Naber M 2004 Time fractional Schrödinger equation *J. Math. Phys.* **45** 3339
[14] Longhi S 2015 Fractional Schrödinger equation in optics *Opt. lett.* **40** 1117
[15] Zhang Y, Liu X, Belić M J, et. al. 2015 Propagation Dynamics of a Light Beam in a Fractional Schrödinger Equation *Phys. Rev. Lett.* **115** 180403
[16] Infeld L and Hull T E 1951 The factorization method *Rev. Mod. Phys.* **23** 21
[17] Andrianov A A, Borisov N V and Ioffe M V 1984 The factorization method and quantum-systems with equivalent energy-spectra *Phys. Lett. A* **105** 19
[18] Mielnik B and Rosas-Ortiz O 2004 Factorization: Little or great algorithm? *J. Phys. A: Math. Gen.* **37** 10007
[19] Olivar F 2014 *Mecánica Cuántica Fraccional: Una Primera Aproximación* (MSc Thesis (in Spanish), Physics Department, Cinvestav, Mexico)
[20] Gutiérrez-Vega J C 2007 Fractionalization of optical beams: I. Planar analysis *Opt. Lett.* **32** 1521
[21] Gutiérrez-Vega J C 2007 Fractionalization of optical beams: II. Elegant Laguerre-Gaussian beams *Opt. Express* **15** 6300

Optimized Binomial Quantum States of Complex Oscillators with Real Spectrum

K D Zelaya and O Rosas-Ortiz

Physics Department, Cinvestav, A.P. 14-740, 07000, México D.F, Mexico

E-mail: kdzelaya@fis.cinvestav.mx, orosas@fis.cinvestav.mx

Abstract. Classical and nonclassical states of quantum complex oscillators with real spectrum are presented. Such states are bi-orthonormal superpositions of $n+1$ energy eigenvectors of the system with binomial-like coefficients. For large values of n these optimized binomial states behave as photon added coherent states when the imaginary part of the potential is cancelled.

1. Introduction

In Ref. [1] a wide family of complex potentials that have the spectrum $\mathcal{E}_n = 2n+1$ of the quantum oscillator $V_{osc}(x) = x^2$ plus the energy eigenvalue $E_0 = -1$ is reported (see Figure 1). The energy eigenvectors $|\psi_k\rangle$ of this new potential form a bi-orthonormal basis of the corresponding space of states. In this communication we study the superposition properties of such a set of vectors. In particular, we construct a packet of $n+1$ adjacent states such that $|\psi_r\rangle$ is the eigenvector of lowest energy in the superposition. That is, the collection $|\psi_r\rangle, |\psi_{r+1}\rangle, \ldots, |\psi_{r+n}\rangle$ is used to tailor a pure state of the system that can be either classical (with nonnegative Wigner function) or nonclassical [2] in the limit of the quantum oscillator (i.e., when the imaginary part of the complex potential is turned off and the appropriate parameters are chosen). The Fourier coefficients c_k associated with the states $|\psi_{r+k}\rangle$ in the superposition are such that $|c_k|^2$ is a binomial distribution delimited by n. We say that this bi-orthogonal superposition is an *optimized binomial state*. The concept of binomial states was introduced in [3] for the first $n+1$ eigenvectors of the quantum oscillator. The soundness of the approximation lies in the fact that such states represent light that is antibunched, sub-poissonian, and squeezed for certain parameter ranges [3] (see also [4]). Here we use the arbitrary collection of vectors described above and show that similar properties are obtained in the limit of the quantum oscillator, even if the generating set of vectors does not correspond to the first $n+1$ energy states. Moreover, we show that the summing of a very large number ($n \to \infty$) of elements in the collection leads to r-photon added coherent states.

In Section 2 the main properties (bi-orthogonality, continuity equation, time-evolution) of the space of states $|\psi_k\rangle$ is revisited. Section 3 deals with the tailoring of the classical and nonclassical states indicated above. Some concluding remarks are given in Section 4.

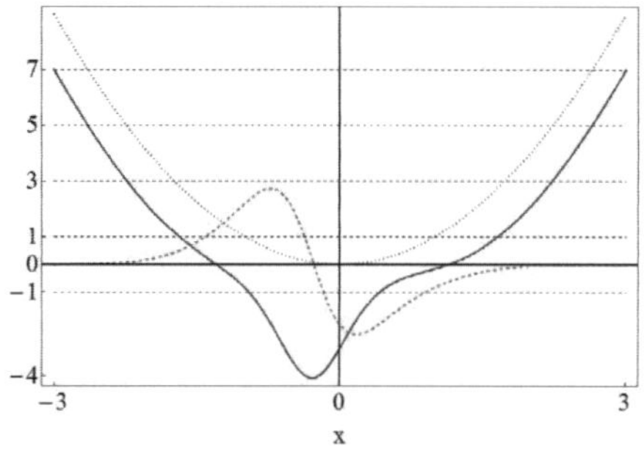

Figure 1. Real (blue, continuous curve) and imaginary (red, dashed curve) parts of the complex oscillator potential $V_\lambda(x)$ defined in (1) for $a = \frac{\pi}{4}$, $b = \frac{\sqrt{\pi}}{2}$ and $c = 1$. The energy eigenvalues $E_n = 2n - 1$, $n = 0, 1, \ldots$, are represented by dotted horizontal lines and the conventional oscillator (black, dotted curve), with eigenvalues $\mathcal{E}_n = 2n + 1 = E_{n+1}$, has been included as a reference.

2. Complex oscillators with real spectrum

Let us consider the family of complex dimensionless potentials

$$V_\lambda(x) = x^2 - 2 - 2\frac{d}{dx}\left[\frac{b + 2a\text{Erf}(x) - i\sqrt{\pi}\lambda}{\sqrt{\pi}\alpha^2(x)}\right], \tag{1}$$

with λ a real parameter and $\alpha(x)$ the real function

$$\alpha(x) = e^{x^2/2}\left[a\text{Erf}^2(x) + b\text{Erf}(x) + c\right]^{1/2}. \tag{2}$$

This family is reduced to the one reported by Mielnik [5] if $\lambda = 0$ and the real parameters a, b and c, are properly chosen [6]. For $\lambda \neq 0$, and any set of non-negative parameters $\{a, b, c > \frac{b^2}{4a}\}$, the potential $V_\lambda(x)$ is exactly solvable [1] with spectrum $E_n = 2n - 1$, $n = 0, 1, 2, \ldots$ (see Figure 1). Indeed, $V_\lambda(x)$ is a complex supersymmetric partner of the quantum oscillator $V_{osc}(x) = x^2$, this last with spectrum $\mathcal{E}_n = 2n + 1$. The set of eigenfunctions is given by

$$\psi_{n+1}(x) = \frac{1}{\sqrt{2(n+1)}}\left[\frac{d}{dx} - \frac{\alpha'(x)}{\alpha(x)} + i\frac{\lambda}{\alpha^2(x)}\right]\varphi_n(x), \quad n \geq 0, \tag{3}$$

and

$$\psi_0(x) = \frac{\kappa_0}{\alpha(x)}\exp\left[i\lambda\int^x \alpha^{-2}(y)dy\right], \tag{4}$$

where κ_0 is a normalization constant and

$$\varphi_n(x) = \frac{e^{-x^2/2}}{\sqrt{2^n n!\sqrt{\pi}}}H_n(x) \tag{5}$$

are the wave-functions of the quantum oscillator.

2.1. Bi-orthogonality

Although (3) and (4) do not form an orthonormal set, it is possible to construct a bi-orthogonal system $\{\psi_n, \overline{\psi}_m\}_{n,m\geq 0}$ by considering that $H_\lambda = -\frac{d^2}{dx^2} + V_\lambda(x)$ and its Hermitian-conjugate $H_\lambda^\dagger \equiv \overline{H}_\lambda$ are formally different. The notion of bi-orthogonality arises when *orthogonality* requires two formally different sets of vectors. That is, if a given set $\{\psi_n\}_{n\geq 0}$ cannot be orthogonal then one can consider the dual set $\{\overline{\psi}_m\}_{m\geq 0}$ to get $(\overline{\psi}_m, \psi_n) = \delta_{m,n}$. As the wave-functions $\varphi_n(x)$ are real we get $\overline{\psi}_n(x) = \psi_n^*(x)$, with z^* the complex-conjugate of $z \in \mathbb{C}$. Therefore, the following bi-orthonormal condition [1] holds

$$(\psi_n^*, \psi_m) = \int_{\mathbb{R}} \psi_n(x)\psi_m(x)dx = \int_{\mathbb{R}} \overline{\psi}_n^*(x)\overline{\psi}_m^*(x)dx = \left(\overline{\psi}_n, \overline{\psi}_m^*\right) = \delta_{n,m}. \tag{6}$$

Any arbitrary state $\phi(x)$ of the complex oscillator, as well as its dual $\overline{\phi}(x)$, can be expressed as a superposition of the fundamental solutions

$$\phi(x) = \sum_{k=0}^{\infty} c_k\psi_k(x), \quad \overline{\phi}(x) = \sum_{k=0}^{\infty} \overline{c}_k\overline{\psi}_k(x), \quad c_k, \overline{c}_k \in \mathbb{C}, \tag{7}$$

where the Fourier coefficients, c_k and $\overline{c}_k$, are determined by the bi-products

$$c_k = \left(\overline{\psi}_k, \phi\right), \quad \overline{c}_k = \left(\psi_k, \overline{\phi}\right). \tag{8}$$

In turn, the normalization condition

$$\left(\overline{\phi}, \phi\right) = \sum_{k=0}^{\infty} \overline{c}_k^* c_k = 1 \tag{9}$$

is fulfilled for the appropriate products $\overline{c}_k^* c_k \in \mathbb{C}$. We shall take $\overline{c}_k = c_k$ for simplicity. On the other hand, given $\phi(x)$ the conventional notions of probability density $\rho = |\phi|^2$ and probability current $J = i\left(\phi\frac{\partial\phi^*}{\partial x} - \phi^*\frac{\partial\phi}{\partial x}\right)$ lead to the continuity equation

$$\frac{\partial J}{\partial x} + \frac{\partial \rho}{\partial t} = 2\mathrm{Im}(V_\lambda). \tag{10}$$

However, using the bi-orthogonal quantities $\rho_B = \overline{\phi}^*\phi$ and $J_B = i\left(\phi\frac{\partial\overline{\phi}^*}{\partial x} - \overline{\phi}^*\frac{\partial\psi}{\partial x}\right)$, one gets

$$\frac{\partial J_B}{\partial x} + \frac{\partial \rho_B}{\partial t} = 0. \tag{11}$$

If $\lambda = 0$ we have $\mathrm{Im}(V_\lambda) = 0$ and $\overline{\psi}_k = \psi_k$, so that ρ_B and J_B coincide with ρ and J respectively. From here, the limit to the quantum oscillator [6] gives $V_{\lambda=0} \to V_{osc} - 2$, and $\psi_k \to \varphi_k$.

2.2. *Time-evolution*

The conventional time-evolution operator $U(t) = e^{-iH_\lambda t}$ is no longer unitary because H_λ is not self-adjoint. To construct the dual of $U(t)$ in proper form, we use the bi-orthogonality introduced in the previous section. Namely, $\overline{U}(t) = e^{-i\overline{H}_\lambda t}$ is such that $\overline{U}^\dagger(t)U(t) = I$ [6]. Then, the time-evolved vectors $|\phi(t)\rangle = U(t)|\phi(0)\rangle$ and $|\overline{\phi}(t)\rangle = \overline{U}(t)|\overline{\phi}(0)\rangle$ are such that

$$\langle\overline{\phi}(t)|\phi(t)\rangle = \langle\overline{\phi}(0)|e^{i\overline{H}_\lambda^\dagger t}e^{-iH_\lambda t}|\phi(0)\rangle = \langle\overline{\phi}(0)|\phi(0)\rangle. \tag{12}$$

That is, the bi-norm of the initial state $|\phi(0)\rangle$ is invariant under time-evolution. Notice that this last result is compatible with the bi-orthogonal continuity equation (11). In position-representation, $\phi(x) = \langle x|\phi\rangle$ and $\overline{\phi}(x) = \langle x|\overline{\phi}\rangle$, one has

$$\langle\overline{\phi}(t)|\phi(t)\rangle = \int_{\mathbb{R}} \overline{\phi}^*(x,t)\phi(x,t)dx = \int_{\mathbb{R}} \rho_B(x,t)dx = 1. \tag{13}$$

3. Optimized binomial states

Let us construct a superposition of $n+1$ adjacent states

$$|\phi_b\rangle = \sum_{k=0}^{n} c_k |\psi_{k+r}\rangle, \tag{14}$$

where the non-negative integer $r \geq 0$ determines eigenvector $|\psi_r\rangle$ of the lowest energy that is included in the packet. If the Fourier coefficients c_k are chosen such that

$$|c_k|^2 = \left(\begin{array}{c} n \\ k \end{array} \right) p^k (1-p)^{n-k}, \quad 0 \leq p \leq 1, \quad k \leq n, \tag{15}$$

then the state (14) is automatically bi-normalized $\langle \overline{\phi}_b | \phi_b \rangle = 1$, with

$$\left(\begin{array}{c} n \\ k \end{array} \right) = \frac{\Gamma(n+1)}{\Gamma(k+1)\Gamma(n-k+1)} \tag{16}$$

the binomial coefficient. Therefore, the probability $|c_k|^2$ of finding the system in the state $|\psi_{k+r}\rangle$ is weighted by the probability p of having success k times in n trials. We call the superposition (14)–(16) an *optimized binomial* state. The energy of such a wave-packet is characterized by the (bi-orthogonal) expectation value

$$\langle H_\lambda \rangle_b \equiv \langle \overline{\phi}_b | H_\lambda | \phi_b \rangle = \sum_{k=0}^{n} \left(\begin{array}{c} n \\ k \end{array} \right) p^k (1-p)^{n-k} (2k-1) = 2np - 1. \tag{17}$$

In turn, the time dependence of the superposition state is encoded in the Fourier coefficients

$$\phi_b(x,t) = \sum_{k=0}^{n} c_k(t) \psi_{k+r}(x), \quad c_k(t) = c_k e^{-2ikt}, \tag{18}$$

where a global phase $e^{-i(2r-1)t}$ has been dropped. As we are going to show, the optimized binomial state (18) is meaningful because any finite set of $n+1$ adjacent energy eigenvectors can be used to tailor either a classical-like or a quantum-like packet.

We use the binomial distribution (15) for two main reasons: (1) this permits the construction of wave-packets having a finite number of elements and (2) its limit for a large number of elements is well known. To be precise, in the limit $n \to \infty$ the expression (15) goes to the normal (Gaussian) distribution if $p = \text{const}$, or to the Poisson distribution if $np = \text{const}$. The former case is avoided in our approach because it implies that the energy of the system increases as $n \to \infty$. In turn, the Poisson distribution is reached by increasing the number of states but preserving the energy of the system. Thus, with $r = 0$ and $n \to \infty$, the optimized binomial states (14) converge to the Glauber coherent states [7] in the quantum oscillator limit. For $r \neq 0$ and the same conditions the superposition (14) goes to an r-photon added coherent state. More details are given in Section 3.1.

The time-evolution of the related bi-orthogonal probability density ρ_B and current J_B is shown in Figure 2 for $n = 30$ and the indicated values of p and r. The center of Re(ρ_B) oscillates back and forth between two *turning points* while Im(ρ_B) exhibits a zero (change of sign) that oscillates with a shorter amplitude. At a given time t, the integration of Im(ρ_B) over all the real line is zero while the summing of Re(ρ_B) gives 1. This last is because the time-variations of ρ_B are compensated by the space-variations of the bi-orthogonal probability current J_B, as indicated in Eq. (11).

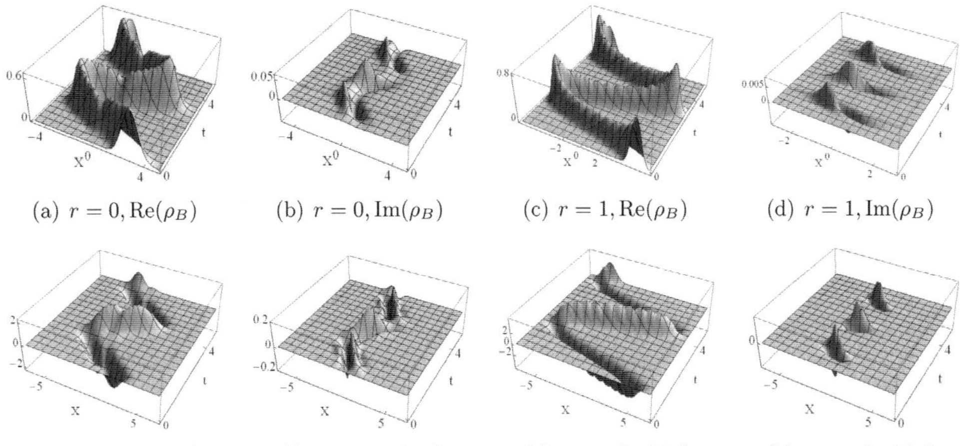

(a) $r = 0, \mathrm{Re}(\rho_B)$ (b) $r = 0, \mathrm{Im}(\rho_B)$ (c) $r = 1, \mathrm{Re}(\rho_B)$ (d) $r = 1, \mathrm{Im}(\rho_B)$

(e) $r = 0, \mathrm{Re}(J_B)$ (f) $r = 0, \mathrm{Im}(J_B)$ (g) $r = 1, \mathrm{Re}(J_B)$ (h) $r = 1, \mathrm{Im}(J_B)$

Figure 2. Time-evolution of the bi-orthogonal probability density ρ_B and current J_B of the optimized binomial state (18) for $n = 30$ and the indicated values of r. The first (last) two columns correspond to a packet of energy $\langle H_\lambda \rangle_b$ equal to 5 (29) and $p = 0.1$ ($p = 0.5$).

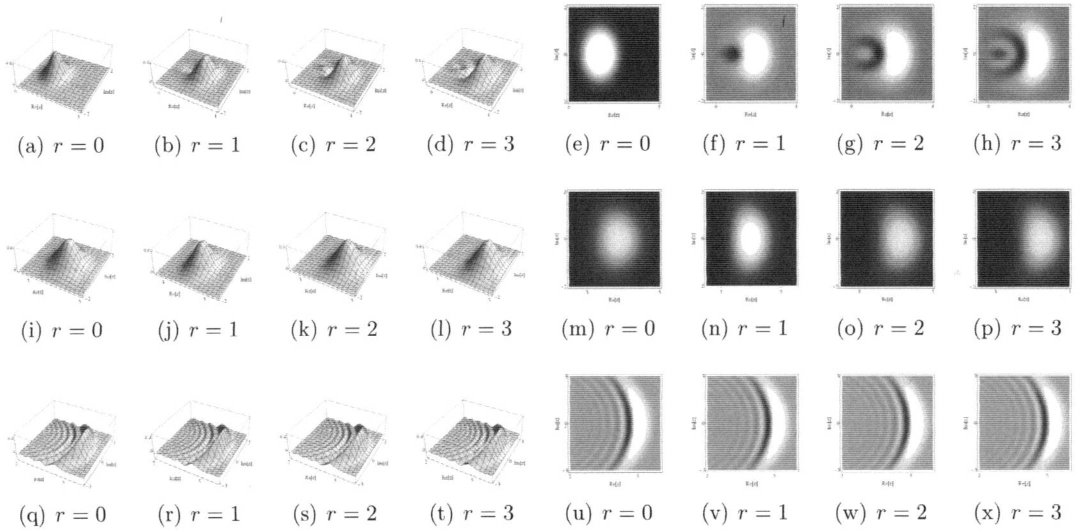

(a) $r = 0$ (b) $r = 1$ (c) $r = 2$ (d) $r = 3$ (e) $r = 0$ (f) $r = 1$ (g) $r = 2$ (h) $r = 3$

(i) $r = 0$ (j) $r = 1$ (k) $r = 2$ (l) $r = 3$ (m) $r = 0$ (n) $r = 1$ (o) $r = 2$ (p) $r = 3$

(q) $r = 0$ (r) $r = 1$ (s) $r = 2$ (t) $r = 3$ (u) $r = 0$ (v) $r = 1$ (w) $r = 2$ (x) $r = 3$

Figure 3. Wigner distribution of the optimized binomial sates $\phi_b(x)$ in the limit of the quantum oscillator for $n = 30$ and the indicated value of r. The last four columns are included as a reference and correspond to the density plot of the first four columns respectively. The probability of success p has been chosen to be $p = 0.1$ (first row), $p = 0.5$ (second row) and $p = 0.9$ (third row).

In general, for $p << 1$ the major contribution to the packet ρ_B is given by the state $|\psi_r\rangle$, so that we can construct a classical-like packet by taking $r = 0$ (i.e., with a nonnegative Wigner function). Any other value of r, with $p << 1$, leads to a nonclassical-like packet. This last is illustrated in Figure 3, see first row, where we have plotted the related Wigner distribution [8] (see also [2]) in the limit of the quantum oscillator. For $p \approx 1$, see third row of Figure 3, we always have a quantum-like packet because the major contribution is given by the state $|\psi_{r+n}\rangle$. Quite interestingly, the situation changes for $p = 0.5$ because the corresponding Wigner distribution is mainly non-negative, no matter the value of r, although this exhibits a squeezing that depends on r (see second row of Figure 3).

3.1. Poisson states

To get a superposition involving a large number of elements we use (7) with

$$c_k = \frac{e^{-|z|^2/2}}{\sqrt{k!}} z^k, \quad z \in \mathbb{C}. \tag{19}$$

That is, $|c_k|^2$ is the Poisson distribution with parameter $|z|^2$. As discussed above, such a distribution can be obtained from (15) by considering $n \to \infty$ such that $np = \text{const}$. On the other hand, the Fourier coefficients (19) together with the Fock states $|\varphi_k\rangle$ of the quantum oscillator produce a Glauber coherent state with $\langle H_{osc}\rangle_z = 2|z|^2 + 1$. We say that the superposition of states defined by (19) is a *Poisson state* ϕ_P. The time-evolution of the related bi-orthogonal probability density and current is depicted in Figure 4. As in the previous case, the integration of $\text{Im}(\rho_B)$ over all the real line is equal to zero, and ρ_B and J_B satisfy the bi-orthogonal continuity equation (11). Notice that in this case the real (imaginary) part of ρ_B exhibits two local maxima (changes of sign) that evolve in time in oscillatory form.

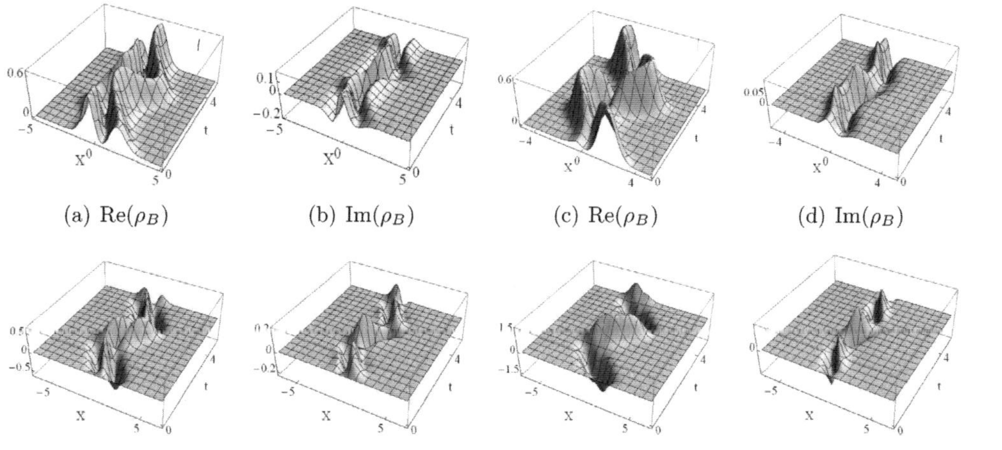

(a) $\text{Re}(\rho_B)$ (b) $\text{Im}(\rho_B)$ (c) $\text{Re}(\rho_B)$ (d) $\text{Im}(\rho_B)$

(e) $\text{Re}(J_B)$ (f) $\text{Im}(J_B)$ (g) $\text{Re}(J_B)$ (h) $\text{Im}(J_B)$

Figure 4. Time-evolution of the bi-orthogonal probability density ρ_B and current J_B of a Poisson state ϕ_P with $|z| = 1$. The first (last) two columns correspond to $r = 0$ ($r = 1$).

In Figure 5 we show the Wigner distribution of the Poisson states ϕ_P in the limit of the quantum oscillator. If $|z| = 0$ the vector ϕ_P coincides with $|\psi_r\rangle$. For $0 < |z| < 1$, see first row of Figure 5, the sensitivity on the value of r is noticeable. That is, the Poisson states ϕ_P behave as r-photon added coherent states [9]. The Wigner distribution is mainly positive for $|z| >> 1$ and exhibits squeezing as r increases.

4. Concluding remarks

The optimized binomial quantum states presented in the previous sections have been constructed as a superposition of $n + 1$ (adjacent) energy eigenvectors of the complex oscillator (1). Such a superposition has been tailored to represent either a classical or a nonclassical state in the limit of the quantum oscillator. Moreover, the superposition goes to an r-photon added coherent state by summing a very large number of energy eigenvectors ($n \to \infty$). On the other hand, the

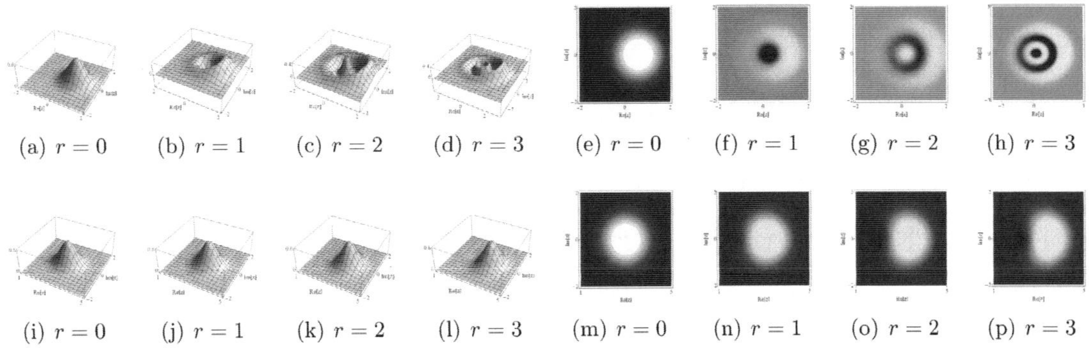

Figure 5. Wigner distribution of the optimized Poisson sates $\phi_P(x)$ in the limit of the quantum oscillator for the indicated value of r. The last four columns are included as a reference and correspond to the density plot of the first four columns respectively. The parameter z has been chosen real and equal to 0.6 (first row) and 3 (second row).

formal construction of the coherent states associated with the complex oscillator (1) involves the algebraic properties of a fundamental set of operators that are not considered in this work, for more information see [6]. However, the results reported here are very close to the ones obtained by using either the Barut-Girardelo or the Perelomov approaches of coherent states [10]. The subject is still in progress, the details will be reported elsewhere.

Acknowledgment

The authors are grateful to the anonymous referee for valuable remarks. The financial support of CONACyT is acknowledged.

References

[1] Rosas-Ortiz O, Castaños O and Schuch D 2015 New supersymmetry-generated complex potentials with real spectra *J. Phys. A: Math. Theor.* **48** 445302
[2] Kenfack A and Zyczkowski K 2004 Negativity of the Wigner function as an indicator of non-classicality *J. Opt. B: Quantum Semiclass. Opt.* **6** 396
[3] Stoler D, Salek B E A and Teich M C 1985 Binomial states of the quantized radial field *Opt. Acta* **32** 345
[4] Aharonov Y, Lerner E C, Huang H W and Knight J M 1973 Oscillator phase states, thermal equilibrium and group representations *J. Math. Phys.* **14** 746
[5] Mielnik B 1984 Factorization method and new potentials with the oscillator spectrum *J. Math. Phys.* **25** 3387
[6] Zelaya K D 2015 *Estados coherentes para Hamiltonianos no Hermitianos* (M.Sc. Thesis (in Spanish), Physics Department, Cinvestav, Mexico)
[7] Glauber R J 2007 *Quantum Theory of Optical Coherence. Selected Papers and Lectures* (Wiley-VCH, Weinheim)
[8] Wigner E P 1932 On the Quantum Correction For Thermodynamic Equilibrium *Phys. Rev.* **40** 749
[9] Agarwal G S and Tara K 1991 Nonclassical properties of states generated by the excitations on a coherent state *Phys. Rev.* A **43** 492
[10] Rosas-Ortiz O and Zelaya K D, in preparation.

Painlevé IV Solutions from Hamiltonians with Equidistant Gapped Spectrum

M I Estrada-Delgado and D J Fernández C

Departamento de Física, Cinvestav A.P. 14-740, 07000 México D.F. Mexico

E-mail: david@fis.cinvestav.mx

Abstract. Supersymmetry transformations are applied to the harmonic oscillator for generating potentials V_k^j whose spectra have a gap with respect to the initial one. The extremal states are found and, as the reduction theorem conditions are satisfied, ensuring that the system has third order ladder operators and it is connected with Painlevé IV (PIV) equation, then solutions to this equation can be generated. An alternative transformation is applied, by adding the levels needed to recover the spectrum of V_k^j. The extremal states are found and, as the reduction theorem is met again, we get also solutions to the PIV equation which will be analysed.

1. Introduction

Supersymmetric quantum mechanics (SUSY QM) is a powerful tool for generating Hamiltonians with known spectra departing from a given initial one. In fact, through this technique it is possible to "add" or "delete" levels at will in the initial spectrum.

On the other hand, the Painlevé IV (PIV) equation has a direct connection with systems ruled by second-order polynomial Heisenberg algebras (PHA) [1]. Particular Hamiltonians having this kind of algebras arise by applying first-order SUSY QM to the harmonic oscillator. Note that the identification of the corresponding extremal states is the key to generate solutions to the PIV equation. Moreover, after a k-order SUSY transformations systems with $(2k+1)$-th order ladder operators will be obtained. In addition, the SUSY partners of the harmonic oscillator that have also third order ladder operators have been recently identified [2]. Through them additional solutions to the PIV equation can be found.

In previous works, when k new levels were created a pattern for the zeros of the non-singular PIV solution (the one arising from the extremal state with lowest energy) was recognized [2, 3]: the number of zeros becomes equal to $2k-1$, where k is the order of the transformation. Here, we will see that the order of the transformation is not what defines the number of zeros, but the number of steps of the finite ladder is indeed what characterizes this quantity.

In this paper, Hamiltonians with added or deleted levels whose spectra are the same (up to energy displacements), are generated through SUSY QM for the harmonic oscillator. The extremal states of those systems are found, and through them solutions to the PIV equation are generated. Finally, a comparison between such PIV solutions is made.

2. Supersymmetric quantum mechanics

In this work the names intertwining technique, factorization method and supersymmetric quantum mechanics are used as synonymous, since they are equivalent methods for generating exactly solvable quantum mechanical potentials, as we will show next.

2.1. *Intertwining technique*

In the k-th order intertwining technique $k+1$ Hamiltonians H_j and $2k$ first-order operators $A_l^{\pm}$ of the form

$$H_j = -\frac{1}{2}\frac{d^2}{dx^2} + V_j(x), \quad A_l^{\pm} = \frac{1}{\sqrt{2}}\left(\mp\frac{d}{dx} + \alpha_l(x,\epsilon_l)\right), \quad j=0,\dots,k, \quad l=1,\dots,k, \tag{1}$$

are intertwined in the way

$$H_i A_i^+ = A_i^+ H_{i-1}, \quad H_{i-1}A_i^- = A_1^- H_i, \qquad i=1,\dots,k, \tag{2}$$

where the initial Hamiltonian H_0 has a known given spectrum. In order to fulfil (2) the superpotentials $\alpha_i(x,\epsilon_i)$, $i = 1,\dots,k$ are determined by solutions to the initial Riccati equation [2]:

$$\alpha_1'(x,\epsilon_j) + \alpha_1^2(x,\epsilon_j) = 2(V_0 - \epsilon_j), \qquad j=1,\dots,k. \tag{3}$$

Equivalently, the substitution $\alpha_1(x,\epsilon_j) = u_j'/u_j$ indicates that we require k solutions to the corresponding stationary Schrödinger equation:

$$H_0 u_j = -\frac{1}{2}u_j'' + V_0 u_j = \epsilon_j u_j, \qquad j=1,\dots,k. \tag{4}$$

The potential V_k is given in terms of them and the initial potential V_0 through [2,3]:

$$V_k(x) = V_0(x) - \sum_{i=1}^{k}\alpha_i'(x,\epsilon_i) = V_0(x) - \left(\ln\{W(u_1,u_2,\dots,u_k)\}\right)'', \tag{5}$$

where $W(u_1,\dots,u_k)$ denotes the Wronskian of the k seed solutions.

2.2. *Factorization*

Due to the intertwining relations (2), the $k+1$ Hamiltonians can be factorized as:

$$\begin{aligned} H_0 &= A_1^- A_1^+ + \epsilon_1, \qquad H_k = A_k^+ A_k^- + \epsilon_k, \\ H_i &= A_i^+ A_i^- + \epsilon_i = A_{i+1}^- A_{i+1}^+ + \epsilon_{i+1}, \quad i=1,\dots,k-1. \end{aligned} \tag{6}$$

Another intertwining relation between H_0 and H_k can be given, involving a k-th order differential intertwining operator which is factorized in terms of the first order ones as follows:

$$H_k B_k^+ = B_k^+ H_0, \qquad B_k^+ = A_k^+ A_{k-1}^+ \dots A_1^+. \tag{7}$$

In addition, the two different products of $B_k^{\pm}$ can be factorized in the following way:

$$B_k^- B_k^+ = \prod_{i=1}^{k}(H_0 - \epsilon_i), \qquad B_k^+ B_k^- = \prod_{i=1}^{k}(H_k - \epsilon_i). \tag{8}$$

2.3. Eigenfunctions

The formal eigenfunction of H_k for the initial eigenvalue E_n is expressed in terms of Wronskians involving the seed solutions $u_1, \ldots, u_k$ and the eigenfunction $\psi_n^{(0)}$ of H_0 which is transformed, and similar relations for those associated to the factorization energies ϵ_i [1]:

$$\psi_n^{(k)} \propto \frac{W(u_1, \ldots, u_k, \psi_n^{(0)})}{W(u_1, \ldots, u_k)}, \quad \psi_{\epsilon_i}^{(k)} \propto \frac{W(u_1, \ldots, \hat{u}_i, \ldots, u_k)}{W(u_1, \ldots, u_k)}, \quad n = 0, 1, \ldots, \quad i = 1, \ldots, k, \tag{9}$$

where $\hat{u}_i$ indicates that the seed solution u_i is excluded of the corresponding Wronskian.

2.4. Supersymmetric quantum mechanics

The SUSY QM introduced by Witten in 1981 is realized through the following operators choice:

$$Q_1 = \frac{Q^+ + Q^-}{\sqrt{2}}, \quad Q_2 = \frac{Q^+ - Q^-}{i\sqrt{2}}, \quad Q^+ = \begin{pmatrix} 0 & B_k^+ \\ 0 & 0 \end{pmatrix}, \quad Q^- = \begin{pmatrix} 0 & 0 \\ B_k^- & 0 \end{pmatrix}. \tag{10}$$

Since the operators $B_k^\pm$ satisfy equations (7-8), it turns out that

$$H_{ss} = \{Q^-, Q^+\} = \begin{pmatrix} B_k^+ B_k^- & 0 \\ 0 & B_k^- B_k^+ \end{pmatrix} = \prod_{i=1}^{k} (H_s^p - \epsilon_i \mathbb{1}), \quad H_s^p = \begin{pmatrix} H_k & 0 \\ 0 & H_0 \end{pmatrix}, \tag{11}$$

where H_k and H_0 in H_s^p are intertwined as in equations (7).

3. Second order polynomial Heisenberg algebra

In order to determine the general potential ruled by a second order PHA, a nonlinear ordinary second-order differential equation, the Painlevé IV equation, needs to be solved. In fact, let us take two Schrödinger Hamiltonians H and H_a such that:

$$HL_1^+ = L_1^+(H_a + 1), \qquad H_a L_2^+ = L_2^+ H \quad \Rightarrow \quad [H, L^+] = L^+, \tag{12}$$

where the ladder operator L^+ is given by [1, 2]:

$$L^+ = L_1^+ L_2^+, \quad L_1^+ = \frac{1}{\sqrt{2}}\left[-\frac{d}{dx} + f(x)\right], \quad L_2^+ = \frac{1}{2}\left[\frac{d^2}{dx^2} + g(x)\frac{d}{dx} + h(x)\right]. \tag{13}$$

The functions $f(x)$, $h(x)$, and $V(x)$ at the end depend on just the function $g(x)$,

$$f = x + g, \quad h = -x^2 + \frac{g'}{2} - \frac{g^2}{2} - 2xg + a, \quad V = \frac{x^2}{2} - \frac{g'}{2} + \frac{g^2}{2} + xg + \varepsilon_1 - \frac{1}{2}, \tag{14}$$

which satisfies the PIV equation:

$$g'' = \frac{(g')^2}{2g} + \frac{3}{2}g^3 + 4xg^2 + 2(x^2 - a)g + \frac{b}{g}, \tag{15}$$

with parameters $a = \varepsilon_2 + \varepsilon_3 - 2\varepsilon_1 - 1$ and $b = -2\Delta^2$, $\Delta = \varepsilon_2 - \varepsilon_3$. The generalized number operator is a polynomial of third degree in H,

$$N(H) = (H - \varepsilon_1)(H - \varepsilon_2)(H - \varepsilon_3). \tag{16}$$

Thus, from an algebraic viewpoint the system will have at most three equidistant ladders, each one starting from ε_1, ε_2 and ε_3. The corresponding extremal states are given by:

$$\psi_{\varepsilon_1} \propto e^{\left(-\frac{x^2}{2}-\int g dx\right)}, \quad \psi_{\varepsilon_2} \propto \left(\frac{g'}{2g} - \frac{g}{2} - \frac{\Delta}{g} - x\right) e^{\int\left(\frac{g'}{2g}+\frac{g}{2}-\frac{\Delta}{g}\right)dx}, \tag{17}$$

$$\psi_{\varepsilon_3} \propto \left(\frac{g'}{2g} - \frac{g}{2} + \frac{\Delta}{g} - x\right) e^{\int\left(\frac{g'}{2g}+\frac{g}{2}+\frac{\Delta}{g}\right)dx}. \tag{18}$$

Therefore, if the extremal states for a system with third order ladder operators are given, then solutions to the PIV equation can be obtained through

$$g(x) = -x - \{\ln(\psi_{\varepsilon_1})\}'. \tag{19}$$

4. SUSY transformations for the harmonic oscillator

The k-th order SUSY partners of the harmonic oscillator are ruled by $2k$-th order PHA (with ladder operators of $(2k+1)$-th order). Moreover, if the seed solutions are connected by the standard annihilation and creation operators $a^{\pm}$ for $k > 1$, then the $(2k+1)$-th order ladder operators will be factorized, leading as well to third order ladder operators for our system. This is the so-called *reduction theorem*, which has been proven and analized in [2].

It has been shown recently that two different SUSY transformations producing a spectral gap of integer size can lead to the same final potential: in one transformation we delete an even number k' of consecutive energy levels to produce a gap of size $k'+1$, while it appears a finite ladder of k steps; in the second, we make a k-th order SUSY transformation which creates the finite ladder with k steps and reproduce the same spectral gap. Thus, up to an energy displacement both spectra are equal, and by adjusting one final parameter both transformations lead indeed to the same final potential. In this paper we will illustrate this result by means of two particular examples; the general situation can be found elsewhere.

The eigenfunctions of the harmonic oscillator Hamiltonian are well known, and the general solution to the stationary Schrödinger equation for $\epsilon \neq E_n, n = 0, 1, \ldots$ is expressed in terms of confluent hypergeometric functions as follows [1, 2, 4]:

$$u(x) = e^{-\frac{x^2}{2}} \left[{}_1F_1\left(\tfrac{1-2\epsilon}{4}, \tfrac{1}{2}; x^2\right) + 2\nu x \frac{\Gamma\left(\frac{3-2\epsilon}{4}\right)}{\Gamma\left(\frac{1-2\epsilon}{4}\right)} {}_1F_1\left(\tfrac{3-2\epsilon}{4}, \tfrac{3}{2}; x^2\right) \right]. \tag{20}$$

Let us make a first-order transformation with factorization energy $\epsilon_1 = -\frac{5}{2}$, the final potential will be non singular for $\nu_1 \in (-1, 1)$, as proved in [2] (some examples of potentials of this family are shown in Figure 1). In particular, for $\nu_1 = 0$ we denote this potential as V_1^{-3}. On the other hand, a second-order transformation is implemented by choosing the first two excited states to generate the SUSY partner potential V_2^1. It turns out that:

$$V_2^1 = \frac{x^2}{2} + 2 - \frac{4}{2x^2+1} + \left(\frac{4x}{2x^2+1}\right)^2 = V_1^{-3} + 3. \tag{21}$$

The eigenfunctions for both potentials turn out to be the same and, after identifying and sorting the extremal states, the corresponding PIV solutions become (see Figure 2):

$$g_{\varepsilon_1}^{1,-3} = g_{\varepsilon_1}^{2,1} = \frac{4x}{2x^2+1} \quad \text{for } a = 3,\ b = -8, \tag{22}$$

$$g_{\varepsilon_2}^{1,-3} = g_{\varepsilon_2}^{2,1} = -\frac{4x^4+3}{4x^5+8x^3+3x} \quad \text{for } a = -6,\ b = -2, \tag{23}$$

$$g_{\varepsilon_3}^{1,-3} = g_{\varepsilon_3}^{2,1} = \frac{8x^5+6x}{1-4x^4} \quad \text{for } a = 0,\ b = -18. \tag{24}$$

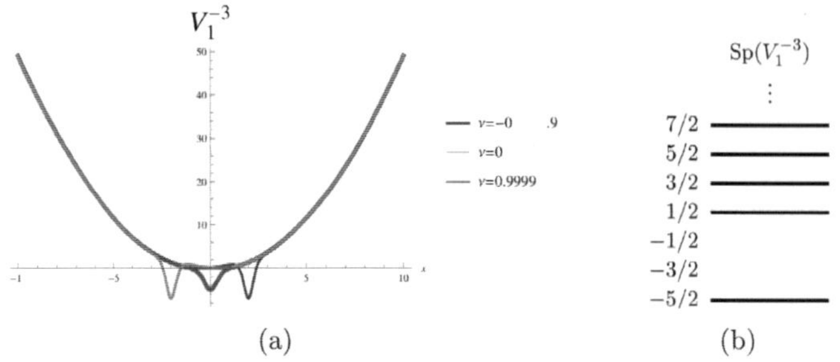

Figure 1. (a) Plot of three potentials belonging to the family of first-order SUSY partners of the oscillator generated by adding a new level at $-\frac{5}{2}$, for different values of ν_1 (for $\nu_1 = 0$ it is called V_1^{-3}); (b) Spectrum associated to this family of potentials, which is also the one of V_2^1 but displaced down by 3.

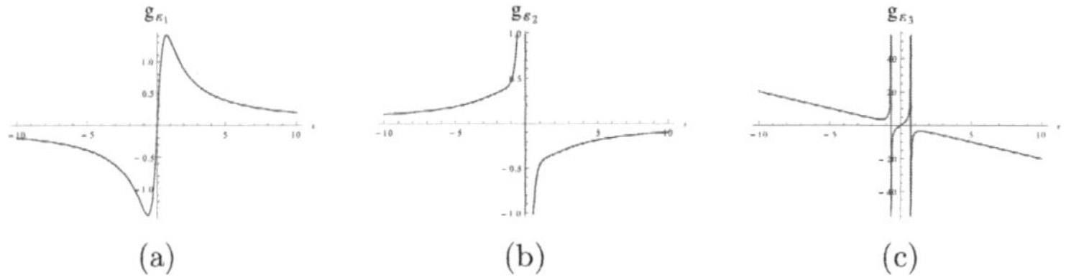

Figure 2. PIV solutions associated to V_2^1 and V_1^{-3}.

Note that the non-singular solution in Figure 2a has one node; by realizing that the finite ladder of V_1^{-3} and V_2^1 has one step, the formula for the number of nodes is now $2k-1$, with $k=1$ being the number of steps of the finite ladder.

The second example appears by doing first a fourth order transformation which deletes four excited state levels starting from the second one and then implementing a second order SUSY to create two levels at $-\frac{9}{2}$ and $-\frac{11}{2}$, leading to (a plot of both potentials is given in Figure 3):

$$\begin{aligned} V_4^2 &= V_2^{-6} + 6 \\ &= \frac{x^2}{2} - \frac{64\left(8x^3+16x\right)x^3 + 8\left(24x^2+16\right)x^4 + 96\left(2x^4+8x^2+15\right)x^2}{8\left(2x^4+8x^2+15\right)x^4+45} \\ &+ \frac{\left(8\left(8x^3+16x\right)x^4 + 32\left(2x^4+8x^2+15\right)x^3\right)^2}{\left(8\left(2x^4+8x^2+15\right)x^4+45\right)^2}. \end{aligned} \tag{25}$$

The corresponding PIV solutions become now (see Figure 4):

$$\begin{aligned} g_{\varepsilon_1}^{2,-6} = g_{\varepsilon_1}^{4,2} &= \frac{8x\left(32x^{10}+208x^8+432x^6+504x^4+90x^2-135\right)}{\left(4x^4+12x^2+3\right)\left(16x^8+64x^6+120x^4+45\right)} \quad \text{for} \quad a=7,\ b=-32, \\ g_{\varepsilon_2}^{2,-6} = g_{\varepsilon_2}^{4,2} &= \frac{32x^3\left(4\left(x^2+3\right)x^2+15\right)}{8\left(2x^4+8x^2+15\right)x^4+45} \\ &- \frac{20x\left(8\left(2x^6+12x^4+27x^2+15\right)x^2+45\right)}{2x^2\left(4\left(4x^6+30x^4+90x^2+75\right)x^2+225\right)-225} \quad \text{for} \quad a=-11,\ b=-8, \\ g_{\varepsilon_3}^{2,-6} = g_{\varepsilon_3}^{4,2} &= -\frac{32\left(4x^4+9\right)x^3}{16x^8+72x^4-135} \\ &+ \frac{32\left(4\left(x^2+3\right)x^2+15\right)x^3}{8\left(2x^4+8x^2+15\right)x^4+45} - 2x - \frac{1}{x} \quad \text{for} \quad a=1,\ b=-72. \end{aligned} \tag{26}$$

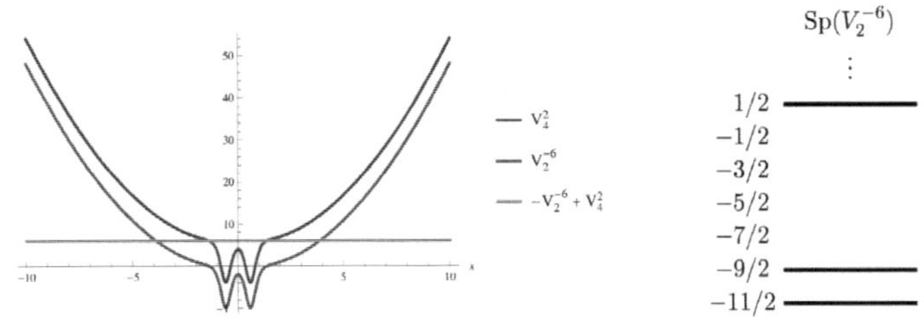

(a) (b)

Figure 3. (a) Plot of the potentials V_4^2 and V_2^{-6} and their difference, which is a constant. (b) The spectrum of V_2^{-6}, which is also the one of V_4^2 but displaced by -6.

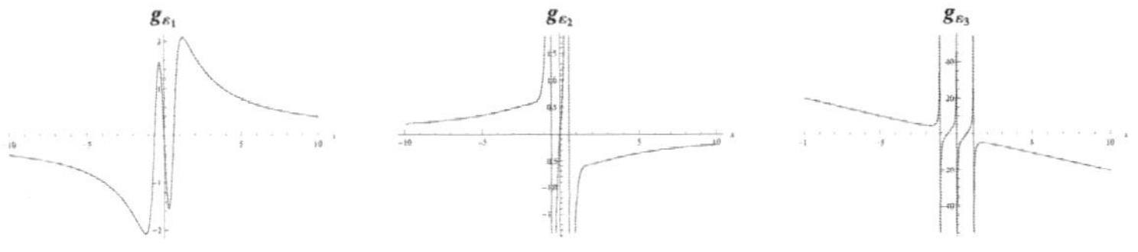

(a) (b) (c)

Figure 4. PIV solutions associated to V_4^2 and V_2^{-6}.

Let us note that, although $g_{\varepsilon_1}^{4,2}$ comes from a fourth order SUSY transformation, it has just three zeros which appear again from the same empirical formula $2k-1$, where the number of steps of the finite ladder is now $k=2$.

5. Conclusions

In this article we have derived some rational solutions to the PIV equation, from potentials whose spectra have a gap of integer size. Since such Hamiltonians appear from two SUSY transformations of different orders, we realized that the number of zeros associated to the non-singular PIV solution which comes from the ground state is equal to $2k-1$, where k is now the number of steps of the finite ladder. A proof of this conjecture is still an open question.

Acknowledgments

The authors acknowledge the support of Conacyt, project 152574.

References

[1] Carballo J M, Fernández D J, Negro J and Nieto L M 2004 Polynomial Heisenberg algebras *J Phys A: Math Gen* **37** 10349

[2] Bermudez D and Fernández D J 2011 Supersymmetric Quantum Mechanics and Painlevé IV equation *SIGMA* **7** 025

[3] Bermudez D and Fernández D J 2013 Solution Hierarchies for the Painlevé IV Equation *Geometric Methods in Physics, Trends in Mathematics* ed P Kielanowski *et al* (Heidelberg: Springer Basel) p 199

[4] Junker G and Roy P 1998 Conditionally exactly solvable potentials: a supersymmetric construction method *Ann Phys* **270** 155177

Isospectral Trigonometric Pöschl-Teller Potentials with Position Dependent Mass Generated by Supersymmetry

C Santiago-Cruz

Unidad Profesional Interdisciplinaria en Ingeniería y Tecnologías Avanzadas del Instituto Politécnico Nacional, Av. Instituto Politécnico Nacional 2580, Col. La Laguna Ticomán, C.P. 07340, México D. F., Mexico

E-mail: csantiagoc0500@ipn.mx

Abstract. In this work a position dependent mass Hamiltonian with the same spectrum of the trigonometric Pöschl-Teller one was constructed by means of the underlying potential algebra. The corresponding wave functions are determined by using the factorization method. A new family of isospectral potentials are constructed by applying a Darboux transformation. An example is presented in order to illustrate the formalism.

1. Introduction

In recent years the study of systems with position-dependent mass (PDM) has gain much interest form both, theoretical and experimental points of views, due to its multiple applications in different areas of physics and engineering [1–8]. The very concept of a PDM system is far from being exhaustively discussed. To this respect, many contributions to the fundamental understanding of the problem have been carried out from different points of view [9–13]. The factorization method [14–16] has been considered also in this context. Its application in the generation of exactly solvable potentials in the classical as well as in the quantum mechanical regimes has been discussed, *e. g.*, in [17–24]. The aim of this work is the construction of position dependent mass Hamiltonians, having the same spectrum of the trigonometric Pöschl-Teller (TPT) potential, by means of the potential algebra of the constant mass case. In section 2 the solution of the constant mass TPT potential by means of the factorization method is presented. In section 3 it is used the constant mass potential algebra to construct position-dependent mass potentials with the TPT spectrum, and the corresponding wave functions are determined. In section 4 these Hamiltonians are used to generate a new family of PDM potentials isospectral to the TPT one by applying a Darboux transformation. Finally, some concluding remarks are presented.

2. The constant mass trigonometric Pöschl-Teller potential revisited

In proper units, the constant mass TPT potential Hamiltonian is

$$\mathcal{H}_\ell = \frac{p^2}{2} + \frac{\ell(\ell-1)}{2\cos^2 x}, \qquad (1)$$

where

$$p = -i\frac{d}{dx}$$

stands for the momentum operator and ℓ is an integer number such that $\ell > 1$. This Hamiltonian can be factorized in the form [25]

$$\mathcal{H}_\ell = a_\ell^+ a_\ell^- + \epsilon_\ell = a_{\ell-1}^- a_{\ell-1}^+ + \epsilon_{\ell-1} \tag{2}$$

with

$$a_\ell^\pm = \frac{1}{\sqrt{2}}\left(\mp ip + \ell \tan x\right), \quad \epsilon_\ell = \frac{1}{2}\ell^2. \tag{3}$$

The factorizations (2) lead to the following intertwining relations

$$a_\ell^- \mathcal{H}_\ell = \mathcal{H}_{\ell+1} a_\ell^-, \quad a_{\ell-1}^+ \mathcal{H}_\ell = \mathcal{H}_{\ell-1} a_{\ell-1}^+, \tag{4}$$

meaning that the operators $a_\ell^\mp$ allow to construct the wave functions of $\mathcal{H}_{\ell-1}$ and $\mathcal{H}_{\ell+1}$ in terms of those of $\mathcal{H}_\ell$. Indeed if $\mathcal{H}_\ell \psi_\ell^n = E_\ell^n \psi_\ell^n$, then $a_{\ell-1}^+ \psi_\ell^n$ is the eigenfunction of $\mathcal{H}_\ell$ with eigenvalue $E_{\ell-1}^{n+1} = E_\ell^n$ and $a_\ell^- \psi_\ell^n$ is the eigenfunction of $\mathcal{H}_\ell$ with eigenvalue $E_{\ell+1}^{n-1} = E_\ell^n$. Thus

$$\psi_{\ell-1}^{n+1} \propto a_{\ell-1}^+ \psi_\ell^n, \qquad \psi_{\ell+1}^{n-1} \propto a_\ell^- \psi_\ell^n. \tag{5}$$

The ground state ψ_ℓ^0 may be constructed as a square integrable function which is annihilated by a_ℓ^-, *i.e.*,

$$a_\ell^- \psi_\ell^0 = 0. \tag{6}$$

In this way, the normalized ground state wave function is

$$\psi_\ell^0(x) = \sqrt{\frac{\ell!}{\pi^{1/2}\Gamma(\ell+\frac{1}{2})}}\,(\cos x)^\ell, \tag{7}$$

and corresponds to the eigenvalue $E_\ell^0 = \epsilon_\ell = \frac{1}{2}\ell^2$. The state ψ_ℓ^n can be obtained by subsequent applications of the corresponding operators a_ℓ^+, indeed

$$\psi_\ell^n(x) = N_\ell^n a_\ell^+ a_{\ell+1}^+ \cdots a_{\ell+n-1}^+ \psi_{\ell+n}^0, \tag{8}$$

with N_ℓ^n a normalization coefficient. Since the operators $a_\ell^\pm$ can be expressed in the form

$$a_\ell^\pm = \mp\frac{1}{\sqrt{2}}(\cos x)^{\mp\ell}\frac{d}{dx}(\cos x)^{\pm\ell}, \tag{9}$$

the equation (8) leads to

$$\psi_\ell^n(x) = \sqrt{\frac{2^{2\ell-1}(\ell+n)n!\,(\Gamma(\ell))^2}{\pi\Gamma(2\ell+n)}}\,(\cos x)^\ell\, C_n^{(\ell)}(\sin x), \tag{10}$$

where $C_n^{(\ell)}(z)$ are the Gegenbauer polynomials. The corresponding spectrum, defined by the intertwining relations (4), is given by

$$Sp(\mathcal{H}_\ell) = \left\{E_\ell^n = E_{\ell+n}^0 = \frac{1}{2}(\ell+n)^2,\ \ell = 2, 3, \ldots,\ n = 0, 1, \ldots\right\}.$$

3. Trigonometric Pöschl-Teller potentials with position-dependent mass

In order to construct TPT potentials with PDM let us consider the Hamiltonian

$$H_\ell = \frac{1}{2} m^a p m^{2b} p m^a + V_\ell, \qquad 2a + 2b = -1, \tag{11}$$

where $m(x)$ is a given function and the potential $V_\ell(x)$ will be determined in such a way that H_ℓ has the same spectrum as the constant mass TPT potential. To this end, let us assume that H_ℓ admit the following factorizations (compare to (2))

$$H_\ell = A_\ell^+ A_\ell^- + \epsilon_\ell = A_{\ell-1}^- A_{\ell-1}^+ + \epsilon_{\ell-1}, \tag{12}$$

in terms of two first order operators of the form

$$A_\ell^+ = \frac{1}{\sqrt{2}} \left(-i m^a p m^b + W_\ell \right), \qquad A_\ell^- = \frac{1}{\sqrt{2}} \left(i m^b p m^a + W_\ell \right), \tag{13}$$

with W_ℓ a function of the position to be determined and $\epsilon_\ell = \frac{1}{2}\ell^2$. These factorizations imply the following set of equations from which W_ℓ and V_ℓ may be constructed

$$-DW_\ell + 4\left(a + \frac{1}{4}\right)(D \ln J)\, W_\ell + W_\ell^2 = 2\left(V_\ell - \epsilon_\ell\right) \tag{14}$$

$$DW_{\ell-1} + 4\left(a + \frac{1}{4}\right)(D \ln J)\, W_{\ell-1} + W_{\ell-1}^2 + 4\left(a + \frac{1}{4}\right)\left(D^2 \ln J\right) = 2\left(V_\ell - \epsilon_{\ell-1}\right), \tag{15}$$

with

$$J(x) = \sqrt{m(x)}, \quad D = \frac{1}{J}\frac{d}{dx}.$$

A first solution to the system (14)-(15) can be set in the form

$$W_\ell = \ell \tan \int_{x_0}^{x} J(r) dr - 2\left(a + \frac{1}{4}\right)(D \ln J) \tag{16}$$

$$V_\ell = \frac{1}{2}\ell(\ell - 1) \sec^2 \int_{x_0}^{x} J(r) dr - 2\left(a + \frac{1}{4}\right)^2 (D \ln J)^2 + \left(a + \frac{1}{4}\right)\left(D^2 \ln J\right), \tag{17}$$

with x_0 an integration constant with length units. As in the constant mass case, the factorization (12) imply the following intertwining relations

$$A_\ell^- H_\ell = H_{\ell+1} A_\ell^-, \qquad A_{\ell-1}^+ H_\ell = H_{\ell-1} A_{\ell-1}^+, \tag{18}$$

meaning that one can determine the eigenfunctions of $H_{\ell+1}$ and $H_{\ell-1}$ in terms of those of H_ℓ, and that $Sp(H_\ell) = Sp(\mathcal{H}_\ell)$. In fact, if $\Psi_\ell^n(x)$ denote the eigenfunctions of H_ℓ, *i.e.*,

$$H_\ell \Psi_\ell^n(x) = E_\ell^n \Psi_\ell^n(x), \tag{19}$$

then, the action of the operators $A_\ell^\pm$ on Ψ_ℓ^n is given by

$$A_\ell^- \Psi_\ell^n = \frac{1}{\sqrt{2}} \sqrt{n\,(2\ell + n)}\, \Psi_{\ell+1}^{n-1}, \qquad A_{\ell-1}^+ \Psi_\ell^n = \frac{1}{\sqrt{2}} \sqrt{(n+1)\,(2\ell + n - 1)}\, \Psi_{\ell-1}^{n+1}. \tag{20}$$

Accordingly, the general wave function of H_ℓ can be written in the form

$$\Psi_\ell^n(x) = \sqrt{\frac{2^n \Gamma(2\ell + n)}{n! \Gamma(2(\ell + n))}} A_\ell^+ A_{\ell+1}^+ \cdots A_{\ell+n-1}^+ \Psi_{\ell+n}^0, \tag{21}$$

where the state Ψ_ℓ^0 is the ground state of the Hamiltonian H_ℓ given by

$$A_\ell^- \Psi_\ell^0 = 0. \tag{22}$$

Now consider the following change of variable

$$y(x) = \int_{x_0}^{x} J(r)dr. \tag{23}$$

It is not difficult to show that [21]

$$A_\ell^\pm J^{1/2} = J^{1/2}\tilde{a}_\ell^\pm \tag{24}$$

with

$$\tilde{a}_\ell^\pm = \frac{1}{\sqrt{2}}\left(\pm D + \ell \tan \int_{x_0}^{x} J dr\right) = \frac{1}{\sqrt{2}}\left(\mp \frac{d}{dy} + \ell \tan y\right). \tag{25}$$

Note that the operators $\tilde{a}_\ell^\pm$ correspond to the factor operators of the constant mass TPT Hamiltonian in the $y-$configuration space. Expressing the ground state Ψ_ℓ^0 in the form

$$\Psi_\ell^0(x) = J^{1/2}(x)\psi_\ell^0(y(x)) \tag{26}$$

then

$$A_\ell^- \Psi_\ell^0 = A_\ell^- J^{1/2}\psi_\ell^0 = J^{1/2}\tilde{a}_\ell^- \psi_\ell^0 = 0.$$

It is clear, thus, that $\psi_\ell^0(y)$ is nothing but the ground state of the constant mass TPT potential evaluated in the variable $y(x)$ defined by the choice of the mass function. Hence

$$\Psi_\ell^0(x) = \sqrt{\frac{\ell!}{\pi^{1/2}\Gamma(\ell + 1/2)}} J^{1/2}(x)\left(\cos \int_{x_0}^{x} J(r)dr\right)^\ell. \tag{27}$$

Now the expression (21), together with (26) and (24) allows the construction of the wave functions Ψ_ℓ^n as

$$\Psi_\ell^n(x) = J^{1/2}\psi_\ell^n(y(x)), \tag{28}$$

where $\psi_\ell^n(y)$ is given by

$$\psi_\ell^n(y) = \sqrt{\frac{2^n\Gamma(2\ell + n)}{n!\Gamma(2(\ell + n))}} \tilde{a}_\ell^+ \tilde{a}_{\ell+1}^+ \cdots \tilde{a}_{\ell+n-1}^+ \psi_{\ell+n}^0(y), \tag{29}$$

from which it is clear that ψ_ℓ^n are the constant mass TPT potential wave functions evaluated in the variable y. Finally, we can immediately write the normalized wave functions for the TPT potential with position-dependent mass as

$$\Psi_\ell^n(x) = \sqrt{\frac{2^{2\ell-1}(\ell + n)n!}{\pi\Gamma(2\ell + 2)}}\Gamma(\ell)J^{1/2}(x)\left(\cos \int_{x_0}^{x} J(r)dr\right)^\ell C_n^{(\ell)}\left(\sin \int_{x_0}^{x} J(r)dr\right). \tag{30}$$

This expression is consistent with the point canonical transformation discussed, *e.g.*, in [11].

In order to illustrate our results, we consider the mass [23]

$$m(x) = \frac{1 - \beta(\lambda x)^2}{1 - (\lambda x)^2}, \tag{31}$$

with $\beta < 0$ a dimensionless parameter and λ a constant with length units. This mass has been used in molecular dynamics to determine the inversion potential of the NH_3 molecule by means of the density functional theory approach (see [4]). Observe that this function has singularities at the points $x = \pm 1/\lambda$, meaning that the choice of λ will be determinant in fixing the domain of V_ℓ. Note also that in the limit as $\lambda \to 0$ we recover the constant mass case with a unit mass $m = 1$. The expression (31) leads to

$$y(x) = \int_0^x \sqrt{\frac{1-\beta(\lambda r)^2}{1-(\lambda r)^2}}dr = \frac{1}{\lambda}E_{\rm int}\left(\arcsin \lambda x, \beta^{1/2}\right), \tag{32}$$

with $E_{\rm int}(\phi, k)$ the incomplete elliptic integral of second kind [23, 26]. The substitution of (31)-(32) in (17) and (30) leads us to the explicit expressions for the PDM TPT potential and its corresponding wave functions. In Figure 1 we show the potential and some wave functions for particular values of the parameters ℓ, a and β.

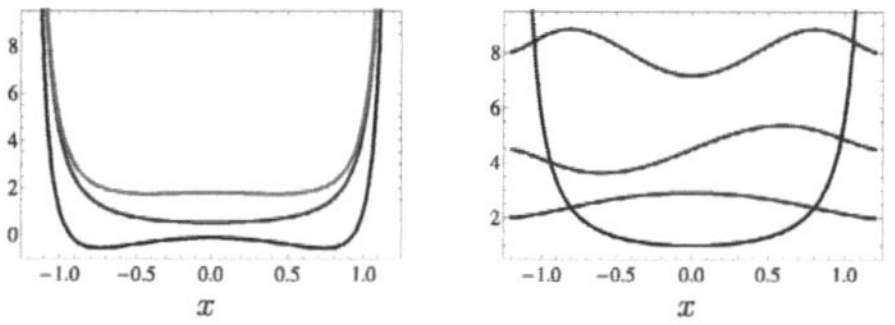

Figure 1. Left: Position-dependent mass TPT potentials $V_\ell(x)$ for the mass (31). In this plot $\ell = 2$, $\lambda = 0.7$, $\beta = -0.3$ and $a = -2$ (black), $a = -1$ (blue) and $a = 1$ (red). Right: Position-dependent mass TPT potential and its first three wave functions for $\ell = 2$, $\lambda = 0.7$, $\beta = -0.3$ and $a = -\frac{1}{4}$.

4. New position-dependent mass Hamiltonians with the TPT potential spectrum

It is well known that, in the constant mass case, the operators $a_\ell^\mp$ factorizing the TPT Hamiltonian (1) are not unique [14]. The same is true for the PDM Hamiltonian (11). Let us assume that

$$H_\ell = B^-_{\ell-1}B^+_{\ell-1} + \epsilon_\ell \, 1, \tag{33}$$

with

$$B^+_\ell = \frac{1}{\sqrt{2}}\left(-im^a p m^b + U_\ell\right), \qquad B^-_\ell = \frac{1}{\sqrt{2}}\left(im^b p m^a + U_\ell\right). \tag{34}$$

The function $U_\ell(x)$ must then fulfill the Riccati equation (15) that may be rewritten as

$$DU_\ell + 4\left(a+\frac{1}{4}\right)(D\ln J)\,U_\ell + U_\ell^2 + 4\left(a+\frac{1}{4}\right)\left(D^2\ln J\right) = 2\left(V_{\ell+1} - \epsilon_\ell\right)$$

where $V_{\ell+1}$ is given by (17). In order to construct the solution let us propose, as usual [14, 27],

$$U_\ell(x) = W_\ell(x) + v(x),$$

where $v(x)$, in turn, must hold

$$Dv(x) + 2\ell v(x)\int_{x_0}^x J(r)dr + v^2(x) = 0,$$

leading to

$$v(x) = D\ln\left(\int_{x_0}^{y(x)} (\cos s)^{2\ell}\,ds + \gamma\right),$$

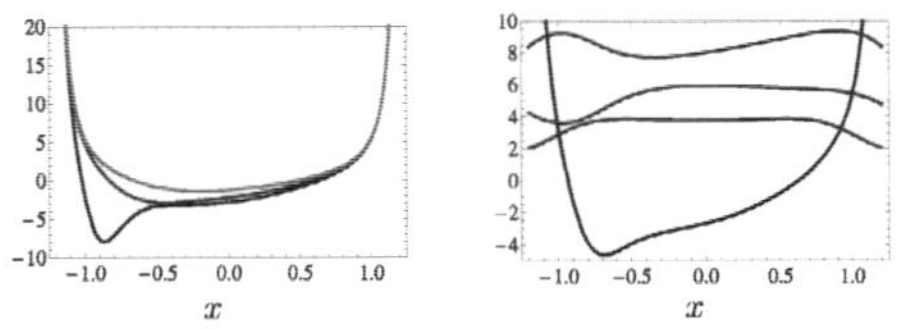

Figure 2. Left: Position-dependent mass supersymmetric TPT potentials $\widetilde{V}_\ell(x,\gamma)$ for the mass (31) with $\ell=2$, $\lambda=0.7$, $\beta=-0.3$, $a=-\frac{1}{4}$ and $\gamma=0.62$ (black), $\gamma=0.9$ (blue), $\gamma=1.2$ (red). Right: Position-dependent mass supersymmetric TPT potential and its first three wave functions for $\ell=2$, $\lambda=0.7$, $\beta=-0.3$ and $a=-\frac{1}{4}$ and $\gamma=0.7$.

with γ an integration constant with the constrain

$$|\gamma| > \max \int_{x_0}^{y(x)} \left(\cos \int_{x_0}^{r} J(s)ds\right)^{2\ell} J(r)dr$$

in order to avoid singularities.

The function U_ℓ can be readily written

$$U_\ell(x) = \ell \tan \int_{x_0}^{x} J(r)dr + D \ln\left[\int_{x_0}^{y(x)} \left(\cos \int_{x_0}^{r} J(s)ds\right)^{2\ell} J(r)dr + \gamma\right] - 2\left(a+\frac{1}{4}\right) D \ln J(x), \tag{35}$$

Following [14, 22], we may construct a new family of PDM Hamiltonians with the spectrum of the TPT one by means of a supersymmetric transformation

$$\widetilde{H}_\ell(\gamma) = B_\ell^+ B_\ell^- + \epsilon_\ell = \frac{1}{2} m^a p m^{2b} p m^a + \widetilde{V}_\ell(x,\gamma) \tag{36}$$

where

$$\widetilde{V}_\ell(x,\gamma) = V_\ell(x) - D^2 \ln\left[\int_{x_0}^{y(x)} \left(\cos \int_{x_0}^{r} J(s)ds\right)^{2\ell} J(r)dr + \gamma\right]. \tag{37}$$

The set of operators $\left\{B_\ell^\pm, H_\ell, \widetilde{H}_\ell\right\}$ fulfill the intertwining relations

$$B_\ell^- \widetilde{H}_\ell = H_{\ell+1} B_\ell^-, \quad B_\ell^+ H_{\ell+1} = \widetilde{H}_\ell B_\ell^+, \tag{38}$$

which means that it is possible to construct the wave functions of $\widetilde{H}_\ell$ in terms of those of H_ℓ. Indeed, if $\widetilde{H}_\ell \Theta_\ell^n = E_\ell^n \Theta_\ell^n$, the wave functions Θ_ℓ^n are related to the eigenfunctions of H_ℓ by

$$\Theta_\ell^n \propto B_\ell^+ \Psi_\ell^{n-1}, \qquad n = 1, 2, \ldots. \tag{39}$$

It may happen that the set $\{\Theta_\ell^n,\ n=1,2,\ldots\}$ do not span the whole space of states of $\widetilde{H}_\ell$. In that case there is an isolated eigenstate $\Theta_\ell^0(x)$, orthogonal to all Θ_ℓ^n $n=1,2,\ldots$, non connected to the set $\{\psi_\ell^n,\ n=0,1,\ldots\}$, given by

$$B_\ell^- \Theta_\ell^0 = 0, \tag{40}$$

and corresponding to the eigenvalue $E_\ell^0 = \epsilon_\ell$. Figure 2 shows some members of the family of potentials $\widetilde{V}_\ell(x,\gamma)$ (Left) and the first wave functions Θ_ℓ^n for $\gamma = 0.7$ (Right), using the mass (31) for some values of the parameters ℓ, a and β.

Concluding remarks

A PDM Hamiltonian having the spectrum of the constant mass TPT one was constructed by means of the factorization method. It was shown that there exist intertwining relations between the constant mass and the PDM factor operators. This fact allows to construct the corresponding PDM wave functions in terms of that of its constant mass counterpart. The method is consistent with the well known point canonical transformation approach. New isospectral Hamiltonians were generated by applying a Darboux transformation and their corresponding eigenfunctions were determined. This method can be generalized to constant mass Hamiltonians with different underlying algebraic structure. Results on the matter are in progress.

Acknowledgements

The financial support of Project SIP 20160527 (IPN Mexico) is acknowledged. The author is grateful to S. Cruz y Cruz for enlightening discussions.

References

[1] Richstone D O and Potter M D 1982 *Astrophys. J* **254** 451
[2] Cveticanin L 1993 *Trans. ASME J. Appl. Mech.* **60** 954
[3] Bastard G 1998 *Wave mechanics applied to semiconductor heterostructures* (Paris: Les Ulis Editions de Physique)
[4] Aquino N, Campoy G and Yee-Madeira H 1998 *Chem. Phys. Lett.* **296** 111
[5] Wolf K B 2004 *Geometric optics on phase space* (Berlin: Springer-Verlag)
[6] Cariñena J F, Perelomov A M, Rañada M and Santander M 2008 *J. Phys. A: Math. Theor.* **41** 085301
[7] Ballesteros A, Enciso A, Herranz J F, Ragnisco O and Riglioni D 2011 *Int. J. Theor. Phys.* **50** 2268
[8] Kordhad R 2011 *Physica* B **406** 3911-3916
[9] Mathiews P M and Lakshmanan M 1974 *Quart. Appl. Math.* **32** 215
[10] Lévy-Leblond J M 1995 *Phys. Rev.* A **52** 1845
[11] Alhaidari A D 2002 *Phys. Rev.* A **66** 042116
[12] Ganguly A, Kuru Ş, Negro J and Nieto L M 2006 *Phys. Lett.* A **360** 228
[13] Musielak Z E 2008 *J. Phys. A: Math. Theor.* **41** 055205
[14] Mielnik B 1984 *J. Math. Phys.* **25** 3387
[15] Mielnik B and Rosas-Ortiz O 2004 *J. Phys. A: Math. Gen.* **37** 10007
[16] Kuru Ş and Negro J 2008 *Ann. Phys.* **323** 413
[17] Milanovic V and Ikonic Z 1999 *J. Phys. A: Math. Gen.* **32** 7001
[18] Plastino A R, Rigo A, Casas M, Gracias F and Plastino A 1999 *Phys. Rev.* A **60** 4318
[19] Quesne C 2006 *Ann. Phys.* **321** 1221
[20] Cruz y Cruz S, Negro J and Nieto L M 2007 *Phys. Lett.* A **369** 400
[21] Cruz y Cruz S and Rosas-Ortiz O 2009 *J. Phys. A: Math. Theor.* **42** 185205
[22] Cruz y Cruz S 2013 Factorization method and the position-dependent mass problem *Trends in Mathematics* (Springer Basel) pp 229-237
[23] Cruz y Cruz S and Rosas-Ortiz O 2013 *SIGMA* **9** 004
[24] Cruz y Cruz S and Santiago-Cruz C 2014 *J. Phys. Conf. Ser.* **538** 012006
[25] Kuru Ş and Negro J 2009 *Ann. Phys.* **324** 2548
[26] Gradshstein I S and Ryzhik I M 2000 *Tables of integrals, series and products* (Academic Press Inc. San Diego CA)
[27] Ince E L 1956 *Ordinary differential equations* (New York: Dover)

Journal of Physics: Conference Series **698** (2016) 012029 doi:10.1088/1742-6596/698/1/012029

Analysis of the Spectral Singularities of Schrödinger Operator with Complex Potential by Means of the SPPS Method

V Barrera Figueroa

Sección de Estudios de Posgrado e Investigación de UPIITA-IPN. Av. Inst. Politécnico Nal. 2580, Col. Barrio la Laguna Ticomán, México D.F., C.P. 07340 MÉXICO

E-mail: vbarreraf@ipn.mx

Abstract. In this work we present an effective way of finding the spectral singularities of the one-dimensional Schrödinger operator with a complex-valued potential defined in the half-axis $[0,\infty)$. The spectral singularities are certain poles in the kernel of the resolvent, which are not eigenvalues of the operator. In this work, the spectral singularities are calculated from the real zeros of $\varkappa(\varrho)=0$, where $\varkappa(\varrho)$ is an analytic function of the complex variable ϱ, which is obtained by means of the Spectral Parameter Power Series Method. This representation is convenient from a numerical point of view since its numerical implementation implies truncating the series up to a M-*th* term. Hence, finding the approximate spectral singularities is equivalent to finding the real roots of a certain polynomial of degree $2M$. In addition, we provide explicit formulas for calculating the eigenvalues of the operator, as well as the eigenfunctions and generalized eigenfunctions associated to both the continuous spectrum and the spectral singularities.

1. Introduction

Let us consider the differential expression $\ell(y) := -y'' + q(x)\,y$, in the half-axis $x \in (0,\infty)$, with the boundary condition at the origin $y(0)=0$, where

$$q(x) := \begin{cases} V(x), & 0 \le x \le a, \\ 0, & a < x < \infty \end{cases} \tag{1}$$

is a complex-valued function called the potential function, and V satisfies certain regularity conditions. By L we denote a differential operator defined by the formula $Ly := \ell(y)$, with the domain $D_L = \left\{ y \in H^2(0,a) \cup H^2(a,\infty) : \ y(0)=0, \ \ell(y) \in \mathscr{L}^2(0,\infty) \right\}$, where $H^2(\alpha,\beta)$ is the Sobolev space [1, Suppl. 2]. Since we are not assuming that $\operatorname{Im} V(x) \equiv 0$ in $[0,a]$, the operator L is non-self-adjoint in D_L. In the work [2] it was addressed a similar differential expression on the whole real axis by considering a real-valued potential function, which leads to a self-adjoint operator without spectral singularities.

In the classical theory of the Sturm-Liouville equation, the spectrum of L is analysed on the basis of certain Volterra integral equations of the second kind representing the solutions of the equation $\ell(y) = \varrho^2 y$, being $\varrho \in \mathbb{C}$ the spectral parameter (see, *e.g.*, [3]). However, in the present work we analyse the spectrum of L from the properties of the solutions of the equation

$\ell(y) = \varrho^2 y$, which are given explicitly in the form of power series of the spectral parameter [4], without using the theory of the integral equations. In addition, we are interested in investigating the spectral singularities of the operator L. The spectral singularities were first discovered by Naimark [5], though the term *spectral singularity* was introduced by Schwartz [6], who defined the spectral singularities for a non-self-adjoint operator in terms of the spectral family of the operator. Comprehensive investigation on the spectral singularities for differential operators has been done by Pavlov [7–9], Lyantse [10], Krein [11] and Langer [12]. The latter investigated the spectral singularities for abstract operators. Nagy [13–15] gave a general notion of the spectral singularities for closed linear operators on a Banach space.

In quantum mechanics, the spectral singularities typically appear in non-Hermitian Hamiltonians having a continuous part of the spectrum [16]. Non-Hermitian Hamiltonians have raised special interest in the literature as complex extensions of quantum mechanics [17]. It is well-known that non-Hermitian Hamiltonians operators with potentials that fulfil the property of $\mathscr{P}\mathscr{T}$-symmetry [18] have real spectra, where $\mathscr{P}$ and $\mathscr{T}$ denote the operations of space reflection and time reversal, respectively. Such symmetry does not by itself guarantee reality of the spectrum, but it does imply that the eigenvalues are either real, or appear as complex-conjugate pairs [19]. Physically, the spectral singularities for some complex potentials have been identified with some resonance phenomena (see, *e.g.*, [20,21]). Hence, identifying the spectral singularities may lead to define their precise physical meaning.

The outline of this work is as follows: in Section 2 we construct some solutions of the equation $\ell(y) = \varrho^2 y$ in $(0, a)$ and in $(0, \infty)$, by means of the SPPS method. Next, in Section 3 we investigate the spectrum of L consisting of the point spectrum and the continuous spectrum, and we identify the spectral singularities. We obtain explicit expressions for calculating the eigenvalues, the spectral singularities and the generalized eigenfunction associated to them. Finally, in Section 4 we provide some concluding remarks.

2. Construction of some solutions of equation $\ell(y) = \varrho^2 y$

Let us introduce the parameter $\lambda \in \mathbb{C}$ admitting the polar representation

$$\lambda := |\lambda| e^{i\theta}, \quad 0 \le \theta < 2\pi, \tag{2}$$

and let $\varrho \in \mathbb{C}$ be a complex number such that $\lambda = \varrho^2$. Thus, ϱ is defined as a branch of $\lambda^{1/2}$ according to the expression

$$\varrho := \sqrt{|\lambda|} e^{i\theta/2}, \quad 0 \le \theta/2 < \pi. \tag{3}$$

Definition (2) specifies a Riemann sheet in the complex λ-plane with a branch cut on the half-axis $[0, \infty)$. Formula (3) implies that the condition $\text{Im}\varrho \ge 0$ holds in each point of this sheet, which is called the proper Riemann sheet [22,23]. Once the condition $\text{Im}\varrho \ge 0$ is invoked, any result with $\text{Im}\lambda^{1/2} < 0$ will violate the requirement of being in the proper Riemann sheet.

Expression (1) leads us to the differential equations

$$-y'' + V(x)\, y = \varrho^2 y, \qquad 0 < x < a, \tag{4a}$$

$$-y'' = \varrho^2 y, \qquad a < x < \infty. \tag{4b}$$

If y satisfies equations (4) simultaneously and if the matching conditions $y(a-0) = y(a+0)$, $y'(a-0) = y'(a+0)$ hold, then y satisfies the equation $\ell(y) = \varrho^2 y$ in $(0, \infty)$. Here we have used the notation $y(a \pm 0) = \lim_{x \to a^\pm} y(x)$ for the one-sided limits. If V were a real-valued function, then equation (4) would be a stationary Schrödinger equation in the proper units, being (4a) the equation describing the bound states, and (4b) the equation describing the scattering states. Thus ϱ^2 would represent the real *eigen*-energies of the bound states and the continuous energy spectrum of the scattering states, respectively. Should V were a complex-valued potential yielding a real spectrum, the physical meaning of ϱ^2 would still be that of energy. However, a complex *eigen*-energy resulting from a complex potential ϱ^2 carries information regarding the lifetime of a decaying *eigen*-state in resonant systems (see, *e.g.*, [24, 25] and references therein).

2.1. A general solution of $\ell(y) = \varrho^2 y$ in the interval $(0, a)$

By y_1 and y_2 we denote two linearly independent solutions of equation (4a) in $(0, a)$, then $y(x) = c_1 y_1(x) + c_2 y_2(x)$ is a general solution of (4a) in $(0, a)$, where c_1, c_2 are arbitrary numbers. Let y_0 be a non-vanishing particular solution of the equation $\ell(y) = 0$ satisfying the regularity conditions $y_0^2, 1/y_0^2 \in \mathscr{C}[0, a]$. According to [4], the series

$$y_1(x) = y_0(x) \sum_{k=0}^{\infty} (-1)^k \varrho^{2k} \widetilde{X}^{(2k)}(x), \qquad y_2(x) = y_0(x) \sum_{k=0}^{\infty} (-1)^k \varrho^{2k} X^{(2k+1)}(x)$$

are two linearly independent solutions of equation (4a) that converge uniformly in $[0, a]$, where the formal powers $\widetilde{X}^{(n)}$ and $X^{(n)}$ are defined by the recursive integration procedure

$$\widetilde{X}^{(n)}(x) := \int_{x_0}^{x} \widetilde{X}^{(n-1)}(s) \left(y_0^2(s)\right)^{(-1)^{n-1}} \mathrm{d}s, \qquad X^{(n)}(x) := \int_{x_0}^{x} X^{(n-1)}(s) \left(y_0^2(s)\right)^{(-1)^{n}} \mathrm{d}s,$$

with the seeds $\widetilde{X}^{(0)} \equiv 1$ and $X^{(0)} \equiv 1$, being x_0 an arbitrary point in $[0, a]$. By choosing $x_0 = a$, the solutions y_1 and y_2 satisfy the initial conditions $y_1(a) = y_0(a)$, $y_1'(a) = y_0'(a)$, and $y_2(a) = 0$, $y_2'(a) = 1/y_0(a)$.

Remark 1. The previous recursive integration procedure yields the estimate $\left|\widetilde{X}^{(2k)}\right| \le c_\varrho^k/(2k)!$, where $c_\varrho := \left|\varrho^2\right| |x-a|^2 \left(\max\left|y_0^2\right|\right)\left(\max\left|1/y_0^2\right|\right)$. Hence, there exist a majorant series such that

$$|y_1(x)| \le |y_0(x)| \sum_{k=0}^{\infty} \frac{c_\varrho^k}{(2k)!}, \qquad 0 \le x \le a.$$

Since the right-hand series converges, then y_1 converges uniformly on $[0, a]$. The uniform convergence of y_2 on $[0, a]$ is proven similarly.

2.2. Solutions of $\ell(y) = \varrho^2 y$ in the half-axis $(0, \infty)$ asymptotically equal to $\exp(\pm i\varrho x)$

The function $d_1 e^{i\varrho x} + d_2 e^{-i\varrho x}$ satisfies equation (4b) in the interval (a, ∞), where d_1, d_2 are arbitrary complex numbers. If the particular solution y_0 is chosen so that $y_0(a) = 1$ and $y_0'(a) = i$, the matching conditions lead to the piecewise continuous function

$$y(x) = \begin{cases} c_1 y_1(x) + c_2 y_2(x), & 0 \le x \le a, \\ d_1 e^{i\varrho x} + d_2 e^{-i\varrho x}, & a \le x < \infty, \end{cases} \tag{5}$$

satisfying $\ell(y) = \varrho^2 y$ in $(0, \infty)$, where $d_1 = \frac{e^{-i\varrho a}}{2i\varrho}\left[c_1 i(\varrho+1) + c_2\right]$, $d_2 = \frac{e^{i\varrho a}}{2i\varrho}\left[c_1 i(\varrho-1) - c_2\right]$.

A solution of the equation $\ell(\varphi) = \varrho^2 \varphi$ in $(0, \infty)$ with the asymptotic behaviour $\varphi(x, \varrho) \sim e^{i\varrho x}$ as $x \to \infty$ is obtained by taking $d_2 = 0$ in expression (5), which implies that $c_2 = c_1 i(\varrho - 1)$. In this case, the sought solution is given by

$$\varphi(x, \varrho) := c_1 \begin{cases} y_1(x) + i(\varrho-1) y_2(x), & 0 \le x \le a, \\ e^{i\varrho(x-a)}, & a \le x < \infty. \end{cases} \tag{6}$$

On the other hand, a solution of the equation $\ell(\psi) = \varrho^2\psi$ in $(0,\infty)$ with the asymptotic behaviour $\psi(x,\varrho) \sim e^{-i\varrho x}$ as $x \to \infty$ is obtained by taking $d_1 = 0$ in (5). This implies that $c_2 = -c_1 i(\varrho + 1)$, which in turn leads us to the solution

$$\psi(x,\varrho) := c_1 \begin{cases} y_1(x) - i(\varrho+1)\, y_2(x), & 0 \le x \le a, \\ e^{-i\varrho(x-a)}, & a \le x < \infty. \end{cases} \tag{7}$$

In expressions (6) and (7) the number c_1 plays the role of a free parameter depending on no boundary condition, so we fix $c_1 = 1$ for simplicity. It follows that the Wronskian $W[\varphi,\psi]$ of the solutions φ and ψ in the whole half-axis $x \in [0,\infty)$ is

$$W[\varphi,\psi](x) := \begin{vmatrix} \varphi(x,\varrho) & \psi(x,\varrho) \\ \varphi'_x(x,\varrho) & \psi'_x(x,\varrho) \end{vmatrix} = -2i\varrho.$$

This implies that φ and ψ are linearly independent in the region $\mathrm{Im}\varrho \ge 0$, $\varrho \ne 0$.

Remark 2. Let us write $\varrho = \alpha + i\beta$. The solutions φ and ψ have the asymptotics $\varphi(x,\varrho) = \mathcal{O}(e^{-\beta x})$, $\psi(x,\varrho) = \mathcal{O}(e^{\beta x})$ as $x \to \infty$. Therefore, in the region $\mathrm{Im}\varrho > 0$ of the complex ϱ-plane, the solutions satisfy $\varphi(x,\cdot) \in \mathscr{L}^2(0,\infty)$, $\psi(x,\cdot) \notin \mathscr{L}^2(0,\infty)$.

3. The spectrum of the operator L

Let us consider the initial value problem

$$\ell(y) = \varrho^2 y, \qquad 0 < x < \infty, \tag{8a}$$

$$y(0) = 0, \qquad y'(0) = 1. \tag{8b}$$

If $y \in D_L$ fulfils this problem, then y also satisfies the equation $Ly = \lambda y$. Since φ and ψ are linearly independent solutions of (8a) for $\varrho \ne 0$, then the function

$$s(x,\varrho) := \frac{\psi(\varrho)\,\varphi(x,\varrho) - \varphi(\varrho)\,\psi(x,\varrho)}{2i\varrho} \tag{9}$$

is a solution of problem (8) in the region $\mathrm{Im}\varrho \ge 0$, $\varrho \ne 0$, where $\varphi(\varrho) := \varphi(0,\varrho) = y_1(0) + i(\varrho-1)\, y_2(0)$, and $\psi(\varrho) := \psi(0,\varrho) = y_1(0) - i(\varrho+1)\, y_2(0)$. Moreover, the function

$$s(x,0) = \begin{cases} \vartheta_1(0)\,\vartheta_2(x) - \vartheta_2(0)\,\vartheta_1(x), & 0 \le x \le a, \\ \vartheta_1(0)(x-a) - \vartheta_2(0), & a \le x < \infty \end{cases} \tag{10}$$

satisfies problem (8) for $\varrho = 0$, where ϑ_1 and ϑ_2 are two linearly independent solutions of $\ell(y) = 0$, which can also be constructed with the SPPS method. On the basis of the solutions $s(x,\varrho)$ and $s(x,0)$ we will investigate the spectrum of the operator L.

3.1. The point spectrum of the operator L

Theorem 3. *The operator L has no positive eigenvalues, moreover,* $\lambda = 0$ *is not an eigenvalue.*

Proof. Let $\lambda \ge 0$ so that $\varrho = \pm\sqrt{\lambda}$ lies in the proper Riemann sheet. The asymptotics of s is

$$s\left(x,\sqrt{\lambda}\right) \sim \begin{cases} \frac{y_1(0)-iy_2(0)}{\sqrt{\lambda}} \sin\sqrt{\lambda}(x-a) - y_2(0)\cos\sqrt{\lambda}(x-a), & \lambda > 0, \\ \vartheta_1(0)(x-a) - \vartheta_2(0), & \lambda = 0, \end{cases}$$

as $x \to \infty$. Hence $s\left(x,\sqrt{\lambda}\right) \notin \mathscr{L}^2(0,\infty)$ for $\lambda \ge 0$. □

Theorem 4. *In order for λ to be an eigenvalue of L it is necessary and sufficient that $\lambda = \varrho^2$, $\mathrm{Im}\varrho > 0$, $\varphi(\varrho) = 0$.*

Proof. According to Remark 2, $\varphi(x,\cdot) \in \mathscr{L}^2(0,\infty)$, $\psi(x,\cdot) \notin \mathscr{L}^2(0,\infty)$ in the region $\mathrm{Im}\varrho > 0$. The solution s defined in (9) will satisfy $s(x,\cdot) \in \mathscr{L}^2(0,\infty)$ in the same region iff $\varphi(\varrho) = 0$. □

Corollary 5. *Let ϱ_n be a zero of the equation $\varkappa(\varrho) = 0$ lying in the region $\mathrm{Im}\varrho > 0$, where $\varkappa$ is an analytic function of the complex variable ϱ defined by its Taylor series*

$$\varkappa(\varrho) := \sum_{k=0}^{\infty} (-1)^k \widetilde{X}^{(2k)}(0)\,\varrho^{2k} + i(\varrho - 1)\sum_{k=0}^{\infty} (-1)^k X^{(2k+1)}(0)\,\varrho^{2k}. \tag{11}$$

Then $\lambda_n = \varrho_n^2$ is an eigenvalue of the operator L.

Proof. The condition $\varphi(\varrho) = 0$ is equivalent to $y_1(0) + i(\varrho - 1)y_2(0) = 0$. The result comes of substituting y_1 and y_2 by their SPPS representations. □

The equation $\varkappa(\varrho) = 0$ is often referred to as the characteristic equation or dispersion relation of the operator L.

Corollary 6. *Let λ_n be an eigenvalue of the operator L, then the function*

$$y_n(x,\lambda_n) := k_n \frac{\psi\left(\lambda_n^{1/2}\right)}{2i\lambda_n^{1/2}}\varphi\left(x,\lambda_n^{1/2}\right), \qquad x \geq 0,$$

is its corresponding eigenfunction, up to a multiplicative constant k_n.

Proof. The result comes by putting $\varphi(\varrho) = 0$ in expression (9). □

3.2. The continuous spectrum of the operator L

Theorem 7 ([26]). *The resolvent R_λ of the operator L is defined by the integral operator*

$$R_\lambda f(x) := \int_0^\infty G(x,\xi;\lambda)\, f(\xi)\,\mathrm{d}\xi, \qquad f \in \mathscr{L}^2(0,\infty),$$

where the kernel is specified by the expression

$$G\left(x,\xi;\varrho^2\right) := \frac{1}{\varphi(\varrho)}\begin{cases}\varphi(x,\varrho)\,s(\xi,\varrho), & 0 < \xi < x,\\ s(x,\varrho)\,\varphi(\xi,\varrho), & x < \xi < \infty.\end{cases}$$

Every number of the form $\lambda = \varrho^2$, $\mathrm{Im}\varrho > 0$, $\varphi(\varrho) \neq 0$, belongs to the resolvent set ρ_L of L.

Theorem 8. *Every number $\lambda \geq 0$ belongs to the continuous spectrum σ_c of the operator L.*

Proof. Let us prove that for $\lambda \geq 0$ the range $\mathscr{R}(L - \lambda I)$ of the operator $L - \lambda I$ is dense in $\mathscr{L}^2(0,\infty)$. It means that the orthogonal complement of $\mathscr{R}(L - \lambda I)$ consists only on the zero element. Indeed, the orthogonal complement of the set $\mathscr{R}(L - \lambda I)$ is the space of solutions u of the equation $L^*u = \lambda u$, where L^* is the adjoint operator of L, which is defined by the differential expression $\ell^* u = -u'' + \overline{q(x)}u$ and the boundary condition $u(0) = 0$. But according to Theorem 3, $\lambda \geq 0$ is not an eigenvalue of the operator L^*, which yields the trivial solution. □

Corollary 9. *Let $\lambda > 0$ belong to the continuous spectrum σ_c of the operator L. Then the bounded, piecewise continuous function*

$$s\left(x, \sqrt{\lambda}\right) = k \begin{cases} y_1(0)\, y_2(x) - y_2(0)\, y_1(x), & 0 \le x \le a, \\ \frac{1}{\sqrt{\lambda}} \left[y_1(0) - i y_2(0)\right] \sin \sqrt{\lambda}\,(x-a) - y_2(0) \cos \sqrt{\lambda}\,(x-a), & a \le x < \infty, \end{cases}$$

is its associated generalized eigenfunction. The generalized eigenfunction associated to $\lambda = 0$ is given by the function (10), up to a multiplicative constant k.

Proof. The assertion follows by substituting φ and ψ in their SPPS form into $s(x, \varrho)$. □

3.3. Spectral singularities

The previous results establish that the complex λ-plane is decomposed as $\mathbb{C} = \rho_L \cup \sigma_p \cup \sigma_c$, *i.e.*, the operator L has no residual spectrum. The continuous spectrum consists of points $\lambda = \varrho^2 \ge 0$ lying in the axis $\mathrm{Im}\varrho = 0$, such that $\varphi(\varrho) \neq 0$. In turn, the eigenvalues are points $\lambda = \varrho^2$ lying in the region $\mathrm{Im}\varrho > 0$, such that $\varphi(\varrho) = 0$. This condition is equivalent to the equation $\varkappa(\varrho) = 0$. A natural question arises regarding the role of the zeros of the equation $\varkappa(\varrho) = 0$ lying in the continuous part of the spectrum of L.

Let us define a singular value of the operator L by a zero of the equation $\varphi(\varrho) = 0$ lying in the region $\mathrm{Im}\varrho \ge 0$, $\varrho \neq 0$. Let ϱ_n be a zero of the equation $\varphi(\varrho) = 0$ lying in the region $\mathrm{Im}\varrho \ge 0$, $\varrho \neq 0$. The non-real singular values indeed correspond to the eigenvalues of L, while the real singular values are the spectral singularities of the operator L. In other words, the point λ_0 is a spectral singularity of the operator L if it is not an eigenvalue of L, and for each λ in the resolvent set ρ_L the kernel $G(x, \xi; \lambda)$ of the resolvent operator $(L - \lambda I)^{-1}$ satisfies $G(x, \xi; \lambda) \to \infty$ as $\lambda \to \lambda_0$ (see, *e.g.*, [27]). The set of the spectral singularities can be identified in the complex λ-plane according to the following result.

Theorem 10. *Let $\alpha_n \in \mathbb{R}$ be a zero of the equation $\varkappa(\alpha) = 0$, where $\varkappa$ is an analytic function defined in (11). Then $\lambda_n = \alpha_n^2$ is a spectral singularity of the operator L.*

Corollary 11. *Let $\alpha_n \neq 0$ be a spectral singularity of the operator L, then its corresponding generalized eigenfunction u_n is a bounded function given by the formula*

$$u_n(x, \alpha_n) := c_n \frac{\psi(\alpha_n)}{2i\alpha_n} \varphi(x, \alpha_n), \qquad x \ge 0,$$

up to a multiplicative constant c_n. If $\alpha_n = 0$ is a spectral singularity, its corresponding eigenfunction is an unbounded function given by expression (10).

Proof. The result comes of applying the condition $\varphi(\alpha) = 0$ in expression (9). □

Theorem 12. *If the potential V is a real-valued function, then the operator L has no spectral singularities.*

Proof. In that case the operator L is self-adjoint in the domain D_L. It is well known that the spectrum of a self-adjoint operator is real, and its residual spectrum is empty [28, p. 182]. Furthermore, all its singular values (eigenvalues) are negative (see, *e.g.*, [1, §2.5], *cf.* [29, §20]) and none of them lie in the continuous part of the spectrum located on $\lambda \ge 0$. □

4. Concluding remarks

Throughout this work we have investigated the spectrum of the operator L by means of the properties of the solutions of the equation $\ell(y) = \varrho^2 y$ in $(0, \infty)$, on the assumption that the potential has compact support on $[0, a]$. In the classical theory of the Sturm-Liouville equation the solutions of the equation $\ell(y) = \varrho^2 y$ in $(0, \infty)$ are represented in the form of Volterra integral equations of the second kind [5, 10, 30–33]. Certain constraints derived from $V \in \mathscr{L}^1(0, \infty)$ are imposed to the potential V in order for the solutions of the integral equations to exist and to be analytic in the domain $\mathrm{Im}\varrho \geq 0$ and $|\varrho| \geq \delta > 0$. If the particular solution y_0 were constructed by means of the SPPS method itself, the only condition imposed to the potential would be $V \in \mathscr{C}[0, a]$. Hence, the present analysis allows the applicability of the theory here presented to a larger class of complex potentials.

Though the theory of the Sturm-Liouville equation $\ell(y) = \varrho^2 y$ is well-known, in the present work we obtained explicit formulas for calculating the eigenvalues, the spectral singularities, and the associated eigenfunctions. Such formulas are given in a SPPS form. From a numerical point of view, the series can be truncated up to a M-*th* term. In this way, the eigenvalues $\lambda_n = \varrho_n^2$, $\mathrm{Im}\varrho_n > 0$, of the operator L can be calculated approximately from the polynomial roots of the equation $\varkappa_M(\varrho) = 0$, where $\varkappa_M$ is the polynomial of order $2M$ defined by

$$\varkappa_M(\varrho) := \sum_{k=0}^{M} (-1)^k \widetilde{X}^{(2k)}(0)\, \varrho^{2k} + i(\varrho - 1) \sum_{k=0}^{M} (-1)^k X^{(2k+1)}(0)\, \varrho^{2k}.$$

Similarly, the spectral singularities $\lambda_n = \alpha_n^2 \geq 0$ can be calculated approximately from the polynomial roots of the equation $\varkappa_M(\alpha) = 0$. The number of approximate eigenvalues (spectral singularities) that can be calculated from the roots of the equation $\varkappa_M(\varrho) = 0$ ($\varkappa_M(\alpha) = 0$) is, in general, lower than M. The remaining spurious roots are due to the truncation of the series $\varkappa(\varrho)$ ($\varkappa(\alpha)$). One would expect that the number of approximate eigenvalues would increase by increasing the number M. However, the upper roots of $\varkappa_M(\varrho) = 0$ ($\varkappa_M(\alpha) = 0$) are numerically more unstable than the lower roots due to small errors in the calculation of the polynomial coefficients. Nonetheless, it is always possible to calculate increasingly higher approximate eigenvalues (spectral singularities) for a fixed M by means of the shifting of the spectral parameter, as is described in [4]. The results presented here show the advantages of the present analysis and the necessity of its further study and numerical deployment, which will be included in a subsequent paper.

References

[1] Berezin F A and Shubin M A 1991 *The Schrödinger Equation* (Dordretch: Kluwer Academic Publ.)
[2] Castillo-Pérez R, Kravchenko V V, Oviedo-Galdeano H and Rabinovich V S 2011 *J. Math. Phys.* **52**, 043522
[3] Marchenko V A 1986 *Sturm-Liouville Operators and Applications* (Operator Theory: Advances and Applications vol 22) ed I. Gohberg (Basel: Birkhuser-Verlag)
[4] Kravchenko V V and Porter R M 2010 *Math. Method Appl. Sci.* **33**, 459
[5] Naimark M A 1954 *Trudy Moskov. Mat. Obshch.* **3**, 181 (In Russian.)
[6] Schwartz J 1960 *Comm. Pure Appl. Math.* **13**, 609
[7] Pavlov B S 1962 *Dokl. Akad. Nauk SSSR* **146**, 1267 (In Russian.) (English Transl. 1962 *Soviet Math. Dokl.* **3**, 1483)
[8] Pavlov B S 1966 On a non-selfadjoint Schrödinger operator I. In: *Probl. Math. Phys. No. 1, Spectral Theory and Wave Processes* ed M S Birman (Leningrad: Izdat. Leningrad Univ.) pp 102–132 (In Russian.) (English Transl. 1967 *Topics in Mathematical Physics* (New York: Consultants Bureau) pp 87-113)
[9] Pavlov B S 1967 On a non-selfadjoint Schrödinger operator II. In: *Probl. Math. Phys. No. 2, Spectral Theory, Diffraction Problems* ed M S Birman (Leningrad: Izdat. Leningrad Univ.) pp 133-157 (In Russian.) (English Transl. 1968 *Topics in Mathematical Physics* (New York: Consultants Bureau) pp 111–134)
[10] Lyantse V È 1964 *Mat. Sb. (N. S.)*, I: **64(106)**, 521; II: **65(107)**, 47 (In Russian.) (English transl.: 1967 *Amer. Math. Soc. Transl.* 2 **60**, 185)

[11] Krein M G and Langer H (= Langer G K) 1963 *Dokl. Akad. Nauk SSSR* **152**, 39 (In Russian.)
[12] Langer H 1982 Spectral functions of definitizable operators in Krein spaces. In: *Functional Analysis* (Lecture Notes in Mathematics vol 948) ed D Butković, H Kraljević and S Kurepa (Berlin: Springer), pp 1–46
[13] Nagy B 1986 *J. Oper. Theo.* **15**, 307
[14] Nagy B 1987 *Acta Math. Hung.* **49**, 51
[15] Nagy B 1988 *Acta Math. Hung.* **51**, 225
[16] Samsonov B F 2005 *J. Phys. A: Math. Gen.* **38** L571
[17] Mostafazadeh A 2010 *Int. J. Geom. Meth. Mod. Phys.* **7**, 1191
[18] Bender C M and Boettcher S 1998 *Phys. Rev. Lett.* **80**, 5243
[19] Dorey P, Dunning C and Tateo R 2002 The ODE/IM correspondence and $\mathscr{PT}$-symmetric quantum mechanics. In: *Statistical Field Theories* (NATO Science Series vol 73) ed A Cappelli and G Mussardo (Netherlands: Springer) pp 13–23
[20] Mehri-Dehnavi H and Mostafazadeh A 2008 *J. Math. Phys.* **49**, 082105
[21] Mostafazadeh A 2009 *Phys. Rev. Let.* **102**, 220402
[22] Dudley D G 1994 *Mathematical Foundations for Electromagnetic Theory* (New York: IEEE Press)
[23] Hanson G W and Yakovlev A B 2002 *Operator Theory for Electromagnetics. An Introduction* (New York: Springer)
[24] Rosas-Ortiz O, Fernández-García N, and Cruz y Cruz S 2008 *AIP Conference Proceedings* **1077**, 31
[25] Sudarshan E C G, Chiu C B, and Gorini V 1978 *Phys. Rev.* D **18**, 2914
[26] Lyantse V È 1968 The non-self-adjoint differential operator of the second order on the half-axis. In: *Linear Differential Operators Part II. Linear Differential Operators in Hilbert Space* ed M A Naimark (London: George G. Harrap & Co., Ltd.), Appendix II
[27] Guseinov G S 2009 *Pramana J. Phys.* **73**, 587
[28] Stakgold I 2000 *Boundary Value Problems of Mathematical Physics. Volume I* (Philadelphia: SIAM)
[29] Vladimirov V S 1971 *Equations of Mathematical Physics* (New York: Marcel Dekker, Inc.)
[30] Levin B Y 1950 *Zap. Nauch. Issled. Inst. Khar. Gos. Univ.* **20,** 83 (In Russian.)
[31] Agranovich Z S and Marchenko V A 1963 *The Inverse Problem in Scattering Theory* (New York: Gordon-Breach)
[32] Levin B Y 1956 *Doklad. Akad. Nauk SSSR* **106**, 187 (In Russian.)
[33] Naimark M A 1968 *Linear Differential Operators Part II. Linear Differential Operators in Hilbert Space* (London: George G. Harrap & Co., Ltd.)

Journal of Physics: Conference Series **698** (2016) 012030 doi:10.1088/1742-6596/698/1/012030

Transmission and Reflection of Wave Packets by Asymmetric Semi-Harmonic Potential Barriers

A Jaimes-Nájera

Physics Department, Cinvestav, A.P. 14-740, México DF 07000, Mexico

E-mail: ajaimes@fis.cinvestav.mx

Abstract. We study the scattering of a particle by a square barrier potential that has a parabolic hole in between. The barrier is parameterized in such a form that the hole can be partially or completely removed. This can be also chosen to be symmetric or asymmetric with respect to the hole. Some expressions for the phase time are given and the time spent by the particle in the interaction zone of the barrier is calculated. It is shown that the related time delay depends on the symmetry operations that one can do on the potential.

1. Introduction

In a previous work we have analyzed the probability of finding a particle in the interaction zone of a scattering process [1]. In particular, we found the intervals of time that are required in order to get a maximum probability in the interaction zone as a lower bound of the duration of the scattering. The simplest example consists of a square barrier (or well) potential as representing the scatterer [2], but this can be also embedded in a semi-harmonic background [3–5]. In this contribution we report our progress in the study of the times involved in scattering under different conditions. To represent the interaction defined by the scatterer we use a square barrier having a parabolic hole in between. This is parameterized such that profiles as symmetry of the potential and the deepness of the hole can be adapted to the necessities of the (possible) experimental setup or to the theoretical approximations. The problem is soundness in areas as solid state physics [6] where the experimental capabilities of manufacturing heterostructures with practically any profile open a wide arena of possible applications (see also [7, 8]). As we are going to show, the time delay is maximum for energies that are close to the transparencies of the scatterer. Evidence of the Hartman effect [9] must be expected for low energies of the incoming particle. The system is sensitive to symmetry transformations so that the time delay of the reflected particles must increase after a change in the configuration of the potential.

The organization of the paper is as follows. In Section 2 we revisit some generalities of the wave-packets in a scattering process. Then, the stationary phase method is applied to calculate the transmission and reflection time delays. In Section 3 we particularize to the potential described above. Final remarks are given in Section 4.

2. Wave packets and time delay

Consider a particle with definite momentum k_0 that is sent from a given point x_0 towards a scatterer at the time $t = 0$. For simplicity, let us assume that the interaction is described by a short-range one-dimensional potential $V(x)$ centered at the origin of the real line. Thus, the potential is equal to zero everywhere except in $(-a_2, a_1)$, with $a_{1,2}$ two nonnegative numbers. The initial quantum state of the particle is given by a momentum distribution $\Lambda(k)$ that is very peaked and centered at $k = k_0$. However, the position of the particle is unknown for $t \geq 0$, so that its position distribution spreads over all the real line. Considering $x_0 > 0$ (i.e., the particle impinges the scatterer from the right), the superposition (proper units assumed)

$$\psi_{inc}(x,t) = \int_{-\infty}^{\infty} dk\, \Lambda(k) e^{i(k(x-x_0)+\omega t)} \tag{1}$$

represents the state of the particle at the right of the interaction zone $(-a_2, a_1)$. The center of the packet moves as a classical particle at the energy $E = \omega = k^2$, with a (group) velocity equal to $v_g(k_0) = d\omega/dk|_{k=k_0} = 2k_0$. The stationary phase condition gives the dynamical law obeyed by the center of the packet

$$\left.\frac{d}{dk}(k(x-x_0)+\omega t)\right|_{k=k_0} = 0 \quad \Rightarrow \quad x = x_0 - v_g t. \tag{2}$$

If $T(k) = |T(k)|e^{-i\gamma(k)}$ and $R(k) = |R(k)|e^{-i\zeta(k)}$ are the complex transmission and reflection amplitudes respectively, the transmitted wave packet can be written as

$$\psi_t(x,t) = \int_{-\infty}^{\infty} dk\, \Lambda(k)|T|e^{-i(k(x-x_0)+\omega t+\gamma)}, \tag{3}$$

where $\gamma(k)$ and $\zeta(k)$ are the corresponding transmission and reflection phase shifts. The center of this new packet obeys the rule $x = x_0 - v_g(t + \partial\gamma/\partial E|_{E=k_0^2})$. Finally, it can be shown [3–5] that the transmission and reflection time delays

$$\Delta\tau_t = -\left.\frac{\partial\gamma}{\partial E}\right|_{E=k_0^2}, \qquad \Delta\tau_r = -\left.\frac{\partial\zeta}{\partial E}\right|_{E=k_0^2}, \tag{4}$$

are in agreement with the notion of phase time introduced by Wigner [10].

3. Scattering by asymmetric semi-harmonic potential barriers

The potential, see Figure 1(a), and the wave functions we are dealing with are defined as follows

$$V(x) = \begin{cases} 0, & x \leq -a_2, \\ V_2, & -a_2 < x < -b_2, \\ x^2, & -b_2 \leq x \leq b_1, \\ V_1, & b_1 < x < a_1, \\ 0, & x \geq a_1, \end{cases} \qquad \psi(x) = \begin{cases} A_5 e^{ikx} + B_5 e^{-ikx}, & x \leq -a_2 \\ A_4 e^{iq_2x} + B_4 e^{-iq_2x}, & -a_2 < x < -b_2 \\ A_3 u_e(x) + B_3 u_o(x), & -b_2 \leq x \leq b_1 \\ A_2 e^{iq_1x} + B_2 e^{-iq_1x}, & b_1 < x < a_1 \\ A_1 e^{ikx} + B_1 e^{-ikx}, & x \geq a_1 \end{cases} \tag{5}$$

where $a_i, b_i > 0$, $i = 1, 2$; $A_j, B_j \in \mathbb{C}$, $j = 1, 2, 3, 4, 5$; $k^2 = E > 0$, $q_1^2 = k^2 - V_1$, $q_2^2 = k^2 - V_2$, and $V_1, V_2 > 0$, while $u_e(x) = e^{-x^2/2}{}_1F_1\left(a, \frac{1}{2}; x^2\right)$ and $u_o(x) = xe^{-x^2/2}{}_1F_1\left(a + \frac{1}{2}, \frac{3}{2}; x^2\right)$ are defined in terms of the confluent hypergeometric function with $a = \frac{1}{4}(1 - k^2)$ [11].

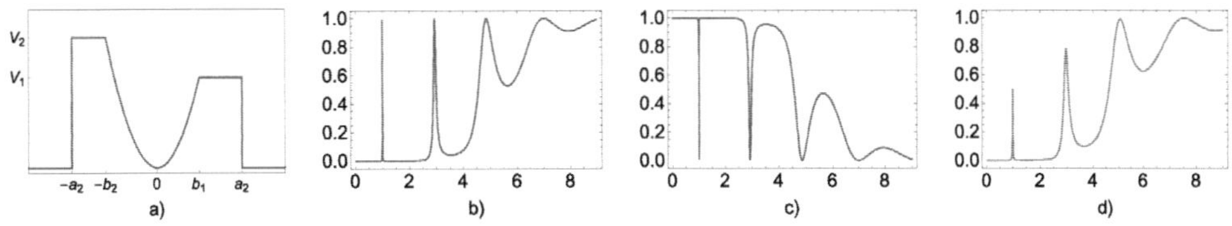

(a) (b) (c) (d)

Figure 1. (a) Schematic representation of the semi-harmonic potential barriers. Transmission (b) and reflection (c) coefficient of the symmetric case as function of the energy for $a_1 = a_2 = 3$, $b_1 = b_2 = 2$, $V_1 = V_2 = b_1^2$. In (d) we have the transmission coefficient for the asymmetric case with $a_1 = 2.5$ and the same values of the other parameters as in (b–c).

After solving the related Schrödinger equation, the scattering amplitudes are of the form

$$T = \frac{-4ikq_1q_2\, W(u_e, u_o)|_{x=b_1}\, e^{-ik(a_2+a_1)}}{t_2 + it_1}, \qquad R = \frac{r_2 + ir_1}{t_2 + it_1}\, e^{-2ika_1}, \tag{6}$$

where

$$t_1 = k(\eta(q_2 + q_1) + \xi(q_2 - q_1)), \qquad t_2 = \alpha(k^2 + q_1q_2) - \beta(k^2 - q_1q_2), \tag{7}$$

$$r_1 = k(\xi(q_2 + q_1) + \eta(q_2 - q_1)), \qquad r_2 = \alpha(k^2 - q_1q_2) - \beta(k^2 + q_1q_2), \tag{8}$$

$$\alpha = (y - q_1q_2z)\cos\varphi_1 + (q_2v_2 + q_1v_1)\sin\varphi_1, \quad \beta = (y + q_1q_2z)\cos\varphi_2 + (v_2q_2 - v_1q_1)\sin\varphi_2, \tag{9}$$

$$\eta = (y - q_1q_2z)\sin\varphi_1 - (v_2q_2 + v_1q_1)\cos\varphi_1, \quad \xi = -(y + q_1q_2z)\sin\varphi_2 + (v_2q_2 - v_1q_1)\cos\varphi_2, \tag{10}$$

$$z = u_e(b_1)u_o(b_2) + u_e(b_2)u_o(b_1), \quad y = \left.\frac{du_e(x)}{dx}\right|_{x=b_1}\left.\frac{du_o(x)}{dx}\right|_{x=b_2} + \left.\frac{du_e(x)}{dx}\right|_{x=b_2}\left.\frac{du_o(x)}{dx}\right|_{x=b_1}, \tag{11}$$

$$v_1 = u_e(b_1)\left.\frac{du_o(x)}{dx}\right|_{x=b_2} + u_o(b_1)\left.\frac{du_e(x)}{dx}\right|_{x=b_2}, \qquad v_2 = u_e(b_2)\left.\frac{du_o(x)}{dx}\right|_{x=b_1} + u_o(b_2)\left.\frac{du_e(x)}{dx}\right|_{x=b_1}, \tag{12}$$

$$\varphi_1 = q_2(b_2 - a_2) + q_1(b_1 - a_1), \qquad \varphi_2 = q_2(b_2 - a_2) - q_1(b_1 - a_1). \tag{13}$$

Here $W(u_e, u_0)$ denotes the Wronskian of u_e and u_0. The above expressions are valid in the cases $E \geq V_j$ ($q_j \in \mathbb{R}$), and $E < V_j$, for $j = 1, 2$, where in the latter ones it is necessary to write $q_j = \pm iq_j^0$, where $q_j^0 = \sqrt{V_j - k^2}$, being indistinct the choice of the sign. Both of the scattering coefficients are invariant under a specular reflection. Indeed, the expressions α, β, η, φ_1, z, y, v_1, v_2, and $W(u_e, u_o)|_{x=b_1}$, are invariant under such a transformation, while $\xi \to -\xi$, $\varphi_2 \to -\varphi_2$. That is, under specular reflection

$$t_1,\ t_2,\ r_2,\ \text{remain invariant, while } r_2 \to -r_2. \tag{14}$$

In Figure 1(b–d) we show the transmission and reflection by a semi-harmonic potential barrier. In the symmetric case, Fig. 1(b), it is clear that the peaks of the transmission are narrower for the transparencies ($|T|^2 = 1$) in the regime of classical confinement ($E < \min(V_1, V_2)$). In turn, the asymmetric case, Figure 1(d), is such that the narrower peaks of the transmission coefficient do not correspond to transparencies ($|T|^2 \neq 1$).

3.1. Phase shift and time delay

To determine the time delay we first calculate the transmission γ and reflection ζ phase shifts. Two main cases of parameters are distinguishable: If $\max(V_1, V_2) < E < \min(V_1, V_2)$ then t_1 and t_2 are real. These last are pure imaginary if $\min(V_1, V_2) < E < \max(V_1, V_2)$. Taking this into account, from (6) we obtain

$$\gamma(k) = k(a_1 + a_2) - \arctan\left(\frac{t_2}{t_1}\right). \tag{15}$$

On the other hand, for the reflection phase shift, if $E < \min(V_1, V_2)$ then $q_j = \pm i q_j^0$, with $q_j^0 = \sqrt{V_j - k^2}$ and $j = 1, 2$. For $V_i < E < V_j$ ($i = 1$, $j = 2$ or $i = 2$, $j = 1$), with $q_i, iq_j \in \mathbb{R}$, it is necessary to write either $q_j = \sqrt{k^2 - V_j}$ for $E > V_j$ or $q_j = i\sqrt{V_j - k^2}$ for $E < V_j$, $j = 1, 2$. Then, for $k > 0$, we have

$$\zeta = 2ka_1 + \arctan\left(\frac{t_1 r_2 - t_2 r_1}{t_1 r_1 + t_2 r_2}\right). \tag{16}$$

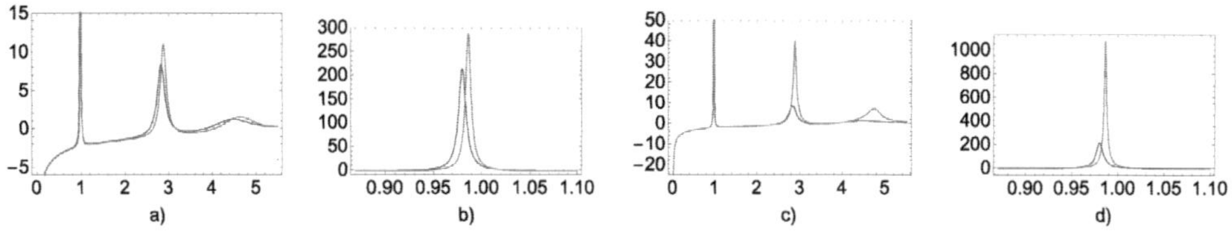

(a) (b) (c) (d)

Figure 2. Transmission time delay $\Delta\tau_T$ (a) and reflection time delay $\Delta\tau_R$ (c) as function of the energy with $a_1 = a_2 = 3$, $b_1 = b_2 = 1.75$, $V_1 = V_2 = b_1^2$ in the symmetric case (blue) and the same parameters except $b_2 = 2$ and $V_2 = b_2^2$ in the asymmetric case (red). Figures (b) and (d) are details of (a) and (b) respectively.

The introduction of (15) and (16) in (4) gives

$$\Delta\tau_T = \frac{W(t_1, t_2)}{(t_1^2 + t_2^2)} - \frac{a_2 + a_1}{2k}, \qquad \Delta\tau_R = \frac{(r_1^2 + r_2^2)W(t_1, t_2) - (t_1^2 + t_2^2)W(r_1, r_2)}{(t_1^2 + t_2^2)(r_1^2 + r_2^2)} - \frac{a_1}{k}. \tag{17}$$

Notice that $\Delta\tau_R = \Delta\tau_T + \Delta\tau_{asy}$, where

$$\Delta\tau_{asy} = -\frac{W(r_1, r_2)}{r_1^2 + r_2^2} - \frac{a_1 - a_2}{2k} \tag{18}$$

accounts for the symmetry breaking in the potential which only depends on r_1, r_2. Indeed, if the potential is symmetric then $\Delta\tau_{asy} = 0$. Moreover, under specular reflection the transmission time delay $\Delta\tau_t$ is preserved but $\Delta\tau_r$ is not, the latter because $\Delta\tau_{asy}$ changes its sign.

Figure 2 shows the behaviour of $\Delta\tau_T$ and $\Delta\tau_T$ in both the symmetric and the asymmetric cases. Note that the peaks of these graphs are very close to the ones of the transmission shown in Figure 1. This last means that closer to a transparency the time delay is maximum. Interestingly,

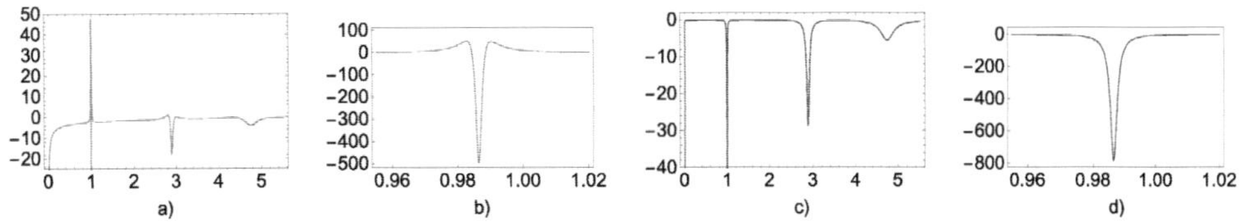

(a) (b) (c) (d)

Figure 3. Reflection time delay $\Delta\tau_R$ (a) as function of the energy with the parameters of the asymmetric case of Figure 2(c), and after specular reflection with $b_1 = 2$, $b_2 = 1.75$, $V_i = b_i^2$, $i = 1, 2$. In (c) it is shown the term of asymmetry $\Delta\tau_{asy}$. Figures (b) and (d) are details of (a) and (c) respectively.

the symmetry breaking enhances the spending of time by the particle in the interaction zone because the peaks of the asymmetric case are higher than the ones in the symmetric case. On the other hand, for low energies we find evidence of the Hartman effect [9] as the time delay is negative.

In Figure 3(a) it is presented the specular reflection of the case shown in Figure 2(c). The change of sign suffered by $\Delta\tau_{asy}$ produces that the peaks of positive time delay are now peaks of negative time delay. That is, the reflected packet is accelerated with respect to a free packet propagating in the same direction and following the same path. Again, the explanation lies on the asymmetric configuration of the barrier. For low energies the incoming packet faces different barriers previous and after the specular reflection. In the first case such a barrier is lower than in the second case, so that tunnelling is most probable and takes less time before than after the symmetry operation. Once the barriers have been interchanged, it is most probable that the incoming packet be reflected so that the process started even before the center of the packet arrives at the interaction zone. This last produces mainly constructive superposition of waves at the right of the scatterer and so the entire reflected packet is formed much earlier than in the case before the symmetry operation.

4. Concluding remarks

The scattering of a particle by a semi-harmonic potential barrier has been studied. The time spent by the particle in the interaction zone has been calculated in terms of the energy derivatives of the phase shift, this last in coincidence with the phase time of Wigner. Then, the time delay has been shown to be maximum close to the transparency energies of the scatterer. As it would be expected, evidence of the Hartman effect has been found for low energies of the projectile. Of particular interest, the system is sensitive to symmetry transformations. That is, the time delay of reflected particles increases after an specular reflection when the asymmetry is such that the first barrier that they face is lower that the second one. The study of the relation between the transparencies and maximum time delay as well as the physical explanation of the peaks in the symmetry breaking term $\Delta\tau_{asy}$ are part of work under progress.

Acknowledgments
I would like to thank Prof. Oscar Rosas-Ortiz for drawing my attention on the subject. The support of CONACyT and Cinvestav is acknowledged.

References

[1] Jaimes-Nájera A 2015 Remarks on the probability of finding a particle in the interaction zone of a scattering process *J. Phys.: Conf. Ser.* **624** 012020
[2] Jaimes A 2012 *Sobre el retraso temporal en la dispersión de un paquete de ondas por una barrera semiarmónica* (MSc. Thesis Cinvestav (In Spanish))

[3] Fernández-García N and Rosas-Ortiz O 2011 Rectangular potentials in a semi-harmonic background: spectrum, resonances and dwell time *SIGMA* **7** 044
[4] Rosas-Ortiz O, Cruz y Cruz S and Fernández-García N 2012 Negative time delay for wave reflection from a one-dimensional semi-harmonic well *Geom. Meth. Phys., Trends in Mathematics* 275-281
[5] Rosas-Ortiz O, Cruz y Cruz S and Fernández-García N 2012 Time delay in the reflection of particles by semi-harmonic wells *J. Phys.: Conf. Ser.* **380** 012018
[6] Gong J, Liang X X and Ban S L 2007 Tunneling time of electronic wave packet through a parabolic quantum well with double barrier *Phys. Stat. Sol.* B **244** 2064
[7] Muga J G, Sala Mayato R and Egusquiza I L 2008 *Time in Quantum Mechanics* (Berlin: Springer)
[8] Muga J G, Ruschhaupt A and del Campo A 2010 *Time in Quantum Mechanics* vol 2 (Berlin: Springer)
[9] Hartman T E 1962 Tunneling of a wave packet *J. Appl. Phys.* **33** 3427
[10] Wigner E P 1955 Lower limit for the energy derivative of the scattering phase shift *Phys. Rev.* **98** 145
[11] Abramowitz M and Stegun I A 1972 *Handbook of Mathematical Functions with Formulas, Graphs, and Mathematical Tables* (AMS-55) (Washington D.C.: National Bureau of Standars)